Physical Science

HOUGHTON MIFFLIN COMPANY • BOSTON
Atlanta
Dallas
Geneva, Illinois
Hopewell, New Jersey
Palo Alto

Physical Science Investigations

Investigations

Charles L. Bickel
Neal D. Eigenfeld
John C. Hogg

Authors

Dr. Charles L. Bickel is Harlan Page Amen Professor and former Chairman of the Science Department at The Phillips Exeter Academy. He has written many science textbooks.

Neal D. Eigenfeld is Science Supervisor of the Milwaukee Public Schools.

John C. Hogg is Harlan Page Amen Professor Emeritus and former Chairman of the Science Department at The Phillips Exeter Academy.

Library of Congress Catalog Card Number: 72–1556

ISBN: 0–395–14091–9

Preface

This textbook deals with topics in physics and chemistry. The authors have taught science classes for many years. They have not presented science as a series of facts and events. Instead, their students have learned by doing and finding out. You will learn in the same way.

Scientific knowledge is obtained by doing experiments. An experiment is performed, observations and conclusions are made, and a little more information is gained. Sometimes scientists suggest a possible explanation for something. Then they do experiments, which support or disprove the explanation. It is the evidence from experiments that determines what is, and what is not, scientific knowledge.

Why not apply this experimental method to the learning of science? Learning from experiments is slower than learning facts by listening to lectures. But learning a lot of facts is not the most important thing you can do in school. By doing experiments you will really understand the basic concepts of physical science. And there are other important advantages.

When you do experiments, you will learn skills. In seeking an answer to a problem, you will work and think independently. Two of the most important skills you can have, not just in science but in all parts of your life, are the ability to make accurate observations and the ability to make reasonable conclusions from your observations. You will develop these skills. You will also learn to measure, classify, and interpret. Each experiment you do will raise questions that you can investigate further.

In this text there are more than 80 experiments. All these experiments have been done many times in the classes of the authors. They all work. Each time you do an experiment you will gain a little more skill and discover something that adds to your knowledge.

Contents

Credits

All illustrations were prepared by George Ulrich, Clementi Associates, Waltham, Massachusetts.

All photos not credited were taken by John T. Urban, Arlington, Massachusetts.

Chapter 1

p. xii	Robert de Gast, Rapho Guillumette Pictures
p. 2	(left) From Donald H. Menzel, *A Field Guide to the Stars and Planets,* Houghton Mifflin Company
	(right) Photograph Courtesy of the Hale Observatories
p. 6	© 1958 United Features Syndicate

Chapter 2

p. 8	Richard P. Rogers
p. 28	Courtesy of the United States Bureau of Standards
p. 29	Jerry Cranham, Rapho Guillumette Pictures
p. 33	(top) Strickler, Monkmeyer Press Photo Service
	(bottom) Courtesy of The Boeing Company
p. 34	Courtesy of United States Steel Corporation

Chapter 3

p. 38	Harold E. Edgerton, Massachusetts Institute of Technology
p. 69	Courtesy of Raymond International, Inc.
p. 75	Elliot Erwitt, Magnum Photos, Inc.

Chapter 4

p. 84	Fred Anderson, Black Star
p. 97	H. James Morrill
p. 100	(left) H. James Morrill
p. 102	H. James Morrill

p. 108	(top) Bruce Roberts, Rapho Guillumette Pictures (bottom) Hal Fay, Design Photographers International
p. 110	H. James Morrill
p. 118	H. James Morrill
p. 121	H. James Morrill

Chapter 5

p. 126	Owen, Black Star
p. 139	H. James Morrill
p. 147	Owen, Black Star
p. 149	Courtesy of the Morton Salt Company, A Division of Morton-Norwich Products, Inc.

Chapter 6

p. 164	Courtesy of Professor Erwin W. Mueller, Pennsylvania State University
p. 168	Courtesy of Carol Goddard
p. 192	(top) Jean Speiser, Black Star (bottom left) Franz Kraus, Design Photographers International (bottom right) Fritz Henle, Monkmeyer Press Photo Service
p. 193	(top left) Courtesy of Ashland Chemical Company (top right) Fundamental Photographs, Inc. (bottom) Courtesy of Freeport Minerals Company

Chapter 7

p. 198	Robert Doisneau, Rapho Guillumette Pictures
p. 204	(bottom) H. James Morrill
p. 215	H. James Morrill
p. 230	James Hubbard, Commonwealth Photos Inc.
p. 234	James Hubbard, Commonwealth Photos Inc.
p. 238	Courtesy of Merck & Co., Inc.

Chapter 8

p. 242	Stock/Boston

Chapter 9

p. 300	Ulrike Welsch
p. 332	Paul Kellogg
p. 335	Courtesy of Carol Goddard
p. 340	Courtesy of Bausch & Lomb
p. 342	Courtesy of Compass Instrument & Optical Company
p. 346	(top) H. James Morrill
p. 347	(top) H. James Morrill

Chapter 1

WHAT IS SCIENCE?

Wondering about the world is part of science.

Summer vacation was almost over, so Tom and Becky spent Saturday on the shore of a large pond. While they were eating lunch a boat came along, loaded with rocks and riding low in the water. The man in the boat dumped the rocks overboard.

"I wonder if the water level goes up when the rocks are dumped in," said Becky. "When I get into a bathtub, the water goes up. Maybe the rocks do the same thing to the pond."

"No, you're wrong," replied Tom. "The boat is higher in the water now that the rocks have been dropped overboard. When the boat was filled with rocks it pushed the water *up* and floated low. So if the boat rides higher, the water level must go *down*."

Becky was willing to think that both she and Tom could be right. "If I am right, the water level goes up. If you are right, it goes down. Maybe we are both right, and maybe it stays the same. How can we find out?"

Tom pointed to a post sticking out of the water. "That post is wet above the level of the pond. So the water level must have gone down, just as I said."

"That doesn't prove anything," Becky answered. "The post is just wet from the waves. Anyway, the pond is so big that even if the water level changes, it might not change enough to measure. We should try it in a bathtub or something like that."

"Yes," said Tom. "My brother has a toy boat and some marbles. Let's try it."

So Becky and Tom decided to try to solve their problem the scientific way. What is the scientific way? What is science?

Becky and Tom did an experiment to find out whether the level of a pond rises, falls, or stays the same when rocks are dumped out of a boat. You will do an experiment too. It should give you an answer to settle the question. In other words, you will gain knowledge by doing an experiment. Scientific knowledge comes from experiments. This way of gaining knowledge applies to science and to no other subject that you will study. In mathematics, history, English, and foreign languages the facts and rules are made by people, so you have to depend on people and books. Science deals with the facts and rules of nature, which you can find out for yourself.

Doing experiments is a partial answer to the question, "What is science?" But there is more to science than experiments. When you know what has happened, then you ask "Why?" A scientist, too, is always asking "Why?" He may suggest a possible reason why, called a **theory** (THEE-uh-ree). Of course a theory is not an experiment. But it is based on the results of experiments, and experiments are done to test it.

Figure 1–1
Left: In a small telescope like those used fifty years ago, the object in the center of this picture is only a blur. **Right:** In a large modern telescope, the object can be seen clearly. It is the Andromeda galaxy, made of billions of stars.

Obviously a scientist cannot learn all he knows about science by doing experiments. He also accepts the experimental findings of other scientists. Otherwise his knowledge would be very limited. In other words, he accepts the background of science and builds upon it. The original experiments are always recorded and published so that if there is any question, an experiment can be repeated under the original conditions. Today scientists have much better equipment than they had fifty years ago. (See Figure 1–1). Observations made with crude equipment are now often found to be incomplete or even misleading. Experimental work not only discovers new knowledge, but also checks old knowledge.

Think about the man in the boat again. You and your teacher will investigate this problem together by doing an experiment. In this way you will get some ideas on how to tackle a problem. You will do most of the later experiments on your own.

In this experiment you will use simple equipment available in the classroom. For example, you will use a graduated cylinder (usually called a graduate), which is the scientist's measuring cup. Water in the graduate will represent the pond and a test tube containing nails will represent the loaded boat. You will float the test tube in the graduate, transfer nails from test tube to graduate, and observe any change in the level of the water in the graduate. One obvious advantage of using a graduate is that the scale on the side shows how high the water is. As a result, measurements by ruler are unnecessary.

experiment 1–1 What Happens to the Water Level?

Materials graduate; test tube; thread; 5 nails; cotton

Procedure Pour water into the graduate until it is about two-thirds filled. Tie a loop of thread around the rim of the test tube. Then place a wad of cotton in the test tube and push it down with a pencil. Slope the tube and place the five nails in it. The wad of cotton prevents the bottom of the test tube from breaking when the nails are dropped in. Put the test tube in the graduate. (See Figure 1–2.) When it is floating freely, read the scale indicating the water level in the graduate. The scale should be read to the nearest half division. Record this reading on the chalkboard.

Figure 1–2
Lower the test tube carefully to avoid splashing.

Lift the test tube out of the graduate by the thread. Remove one nail from the test tube and drop it into the graduate. Replace the test tube in the graduate and read the water level. Again record the reading on the chalkboard.

Continue transferring nails from test tube to graduate until all the nails have been transferred. Read and record the water level after each operation. Copy all readings in your notebook.

The readings are called the data. **Data** (DAY-tuh) are measurements taken and recorded in an experiment. Your data give enough information to tell what happened to the water level of the pond when rocks from the boat were thrown overboard.

Questions

1. What was your original guess about the water level in the pond?

2. Do the data support or oppose your guess?

3. How do the different scale readings compare with each other? Is this what you expect?

4. Would the water level in the pond go up, go down, or stay the same?

5. Why does the water level behave as it does?

Observation and Inference

In Experiment 1–1 you made several observations. An **observation** (ob-zer-VAY-shun) is something you actually see. You observed the change in the water level of the graduate. However, when you answered Question 4 you were not making an observation, but instead you were making an inference. An **inference** (IN-fer-ens) is a judgment based on an observation. Having observed what happens in the graduate, you are able to infer (in-FER) what will happen to the water level in the pond. Making inferences is an important part of scientific procedure. Having made observations in a particular case, a scientist can make inferences about similar cases, so further experiments are unnecessary.

Like scientists, you will learn by doing. In each chapter of this book there are suggestions to guide you in doing several experiments. Also in each chapter there is reading material, called Perspectives. This material will not answer questions raised by the experiments, but it will give you a better understanding of the problems you solve experimentally.

In this book many questions will be asked about the experiments you perform. They may concern an observation, or an inference, or a computation from data. Or they may even involve comparing new information with information you gathered from earlier experiments, in some cases months earlier. For this reason, as well as for reference, it is important that you keep a notebook. For each experiment, record what you do and the results you get.

QUESTIONS

A

1. What is the most important part of science?

2. A scientist suggests why something happened the way it did. What is his suggestion called?

3. Why do scientists repeat experiments that were performed in the past?

4. Does a scientist obtain all his information by doing experiments? Explain.

5. What is the difference between an observation and an inference?

B *The statements numbered 6–15 are about Experiment 1–1, and about the story of the man in the boat. Assume that you actually watched the man. Identify the statements which are observations and those which are inferences.*

6. The man in the boat dropped rocks into the water.

7. The rocks sank to the bottom of the pond.

8. The man placed the rocks in the boat because he wanted to clear the shore area.

9. The man emptied the boat because he wanted to make it lighter.

10. The level of the water rose when the test tube was placed in the graduate.

11. The cotton acted as a cushion to prevent the nails from breaking the test tube.

12. The test tube floated because it was "lighter" than the water.

13. The nails sank to the bottom of the graduate when they were dropped into the water.

14. The boat could be rowed more easily after the rocks were dropped overboard.

15. No cotton is necessary at the bottom of the graduate because water cushions the nails when they are dropped into the graduate.

16. For every inference you listed in Questions 6 through 15, explain why you listed it as an inference, and not as an observation.

Figure 1–3
Questions 17–23 are about this comic strip.

The comic strip (Figure 1–3) is about Charlie Brown's friend, Linus. Indicate which of statements 17–23 are observations and which are inferences.

17. Linus is looking at the puddle of water.

18. Linus is thinking as he studies the puddle.

19. Linus stomps on the puddle angrily.

20. Some water evaporated from the puddle.

21. Linus produced more than six small puddles by stomping on the large puddle.

22. None of the puddles produced are the same size as the original puddle.

23. If all of the small puddles were combined, they would equal the size of the original puddle.

Chapter 2

MEASUREMENT OF MASS AND LENGTH

A unit of measurement is whatever people agree to use. Students at M.I.T. measured the length of a bridge in units of the height of a student named Smoot.

Why do you measure? Does measurement really affect you? Imagine building a house and having to guess whether the opposite walls are the same length. Imagine buying ten apples that are the same price no matter how big they are, and having to take the ones the grocer chooses. Imagine shopping for clothes without knowing your size.

Measurements are important in communicating with other people. If you say that the length of a pencil is six inches, people in the United States will understand what you mean because they know what is meant by a length of one inch. But people in other countries probably will not understand you, because they do not use inches. Before people can communicate measurements, they have to agree on standards.

Even in the Bible there are references to measurement. The Old Testament states that Noah built an ark, 300 cubits in length, 50 cubits in breadth, and 30 cubits in height. A cubit was the length of a man's forearm, measured from his elbow to the tip of his middle finger. This was a convenient way of measuring, since a ruler was not needed. But a length measured this way depended upon the size of the man doing the measuring. If a tall man and a short one measured the length of the same wall, they would get different answers.

We in the United States inherited such terms as inch, foot, and yard from the early English settlers. The original lengths represented by these terms were just as inexact as Noah's cubits. How did these terms come into use?

In the 11th and 12th centuries, life in England was primitive. There were almost no roads, and villages were isolated. Neighbors traded among themselves, so there was little need for standards of length. However, when roads were built the situation changed. Handmade articles could be exchanged by the citizens of different towns. But it was impossible for craftsmen in different towns to share in a task (such as making different parts of a machine) until there was some recognized standard of length. Otherwise the parts would not fit. To solve this problem the king chose a certain length as a standard, and it became law.

One of the oldest units of length was the yard. Its length was fixed by an English king in the 12th century. The length he chose was the distance from his nose to the tip of his outstretched arm. The yard was useful in cutting boards for large buildings, but it was too long for some purposes. Another 12th-century king gave approval to a smaller unit of length called an inch. He ordered that an inch was to be the length of three grains of barley laid end to end. Still later came the foot, shorter than the yard and longer than the inch.

These rough standards were not related to each other. For exact measurements, two conditions are essential. There must be a definite and exact standard of length, and this standard must be divisible into definite and exact parts. In England the yard became the standard of length. A bronze bar was cut to an agreed length and kept in a government office in London. With the length of a yard known exactly, a foot could be made an exact length. One foot was made equal to one third of a yard. This length of a foot was considerably longer than the length of the king's foot. An inch also became an exact length. It was made equal to one twelfth of a foot.

This table of English measurements was set up in the 16th century. Other countries also had their own standards of measurement. Such a variety of standards greatly complicated the exchange of goods between nations. To relieve this situation the French government, in 1790, appointed a committee to study the old units and, if possible, to devise a more useful set of units for both length and mass. The committee recommended an entirely new system, which was legally established about 1800.

The basic unit of length in this system is the meter, which was defined as one ten-millionth of the distance from the equator to the North Pole. This system is called the **metric system** or the International System of Units (SI). The great advantage of the metric system is that it is a decimal system, like United States money. This means that to change from one unit to another, you simply move the decimal point. Multiplication and fractions are not needed.

More than 90 percent of the people in the world now use the metric system. Practically all countries except the United States use it. Scientists all over the world use it. Many companies in the United States use the metric system along with the old system. This makes their products more attractive to other countries, and makes it easier for buyers to compare prices. Probably the United States will adopt the metric system soon.

MASS

Mass and Units of Mass

The whole earth is made up of matter. Rock is matter, water is matter, and even air is matter. Matter may be a solid or a liquid or a gas. However, sound is not matter, nor is light. The amount of matter in an object is called the **mass** of that object. So mass is one of the properties of matter. (A **property** of matter is a feature that identifies the matter.) In many of the experiments described in this book you will be concerned with the mass of an object.

Before you can answer the question "How much?" there must be agreement on the standard of mass. In the metric system the standard of mass is the **kilogram** (KIL-oh-gram), abbreviated **kg.** How much is a kilogram? Roughly, a quart of water has a mass of one kilogram, and the mass of this book is about half a kilogram. However, because the masses in most of the experiments in this book are far less than one kilogram, it will be convenient to use a unit that is one thousandth of a kilogram. This unit is called a **gram,** abbreviated **g.** A dime has a mass of about 2.5 grams (2.5 g) and a nickel has a mass of about 5 grams (5 g).

How Is Mass Measured?

An **equal-arm balance,** sometimes called a double-pan balance, is used to find the mass of an object. As shown in Figure 2–1, it consists of two pans, one at each end of a bar that is balanced at its midpoint. The object is placed on the left-hand pan. If known standard masses are placed on the right-hand pan until the bar is balanced, the sum of the known masses is then equal to the

Figure 2–1 (right)
You will probably use an equal-arm balance like this one in your experiments.

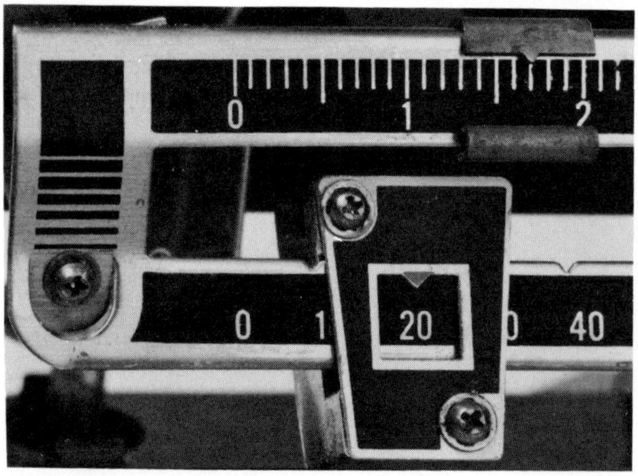

Figure 2–2 (below)
What mass is represented by the riders?

Figure 2–3 (below right)
Before using a balance, put the riders on zero and make sure it balances.

mass of the object. For example, if the known masses are 10 g, 5 g, 2 g, and 1 g, the mass of the object is 18 g.

However, it is more convenient to use the movable riders on the two beams than known masses on the right pan. Notice that there are two beams, each with a scale. The scale on the upper beam goes from 0 to 10 g. Each length representing one gram is subdivided into 10 parts. The scale on the lower beam goes from 0 to 100 g. The way to measure a mass is shown in Figure 2–2, which shows part of the left-hand section of the beams.

Notice that the lower beam is notched and that the rider rests on the notch marked 20 g. The rider on the upper beam rests between the 1-g mark and the 2-g mark. Actually, the center mark on the rider is on the seventh line to the right of the 1-g mark, so the reading of the upper rider is 1.7 g. Thus the two rider readings represent a mass of 20.0 g + 1.7 g, or 21.7 g.

The smallest division on the scale of the equal-arm balance represents one tenth of a gram (0.1 g). Therefore, measurements of mass can be made only to the nearest 0.1 g on this balance.

When both riders rest on the zero marks of the beams, as shown in Figure 2–3, the balance should be "balanced." That is, the pointer on the top of the balance should point to the middle of the scale behind it, or it should swing an equal distance on either side of the middle of the scale. (If the empty balance does not "balance," ask your teacher to help you adjust it.)

experiment 2–1 The Masses of Regular Solids

Materials

4 regular solids; equal-arm balance

Procedure

First check the empty balance and, if necessary, adjust it. Then place a solid on the left pan of the balance. Move the lower rider to the *right,* one notch at a time, until the left pan moves up. Then move the rider one notch to the *left* so that the left pan moves down. Next, slide the upper rider to the right until the balance is "balanced." The sum of the beam readings is the mass of the solid. Enter the separate beam readings in your notebook and then add them. For example, Mass of rectangular solid = 80.0 + 7.6 = 87.6 g.

Following the same procedure, find the masses of the other solids. In each case, enter in your notebook the name of the solid, the separate beam readings, and the sum of the readings (that is, the mass of the solid).

LENGTH AND VOLUME

Unit of Length

Another property of matter is that it occupies space. The space occupied by a solid is called its **volume** (VOL-yum). If the solid is **regular,** that is, if it has sides that can be measured, its volume can be computed from these measured lengths. For example, to find the volume of a rectangular solid, three lengths must first be measured: the length, the width, and the thickness. The unit of length in the metric system is the **meter** (MEE-ter), a distance a little longer than a yard. The abbreviation for meter is **m.** For our experiments, a meter is too long. It is much more convenient to use one hundredth of a meter as a unit of length. This unit is called a **centimeter** (SEN-ti-mee-ter). The abbreviation for centimeter is **cm.**

A ruler is the simplest device for measuring length. Obtain a ruler and notice that there are two scales, one on each edge, as shown in Figure 2–4. One scale is numbered in inches and the other in centimeters. You will use the centimeter scale only. Notice that the smallest division on the centimeter scale is one tenth of a centimeter (written as 0.1 cm). When using the ruler, estimate lengths to the nearest 0.1 cm. Thus the length of a page of this book, as measured by the ruler, is 23.2 cm.

If a length comes out exactly seven centimeters, record it as 7.0 cm, not 7 cm. The reading 7.0 cm shows that you measured it to the nearest 0.1 cm. If you recorded it as 7 cm, a person looking at your notebook would think that the line was measured only to the nearest centimeter.

Figure 2–4
With a ruler like this, you can make measurements that can be understood almost anywhere.

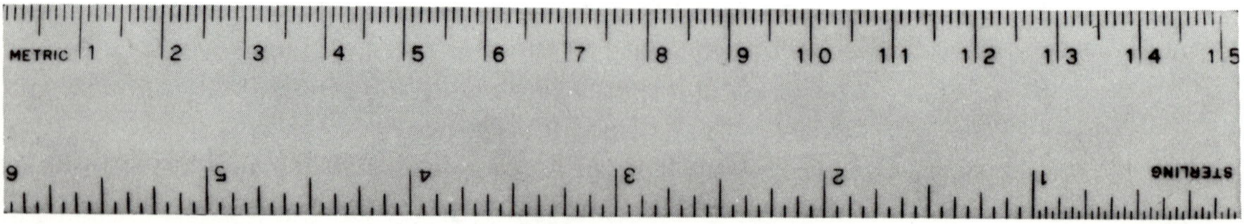

experiment 2–2

The Volumes of Regular Solids

Materials

the same solids you used in Experiment 2–1; centimeter ruler

Procedure

A. Rectangular and Cubical Solids

Probably most of your solids are rectangles or cubes. Choose one and, in your notebook, make a table for your data like Figure 2–5.

length of solid = ? cm
width of solid = ? cm
thickness of solid = ? cm

Figure 2–5
In your notebook, make a table like this for your data.

Using the ruler, measure the length of the solid and record the value in the table. The ruler should rest on the object along the length to be measured. Place the zero mark of the centimeter scale at the left edge of the object. Then read the marking of the ruler that is closest to the right edge of the object. After you have recorded the reading in your notebook, make a second reading along another lengthwise edge of the object. If these two readings do not agree, make a third reading.

Now measure the width and thickness of the solid in the same manner, and record them in the table.

The volume of a regular object is the product of its dimensions, and it can be calculated as follows.

$$\text{Volume} = \text{length} \times \text{width} \times \text{thickness}$$

Example: Consider an object whose dimensions are: length = 8.6 cm, width = 5.0 cm, thickness = 3.3 cm.

$$\begin{aligned}\text{Volume} &= \text{length} \times \text{width} \times \text{thickness} \\ &= 8.6\text{ cm} \times 5.0\text{ cm} \times 3.3\text{ cm} \\ &= 142\text{ cm}^3\end{aligned}$$

The "exact" answer is 141.9 cm^3, which is to four figures. However, this answer is rounded off to the nearest cubic centimeter because you cannot measure to four figures. Also notice that the units are attached to numbers and that units are multiplied like numbers. That is, cm × cm × cm = centimeters cubed, or more simply, cm^3.

Now measure the other rectangular and cubical objects in the same way. Record their dimensions in your notebook and calculate their volumes.

B. Triangular Solids

You may have a triangular solid that looks like half of a rectangular solid. Measure its length, width, and thickness. You do not need to measure the long side of the triangle. Calculate the volume of the solid, using the relationship

$$\text{Volume} = \tfrac{1}{2}(\text{length} \times \text{width} \times \text{thickness})$$

Example: Consider a triangular object whose dimensions are: length = 9.3 cm, width = 4.5 cm, thickness = 2.3 cm.

$$\begin{aligned}\text{Volume} &= \tfrac{1}{2}(\text{length} \times \text{width} \times \text{thickness})\\ &= \tfrac{1}{2}(9.3\text{ cm} \times 4.5\text{ cm} \times 2.3\text{ cm})\\ &= \tfrac{1}{2}(96\text{ cm}^3) = 48\text{ cm}^3\end{aligned}$$

C. Cylindrical Solids

If one of your solids is a cylinder, measure and record its height and diameter. Then calculate its volume, using the relationship

$$\text{Volume} = \pi \times \frac{\text{diameter}^2}{4} \times \text{height}$$

$$V = \pi \times \frac{d^2}{4} \times h$$

where π (called *pi*) is a constant and has the value 3.14, d is the diameter of the cylinder, h is its height, and V is the volume of the cylinder. Note that the unit for d^2 is cm^2, and that the unit for $d^2 \times h$ is cm^3, which is a unit of volume.

Example: Suppose that a cylinder is 12.6 cm long and 3.7 cm in diameter. The volume is calculated as follows.

$$\begin{aligned}V &= \pi \times \frac{d^2}{4} \times h\\ &= 3.14 \times \frac{(3.7\text{ cm})^2}{4} \times 12.6\text{ cm} = 135\text{ cm}^3\end{aligned}$$

Questions

1. How does the diameter of your cylinder compare with the diameters found by other students?
2. Why is the diameter of a cylinder a difficult measurement to make?

The Slide Rule

The computations you made in Experiment 2–2 probably took a long time. You can save much time by making such computations on a slide rule, and you are strongly urged to do so. Both multiplication and division can be done on a slide rule. Answers by slide rule are given to three figures, which is accurate enough for your experimental data.

The way to use a slide rule is described in Appendix 1.

MASS/VOLUME RATIOS

experiment 2–3

Mass/Volume Ratios of Regular Solids

Procedure

In Experiment 2–1 you used four objects of the same material but of different shapes, and you found their masses. In Experiment 2–2 you found the volumes of these same four objects. Is the mass of each object related to its volume? You can answer this question by using the data you have collected.

Construct a table like that in Figure 2–6. Enter the mass and volume for the four objects studied in Experiments 2–1 and 2–2.

Figure 2–6
Data table for Experiment 2–3.

Object	Mass (M)	Volume (V)	M/V
rectangular	? g	? cm^3	? g/cm^3
square flat	? g	? cm^3	? g/cm^3
cubical	? g	? cm^3	? g/cm^3
cylindrical	? g	? cm^3	? g/cm^3

Then compute the mass/volume ratio for each object and enter your answer in the table. Your computed values should include both a number and units.

Example: Suppose the mass of an object is 80.0 g and the volume of the object is 29.0 cm^3. The mass/volume ratio is

$$\frac{M}{V} = \frac{80.0 \text{ g}}{29.0 \text{ cm}^3} = 2.76 \text{ g/cm}^3$$

which is read as "two point seven six grams per cubic centimeter."

experiment 2–4

A Graph of the Mass and Volume Values of the Regular Solids

Materials

graph paper

Procedure

On a piece of graph paper draw a vertical (up and down) line about two or three centimeters from the left-hand margin of the

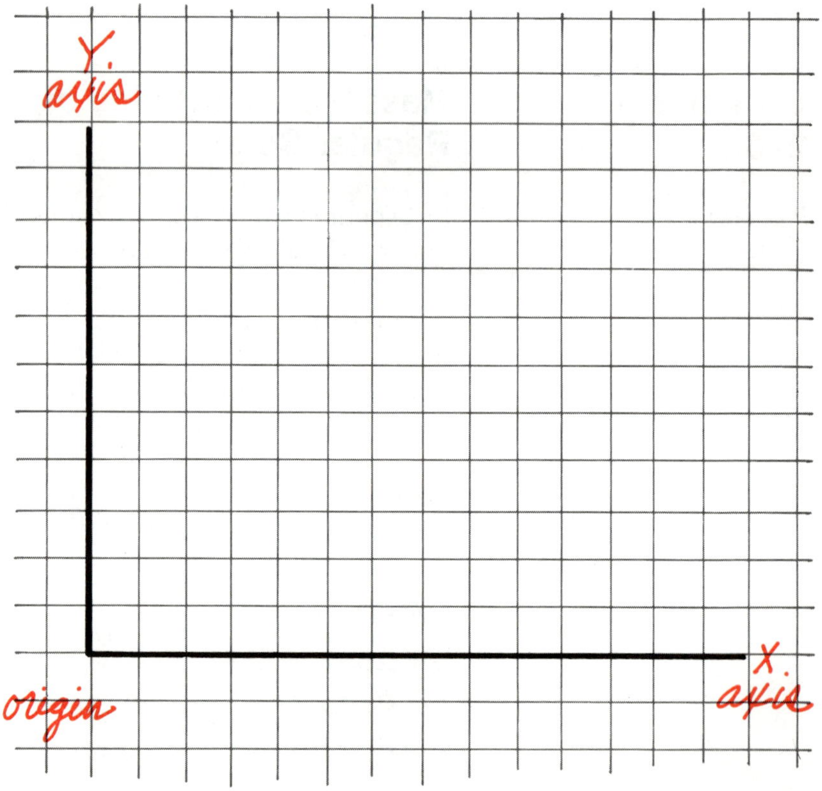

Figure 2–7
Start your graph by drawing the axes.

graph paper to within two or three centimeters of the bottom of the page, as in Figure 2–7. This vertical line is called the **Y axis.** Now draw a horizontal (across) line about two or three centimeters from the bottom of the page so that it crosses the vertical line. This horizontal line is called the **X axis.** The point where the two lines cross is called the **origin** (OR-uh-jin).

Next, set up a scale along the Y axis for your experimental mass values. For example, if your largest mass value is 93 g, make the scale go from 0 to 100 g. (See Figure 2–8.) Also set up a scale along the X axis for your experimental volume values. For example, if your largest volume is 56 cm^3, make the scale go from 0 to 60 cm^3.

Suppose that your mass value for the rectangular object is 75 g and that the volume of this object is 50 cm^3. This pair of data should be entered on the graph as a point. This point is where an imaginary line "drawn" (the line is not actually drawn) upward from the 50 cm^3 mark on the X axis crosses an imaginary line "drawn" to the right from the 75-g mark on the Y axis. Draw

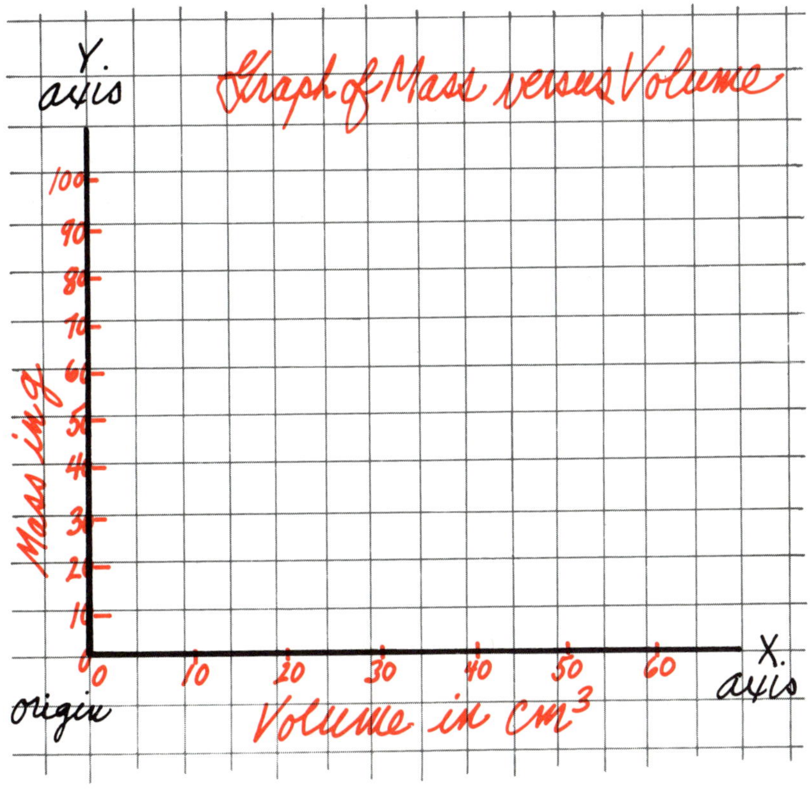

Figure 2–8
Next, add the scales.

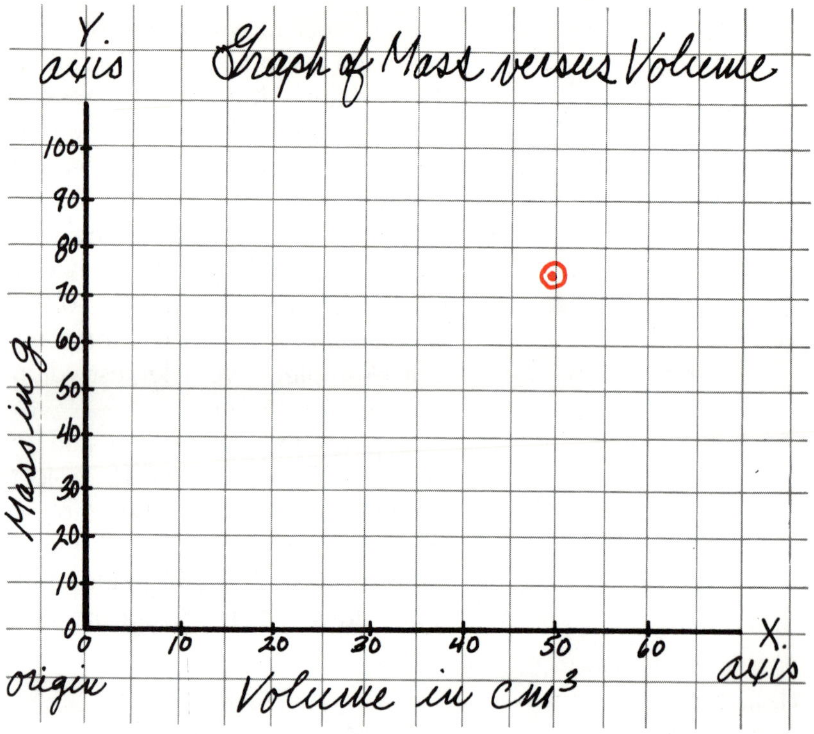

Figure 2–9
Place each point carefully and check it.

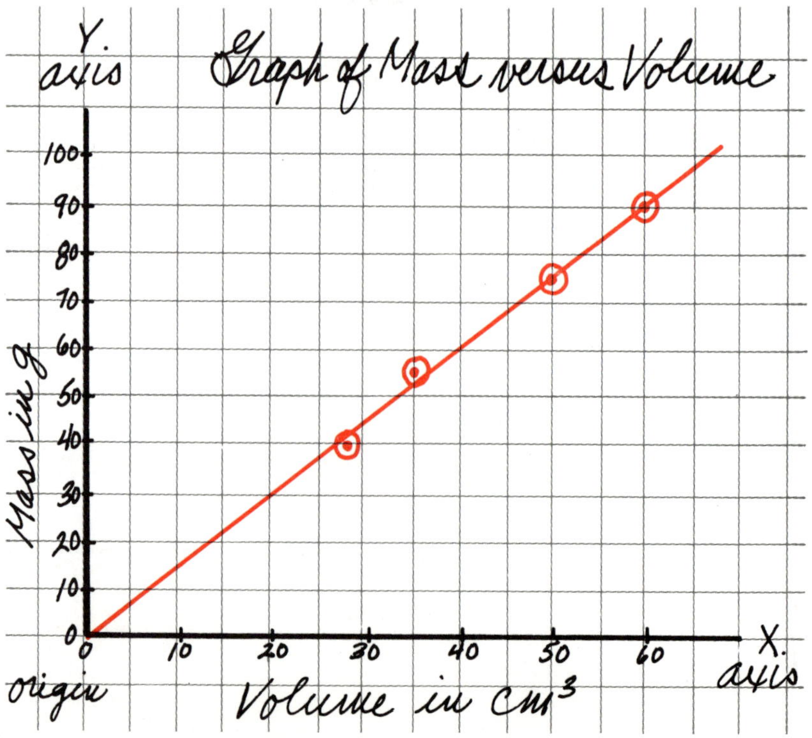

Figure 2–10
Finally, draw a line through the points.

a small circle around the point. The point is said to have the *coordinate values* of 75 g and 50 cm^3. (The values along the X axis are called the X coordinates, and the values along the Y axis are called the Y coordinates.) The graph, at this stage, should look like Figure 2–9.

Now enter the other three pairs of data as circled points. Since the mass of zero volume of material is zero, the origin should be included as a fifth point.

Draw a line that most nearly passes through the points. This is the graphed line. The graphed line may pass slightly below some points and slightly above others, as shown in Figure 2–10. This is because your experimental values are not quite accurate. However, you know that if the mass of the material is zero, the volume is also zero. Therefore, the graphed line should pass through the origin.

experiment 2–5 To Find the Mass of an Object from the Graph

Materials regular solid object; centimeter ruler; equal-arm balance

Procedure Obtain a regular solid object from your teacher. (The object should be made of the same material as the four regular solid objects already studied.) Record the dimensions of the object and then compute its volume.

To find the mass of the object, use the graph that you prepared in Experiment 2–4. To do this, locate the calculated volume value of the object on the graphed line, and then read off and record the corresponding value of the mass.

Then find the mass of the object by means of the equal-arm balance and record this value.

One way of comparing two values is to calculate the percentage difference between them. To do this, use the relationship

$$\text{percentage difference} = \frac{\text{the difference between the values}}{\text{the average of the values}} \times 100\%$$

Example: Suppose that the mass found from the graph is 47.6 g and the mass obtained by balance is 46.3 g.

$$\text{percentage difference} = \frac{47.6\text{ g} - 46.3\text{ g}}{\frac{1}{2}(47.6\text{ g} + 46.3\text{ g})} \times 100\%$$

$$= \frac{1.3\text{ g}}{47.0\text{ g}} \times 100\% = 2.77\% \text{ or } 2.8\%$$

Questions

1. How does the mass obtained from the graph compare with the mass found by balance?
2. Compute your percentage difference for the two mass values.
3. Is your percentage difference reasonable? Explain.

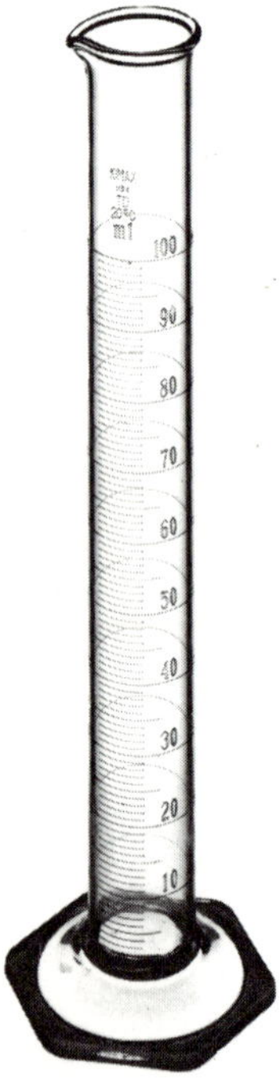

Figure 2–11
A graduate is used to measure the volume of a liquid. This one holds 100 ml.

Density

You have now found the mass/volume ratios of four regular solids. All four solids are made of the same material. The mass/volume ratio of a given material is called its **density** (DEN-suh-tee). Density is usually defined as *the mass of a unit volume of a substance.*

This definition of density can be expressed mathematically as

$$\text{Density} = \frac{\text{Mass}}{\text{Volume}}$$

or, in symbols,

$$D = \frac{M}{V}$$

From this equation you can see that if any two of the factors are known, the third can be computed.

Example: What is the volume of 1000 g of platinum? You can solve this problem if you know the density of platinum. The density of platinum is 21.4 g/cm³. Thus the two known factors are mass and density.

$$V = \frac{M}{D} = \frac{1000 \text{ g}}{21.4 \text{ g/cm}^3} = 46.8 \text{ cm}^3$$

Notice that the gram units (g) in numerator and denominator cancel out.

experiment 2–6

To Compute the Mass of a Regular Solid from Its Density and Its Dimensions

Laboratory Test

Materials regular solid object; centimeter ruler; equal-arm balance

Procedure Obtain a regular solid object from your teacher. Measure and record the dimensions of the object and calculate its volume.

Now ask your teacher for the density of the substance and calculate the mass of the object.

Then find the mass of the object by means of an equal-arm balance and record this value.

Questions

1. Compute the percentage difference between the mass you calculated and the mass you found by the balance.

2. Is this percentage difference reasonable? Explain.

LIQUIDS

Volume of a Liquid

You have now determined the density of several regular solid objects. Can the same experimental method be used to find the density of a liquid? Since a liquid must be confined in a container, you clearly cannot measure the volume of a liquid by the method used for regular solids.

experiment 2–7

The Level of a Liquid in a Graduated Cylinder

Materials

100-ml graduate

Procedure

One way to find the volume of a liquid is to use a graduated cylinder or, as it is usually called, a graduate. Figure 2–11 shows a graduate marked in **milliliters** (MIL-uh-lee-ters) abbreviated **ml.** A milliliter is the same as a cubic centimeter (cm^3).

In Chapter 1 you read a graduate in scale divisions. These same scale divisions are now read as milliliters.

Pour some water into a 100-ml graduate and notice that the surface of the water is slightly curved. The curved free surface of a liquid is called a **meniscus** (men-NIS-kus). A meniscus is shown in Figure 2–12. To read the volume of the water, raise or lower the graduate until your eye is on the same level as the meniscus, as shown in Figure 2–13. Then read the *bottom* of

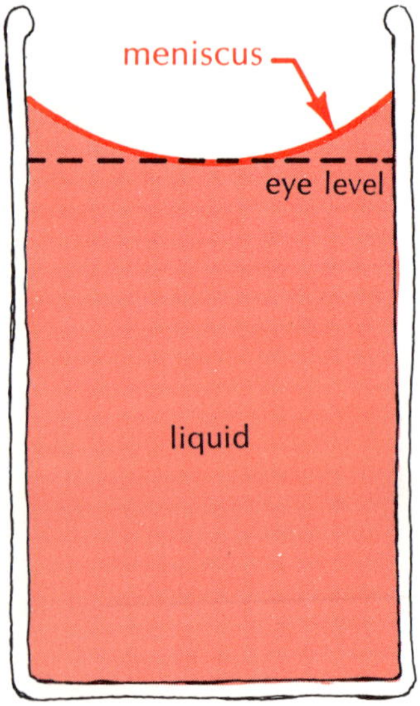

Figure 2–12 (above)
Why should you read the bottom of the meniscus?

Figure 2–13 (right)
Raise the graduate or lower your eye.

the meniscus. Read the volume of water in the graduate and record it in your notebook as follows:

Volume of water in graduate = ? ml

Now ask your teacher to check your reading.

Questions

1. What reading do you get if your eye is (a) below the level of the meniscus, (b) above the level of the meniscus?

experiment 2–8 The Density of Water

Materials

100-ml graduate; equal-arm balance

Procedure

Find the mass of a 100-ml graduate and record the value in your notebook. Pour water into the graduate until it is about one-quarter filled. Read the volume (estimate the reading to the nearest 0.5 ml) and record it. Then find the mass of the graduate plus the water, and record the mass value.

By subtraction, compute the mass of the water in the graduate. Then calculate the density of water.

The data should be recorded neatly as in Figure 2–14.

Figure 2–14
Data table for Experiment 2–8.

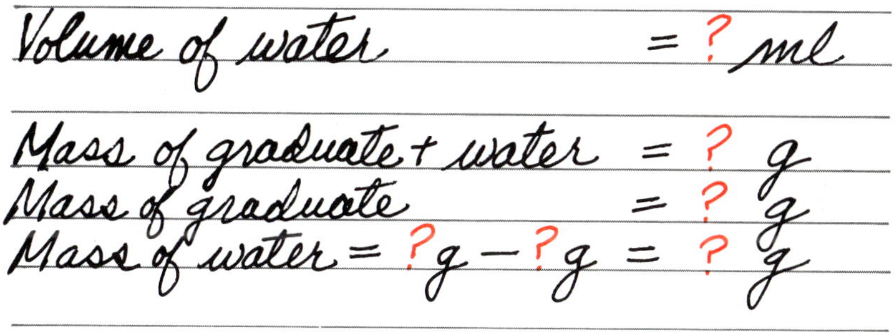

Calculate the density of water as follows:

$$D \text{ (of water)} = \frac{M}{V} = \frac{?\text{ g}}{?\text{ ml}} = ?\text{ g/ml}$$

Now repeat the above procedure, using about 50 ml of water, then about 75 ml of water, and finally about 100 ml of water. Compute the density of water in each case and then find the average value for the density.

The density of water is an important physical constant. It is used as a standard of reference to which the densities of other substances are compared.

experiment 2–9

The Density of Automobile Radiator Antifreeze

Materials

100-ml graduate; equal-arm balance; sample of antifreeze

Procedure

Most commercial permanent antifreeze liquids are a chemical called ethylene glycol (ETH-uh-leen GLY-kawl). Non-permanent antifreeze liquids are usually a chemical called methanol (METH-uh-nawl).

Find the mass of a *dry* 100-ml graduate and record the mass value in your notebook. Ask your teacher for a sample of antifreeze and pour it into the graduate. Obtain the required data (as in Experiment 2–8) and then compute the density of the liquid.

Do not pour the antifreeze down the sink. Return it to your teacher for storage.

Ask your teacher for the density values of ethylene glycol and methanol, and record these values in your notebook.

Questions

1. From the computed density, state the kind of antifreeze you used, ethylene glycol or methanol.

2. Was your sample of antifreeze pure? In answering this question, remember that the density of a substance is a characteristic property of that substance. The commercial antifreeze may contain water. (Water is much cheaper than either ethylene glycol or methanol.)

3. Comparing the density of pure antifreeze and the density of antifreeze "contaminated" with water, which would be nearer the density of water? Why?

IRREGULAR SOLIDS

Volume of an Irregular Solid

As you know, the volume of a regular solid can be computed from its measured dimensions. But how can the volume of an irregular solid be found? If the solid does not dissolve in water, its volume can be found by measuring the volume of water it displaces, as in the following experiment.

experiment 2–10 The Volume of a Sample of Gravel

Materials

100-ml graduate; pea stone gravel

Procedure

Half fill a 100-ml graduate with water. Read and record the volume. Slide the sample of gravel into the graduate. Air bubbles that cling to the surface of the gravel must be released. To do this, hold the graduate in a vertical position and twist your hand sharply to right and left until the air bubbles are dislodged and escape. See Figure 2–15 on page 28.

Now read and record the volume of water plus gravel. Clearly the volume of the gravel is the difference between the two volume readings.

Pour the gravel into the container provided by your teacher. Do *not* dump it into the sink.

Questions

1. If you wish to find the density of the gravel, what additional information would you need?

2. How would you get the information in Question 1?

3. How would you compute the density of the gravel?

experiment 2–11 The Density of Marble

Laboratory Test

Materials

100-ml graduate; marble chips; equal-arm balance

Procedure

Obtain a sample of marble chips from your teacher. Following the same procedure as in Experiment 2–10, find the density of the marble.

PERSPECTIVES

perspective 2–1 Mass

In much of their work scientists are concerned with the masses of substances. To understand each other's scientific literature and to exchange ideas, scientists in all countries must use the same standard of mass. An international committee of scientists selected the kilogram as the standard unit of mass.

The standard kilogram is a block of metal. Kilogram masses everywhere in the world are compared with it. It is kept in a laboratory near Paris, and carefully protected against wear as shown in Figure 2–16. A number of exact copies of this standard mass have been made and sent to other countries for purposes of reference. One of these copies is in the National Bureau of Standards in Washington, D.C.

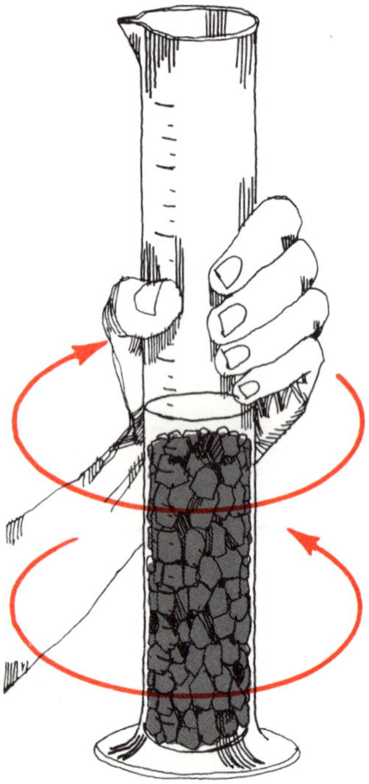

The kilogram is the unit of mass in the metric system. As you probably know, metric units are based on decimals. The great advantage of this system is that you can convert a larger unit to a smaller unit merely by shifting the decimal point. A kilogram is too large a unit for most purposes such as measuring small quantities of chemicals. It is convenient to divide a kilogram, abbreviated kg, into 1000 parts, each of which is called a gram, abbreviated g. Thus, the relationship between grams and kilograms is

$$1000 \text{ grams} = 1 \text{ kilogram}$$

or

$$1000 \text{ g} = 1 \text{ kg}$$

Thus, to convert kilograms to grams, shift the decimal point three places to the right as in

$$1.25 \text{ kg} = 1250 \text{ g}$$

Conversely, to convert grams to kilograms, merely shift the decimal point three places to the left as in

$$45.6 \text{ g} = 0.0456 \text{ kg}$$

Figure 2–15 (above)
To release air bubbles from the gravel, twist the graduate sharply but do not shake it.

Figure 2–16 (below)
The block of metal under the two glass jars is the standard kilogram. Behind it is the standard meter bar.

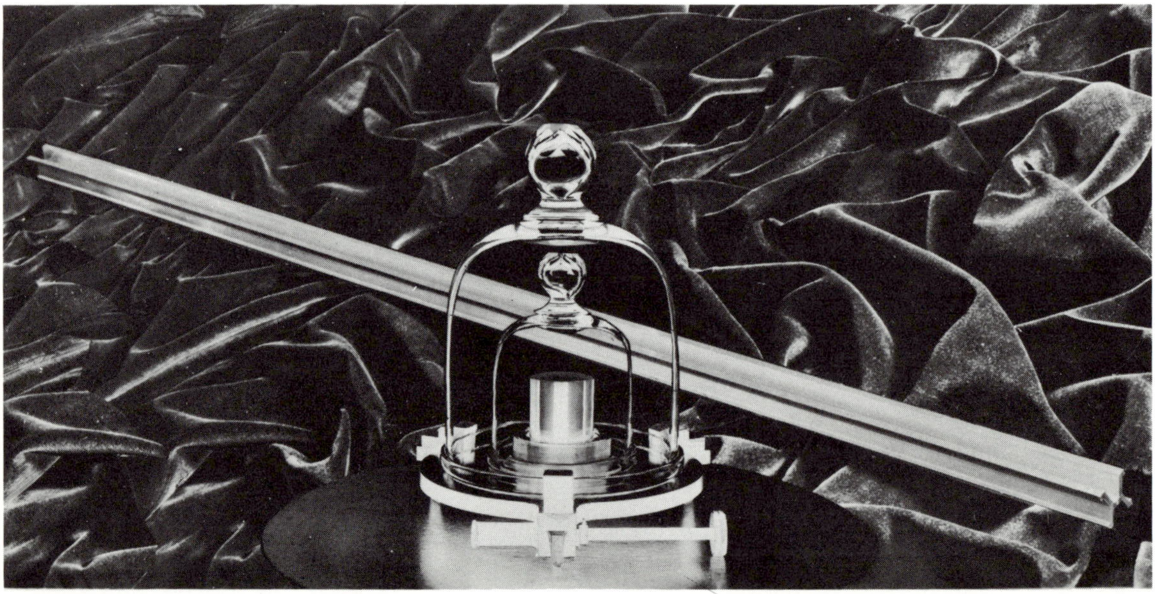

Figure 2–17
These athletes are running a 100-yard dash. How much farther would they have to run in a 100-meter dash?

perspective 2–2

Length

As you know, lengths in the United States are measured in inches, in feet, in yards, and in miles. However, in nearly all other countries they are measured in meters. A meter is a little longer than a yard, about three inches longer. In track meets in the United States there is a 100-yard dash (for which the record is 9.1 seconds). But in the Olympics, where athletes from all countries compete, there is a 100-meter dash (for which the record is 9.9 seconds). See Figure 2–17.

The meter is the metric unit of length. The standard meter bar is kept in the same laboratory in Paris as the standard kilogram mass. And, as with the kilogram, exact copies of this standard have been made. One is kept at the Bureau of Standards in Washington, D.C.

A smaller unit, one hundredth of a meter in length, is called a centimeter. Hence

$$1 \text{ meter} = 100 \text{ centimeters}$$

or

$$1 \text{ m} = 100 \text{ cm}$$

This can also be written

$$1 \text{ cm} = 0.01 \text{ m}$$

Thus, to change a length in meters to centimeters, the decimal point is moved two places to the right, as in

$$1.37 \text{ m} = 137 \text{ cm}$$

Conversely, to change centimeters to meters, the decimal point is moved two places to the left, as in

$$85.2 \text{ cm} = 0.852 \text{ m}$$

A ruler is the simplest device for measuring lengths. Many ordinary "foot" rulers are graduated in centimeters along one edge and in inches along the other, a reminder that there are two different systems for measuring length. A comparison of these scales (Figure 2–4) shows that a centimeter is a little less than half an inch.

A larger unit of length is the **kilometer** (kih-LOM-ih-ter), abbreviated **km.**

$$1 \text{ km} = 1000 \text{ m}$$
$$1 \text{ m} = 0.001 \text{ km}$$

perspective 2–3 Volume of a Solid

A line has length, but a solid occupies space. The space occupied by a solid is called its volume. A regular solid has three dimensions (length, width, and thickness) and its volume is found by multiplying these three dimensions. The volume of a solid is expressed in cubic centimeters, abbreviated cm^3. If the solid is not regular in shape (that is, if its three dimensions cannot be easily measured), its volume can be found by measuring the volume of water it displaces in a graduated cylinder.

There is no standard of volume such as a standard cubic meter or a standard cubic centimeter. This is because volume is a product of three lengths, all of which are standardized. Therefore a standard of volume is unnecessary.

perspective 2–4

Volume of a Liquid

The unit of liquid volume in the metric system is called the **liter** (LEE-ter). One liter, abbreviated **l,** is the volume occupied by one kilogram of water. Thus a liter of water has a mass of one kilogram, which is the standard mass.

A liter is a little larger than a quart. That is, a liter of milk bought in France is a little more than a quart of milk bought in the United States. For many purposes scientists find that a convenient liquid volume is one thousandth of a liter, called a milliliter (abbreviated ml). Hence,

$$1 \text{ liter} = 1000 \text{ milliliters}$$

or

$$1 \text{ l} = 1000 \text{ ml}$$

The volumes of 1 cm^3 and 1 ml are equal.

perspective 2–5

Density

Both mass and volume are properties of matter that can easily be measured. Clearly, different objects made of the same material may have different masses and also different volumes. For example, one piece of aluminum could have a mass of 85 g and a volume of 32 cm^3, and another piece could have a mass of 6.7 g and a volume of 2.5 cm^3.

The ratio of mass to volume for a material is called its density. Thus, the density of a substance is computed from its mass and volume. The units of density are grams per cubic centimeter (g/cm^3) or grams per milliliter (g/ml), depending upon whether the substance is a solid or a liquid.

Again it should be emphasized that density is independent of the amount of mass or volume. Whatever the mass or volume of a piece of aluminum, for example, its density always has the same value. That is why density is so important. Mass is a variable property of matter and so is volume. But density is a characteristic property of matter that does not vary, or varies only very slightly with change in temperature. The volume of a solid or liquid increases slightly when a given mass of it is heated. As a result, the density decreases slightly as the temperature rises. However, the density changes are

Figure 2–18
The densities of common solids and liquids.

Solids		*Liquids*	
Substance	*Density in g/cm³*	*Substance*	*Density in g/ml*
aluminum	2.7	benzene	0.88
copper	8.9	ether	0.71
cork	0.2	ethyl alcohol	0.79
gold	19.3	gasoline	0.70
ice	0.92	glycerine	1.26
iron	7.9	mercury	13.6
magnesium	1.7	sulfuric acid	1.84
lead	11.3	water	1.00
oak	0.8		
pine	0.6		
platinum	21.4		
silver	10.5		
sodium	0.97		
tin	7.3		
uranium	18.7		
zinc	7.1		

so slight that for ordinary purposes they can be disregarded. For this reason it is possible to have a table of densities of various solids and liquids. (See Figure 2–18.)

It should now be apparent that knowing the density of a metal helps to identify it. *Example:* Suppose you find that the mass of a metal is 25.4 g and that its volume is 9.4 cm³. Obviously neither the mass (25.4 g) nor the volume (9.4 cm³) tells us anything about the nature of the metal. However, by computation the density of the metal is

$$\frac{25.4 \text{ g}}{9.4 \text{ cm}^3} = 2.7 \text{ g/cm}^3$$

A glance at the table shows that the metal could be aluminum. It certainly could not be platinum or iron. But it might be an alloy (a solid mixture) of magnesium and tin. In other words, additional tests would be required to finally establish the identity of the metal.

perspective 2–6 Iron and Aluminum

Iron and aluminum are the most plentiful metals in the earth's crust. Aluminum is about twice as plentiful as iron. Both metals are used on a vast scale in construction work. Skyscrapers (Figure 2–19) are supported by a framework of steel, which is about 90 percent iron, and wooden houses are held together by iron nails. Airplanes are built of aluminum alloys (Figure 2–20) since aluminum is a lightweight metal. As you see from the density table, the density of aluminum is only about one-third the density of iron.

Figure 2–19
Skyscrapers are supported by steel, which is strong and inexpensive.

Figure 2–20
Because of its low density, aluminum is used in airplanes.

The two metals have very different histories. Iron has been in use for thousands of years. The famous Damascus swords made the Syrian army almost unbeatable as early as 1500 B.C. Moreover, in the Bible there are frequent references to iron tools such as axes and saws and nails. For instance, in the Book of Chronicles, written about 1000 B.C., there is a reference to King David's plan for building a temple. It reads, "And David prepared iron in abundance for the nails for the doors of the gates, and for the joinings." On the other hand, there is no mention of aluminum in the Bible. In spite of its abundance, aluminum was not discovered until the 19th century. Iron has been known for more than 3000 years, but aluminum for less than 200 years. Why was iron discovered so early and aluminum so late?

Figure 2–21
Molten iron from a blast furnace can be poured.

Neither iron nor aluminum occurs as a metal in nature. Instead, they occur in minerals called ores. The metals had to be extracted from their ores. The ancient metal workers found it was fairly easy to extract iron from iron ore. All they had to do was to heat iron ore with charcoal. When the temperature was high enough a chemical reaction took place. The iron in the ore did not combine with the charcoal but the other substances did. In this way iron was separated from the ore and could be poured off as molten iron. Indeed, iron is still obtained in this way (Figure 2–21).

However, charcoal would not release aluminum from aluminum ore no matter how strongly the mixture was heated. So an entirely new method had to be devised. When electricity became available, it was found that aluminum could be released from its ore by an electric current. Powerful electric currents are produced by electric generators. The generator was not invented until 1840. In 1886 an American chemist named Charles Martin Hall discovered the electrical process by which aluminum could be extracted from its ore. The year 1886 marks the birth of aluminum and the beginning of the vast aluminum industry.

QUESTIONS

A

1. Put the ten words below into three lists: "Mass," "Length," and "Volume." A word may be placed in more than one list.

liter	cubic centimeter
kilogram	density
meter	milliliter
cylindrical solid	graduated cylinder
ruler	equal-arm balance

2. How would you change a given number of meters to centimeters?

3. How would you change a given number of grams to kilograms?

4. What is the word that expresses the constant mass/volume ratio of a given substance?

5. There are standard units of mass and length, but no standard unit of volume. Why not?

6. How would you change a given number of cubic centimeters to liters?

7. How many cubic centimeters of alcohol can be placed in a container whose capacity is 0.500 l?

B

8. Two 20-g masses are on the right-hand pan of a double pan balance. One rider of the balance is on the 10-g mark and the other rider is on the 3-g mark. A beaker has a mass of 68 g. Is the left-hand pan of the balance up or down (a) before the beaker is placed on the left-hand pan of the balance, (b) after the beaker is placed on the left-hand pan of the balance?

9. When the volumes and lengths of two regular solid objects are compared, one of the lengths of the first object is greater than any of the lengths of the second object, yet the volume of the first object is smaller than that of the second object. How can this be explained?

10. The density of lead is 11.3 g/cm^3. If the density of an unknown solid is found to be 11.3 g/cm^3, what can you say about the solid?

11. 20 ml of ethyl alcohol are mixed with 100 ml of ethylene glycol. After thorough mixing, the density of the mixture is determined. Is the resulting density greater, the same, or less than the density of the alcohol? Explain.

12. In Figure 2–22, A and B are the mass-volume graphs of two solids. Which has the greater density, A or B? Why?

13. In Figure 2–23, A and B are the mass-density (not mass-volume) graphs of two metals. Taking a mass of 80 g of each metal, which one has the greater volume? Why?

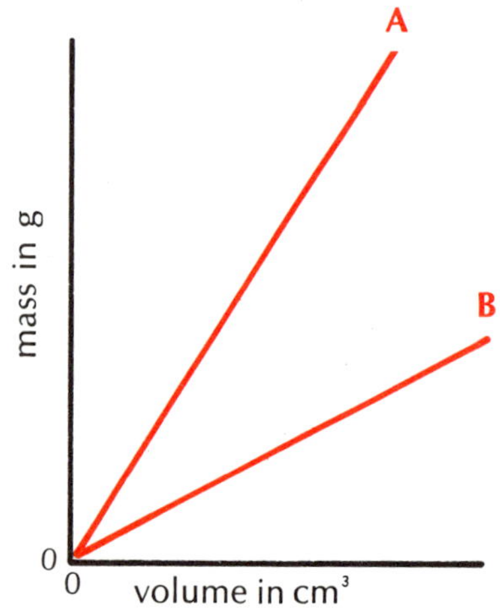

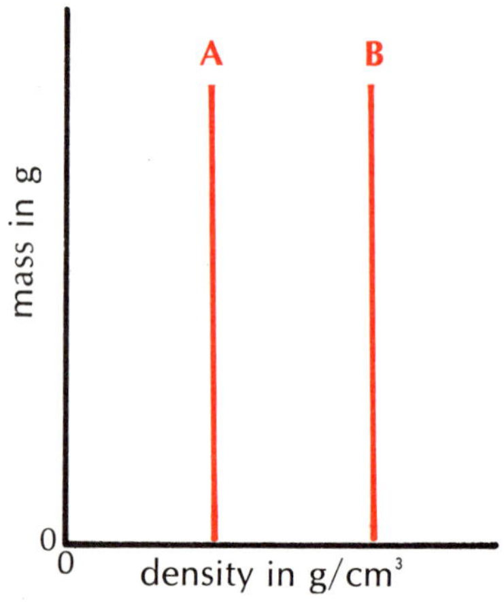

Figure 2–22 (left) Graph for Question 12.

Figure 2–23 (right) Graph for Question 13.

PROBLEMS

1. Calculate the density of a rock if a 25.0 cm^3 sample of it has a mass of 70.0 g.

2. The density of iron is 7.9 g/cm^3. What is the mass of a cube of iron that is 2.5 cm on an edge?

3. The density of aluminum is 2.7 g/cm^3. What volume does 100 g of aluminum occupy?

4. What is the mass in grams of 500 ml of water?

5. Find the volume of 10.0 g of gasoline whose density is 0.70 g/ml.

6. What mass of ether can be held in a 130-ml flask? The density of ether is 0.71 g/ml.

7. 100 ml of water are required to fill a bottle. 125 g of another liquid will fill the same bottle. Compute the density of the other liquid.

8. (a) From the graph you made in Experiment 2–4, estimate the mass of 43 cm^3 of the substance. (b) Can you estimate the mass of 100 cm^3 of the substance? If so, what is your estimated value?

9. The edge of a cube of aluminum is 5.0 cm. The base of a rectangular block of aluminum is 8.0 cm by 4.0 cm. What is the height of the rectangular block if its volume is the same as the volume of the cube?

10. Compare the volumes of pieces of aluminum and iron, both of which have a mass of 100 g.

11. Compare the masses of pieces of aluminum and iron, both of which have a volume of 100 cm^3.

12. A cylinder has a height of 10.0 cm and a diameter of 4.0 cm. A rectangular solid has a height of 10.0 cm, a length of 6.0 cm, and a width of 7.0 cm. Which of the solids has the greater volume?

13. A block of metal has a mass of 90 g and a volume of 12.5 cm^3. What metal might it be?

Chapter 3

FORCE, WORK, AND ENERGY

When a golf club strikes a golf ball, it exerts a force that changes the shape of the ball.

Sixty years ago there were no paved or graveled roads in the farming communities of northern Missouri. In rainy weather the rich black soil turned into a sea of mud that became badly rutted when the rain stopped and the soil began to dry out. The townships did not have equipment for keeping the road surface smooth, but farmers were paid by the township to "drag" the roads near their land. They received 25 cents per hour for a two-horse team and 35 cents per hour for a three-horse team.

The horse-pulled road drag was a heavy wooden device, about eight feet long and three feet wide. It scraped the tops off the ruts and filled in the hollows.

One morning a ten-year-old Missouri farm boy asked his father for permission to "drag" the road (with the understanding that the payment by the township would go to the boy instead of to his father). The father agreed to this arrangement.

The boy got three horses (Lady, Queen, and Topsy) from the pasture, curried and fed them, and harnessed them. He then assembled the three-horse hitch, attached it to the drag, and hitched the horses to it. Lady and Queen were attached to the long arm of the hitch and Topsy to the short arm. The arrangement is shown in Figure 3–1.

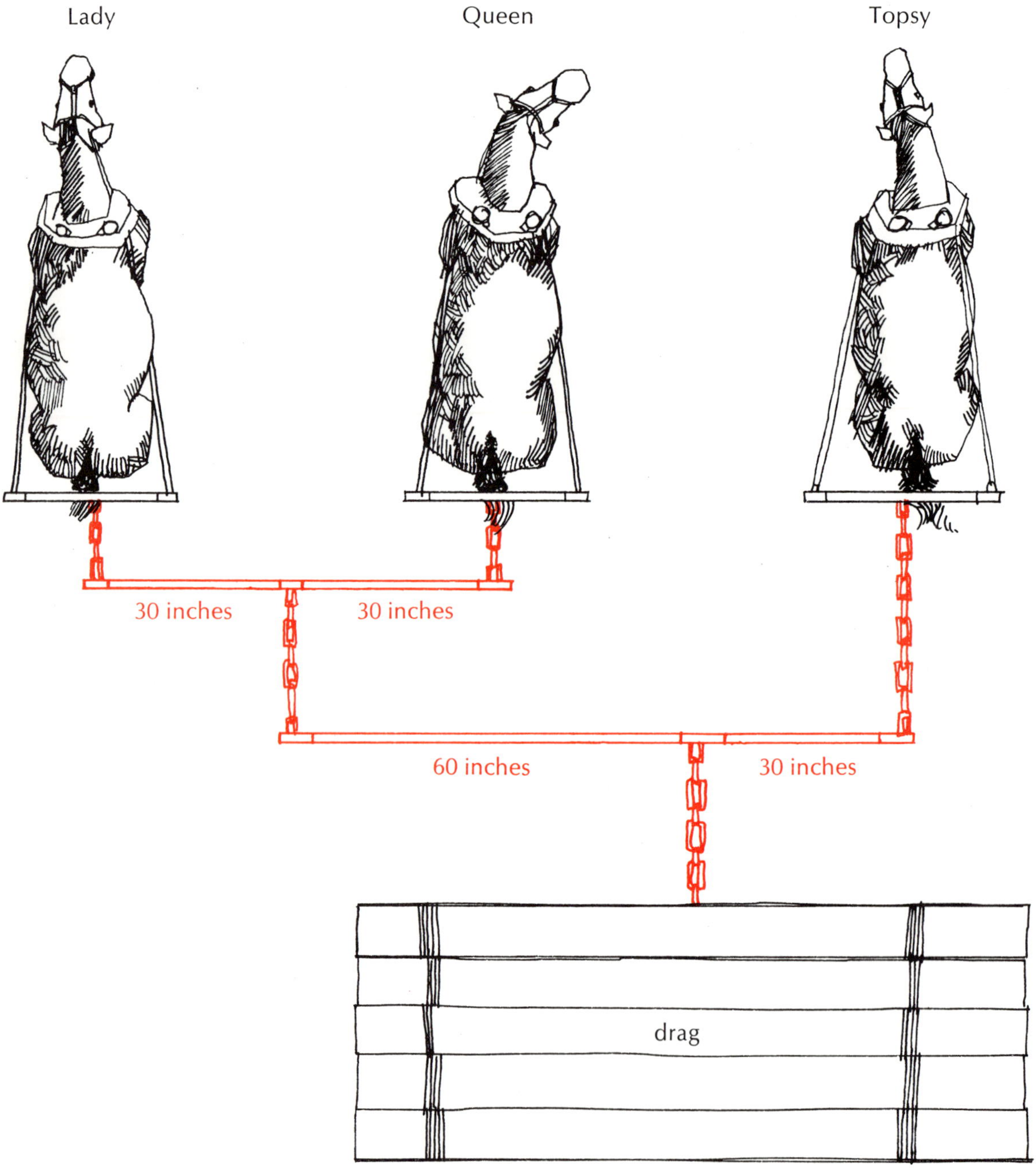

Figure 3–1
The horses were hitched to the road drag like this. What was wrong?

After going about 200 yards, the boy noticed that Topsy was sweating and puffing, but that Lady and Queen were fresh and lively. He stopped to rest Topsy and to think about the problem. What was wrong?

About ten minutes later the boy thought of a solution and put it into operation. It worked, and Lady and Queen did their fair share of the work. What was his solution?

When you have done the experiments in this chapter, you too should be able to figure out why Topsy was so tired.

FORCE

What Is Force?

If you push on a lawn mower hard enough, it moves. If you pull on an open door, it moves too. A push or pull is called a **force.** A force applied to an object usually causes it to move. However, pushes and pulls are not the only kinds of force. There are many others. For instance, if weights are hung on a wire, the wire stretches. This is because the weights pull on it. When the weights are removed, the wire becomes shorter again, as it was before the weights were put on. Why? There must be a force in the wire when it is stretched, causing it to move in a direction opposite to the direction the weights are pulling. This force is called **elasticity** (ee-las-TIS-ih-tee). The force of elasticity pulls the wire back to its original position when the stretching force (the weights) is removed. In the following experiment you will observe the amount of force needed to stretch a spring.

experiment 3–1 The Force to Stretch a Spring

Materials meter stick; tall stand; 2 clamps; spring; weight support; 3 sets of 5 large washers

Procedure Clamp a meter stick to a stand with the 100-cm mark of the stick on the base of the stand. Attach a second clamp to the stand

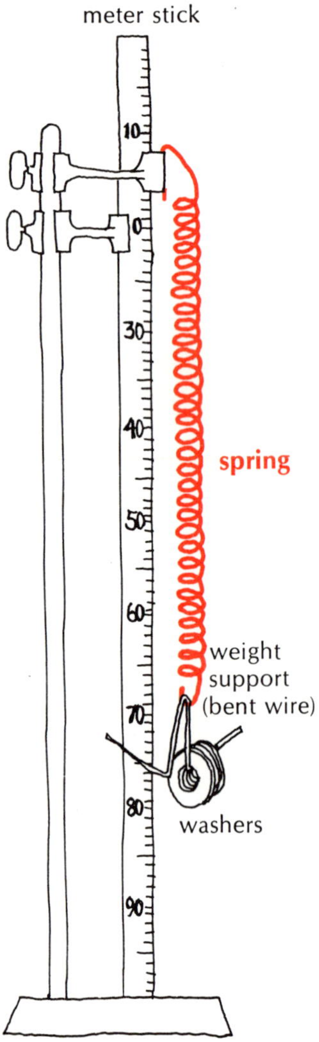

Figure 3–2
Wait until the spring stops moving before making a reading.

and hang a spring from this clamp. Now attach a weight support to the spring. The arrangement is shown in Figure 3–2. Note the reading of the meter stick opposite the bottom of the spring and record it as the first reading of the spring.

Suspend a set of five washers from the weight support attached to the spring. Record the meter stick reading at the bottom of the spring. Repeat this procedure, adding washers in groups of five until fifteen washers are suspended from the spring. Then remove the washers in groups of five and record the meter stick reading each time. Record your data in a table like that shown in Figure 3–3.

Construct a graph like that in Figure 3–4. The vertical axis shows the increase in length in centimeters, while the horizontal axis shows the number of washers. The origin of the graph should have a value of zero on both axes.

Number of Washers	Reading of Meter Stick	Total Change in Length of Spring
0	?	?
5	?	?
10	?	?
15	?	?
10	?	?
5	?	?
0	?	?

Figure 3–3
Data table for Experiment 3–1.

Questions

1. (a) Describe the line of the graph. Is it curved or straight? (b) What does the line of the graph indicate?

2. Compare the increase in length when washers are added with the decrease in length when washers are removed.

LEVERS, GRAVITY, AND WEIGHT

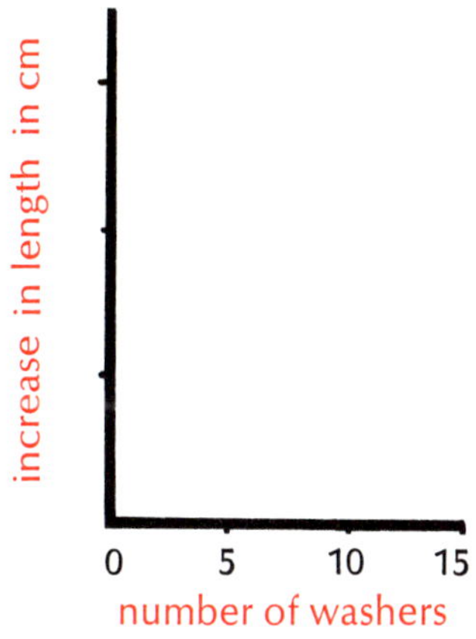

Figure 3–4
Graph axes for Experiment 3–1.

The Lever

Even in earliest times men were faced with the problem of moving large rocks or lifting heavy loads. So they invented simple machines. A **machine** (mah-SHEEN) is a device that makes it easier to do work. Some machines help a person work faster. Others make it possible to move a heavy load by exerting only a small force.

One of the first machines was the **lever** (LEV-ur). A lever is simply a bar that can rotate about a point of support, usually called a **fulcrum** (FULL-krum). In the next experiment you will use a meter stick as a lever.

experiment 3–2

The Center of Gravity of a Lever

Materials

meter stick; loop of thread; stand; clamp

Procedure

Place a loop of thread around a meter stick and attach the loop to a clamp on a stand, as shown in Figure 3–5. Hang the stick so that you can see the centimeter scale from the front. Carefully move the thread back and forth until you find a position where the stick hangs level without moving. This position is called the center of gravity. The **center of gravity** of a body is the point where all the weight of a body appears to be concentrated. It is the point of balance.

Record the position of the center of gravity of the meter stick.

Questions

1. Is the center of gravity exactly where you expected it to be?
2. If not, why not?

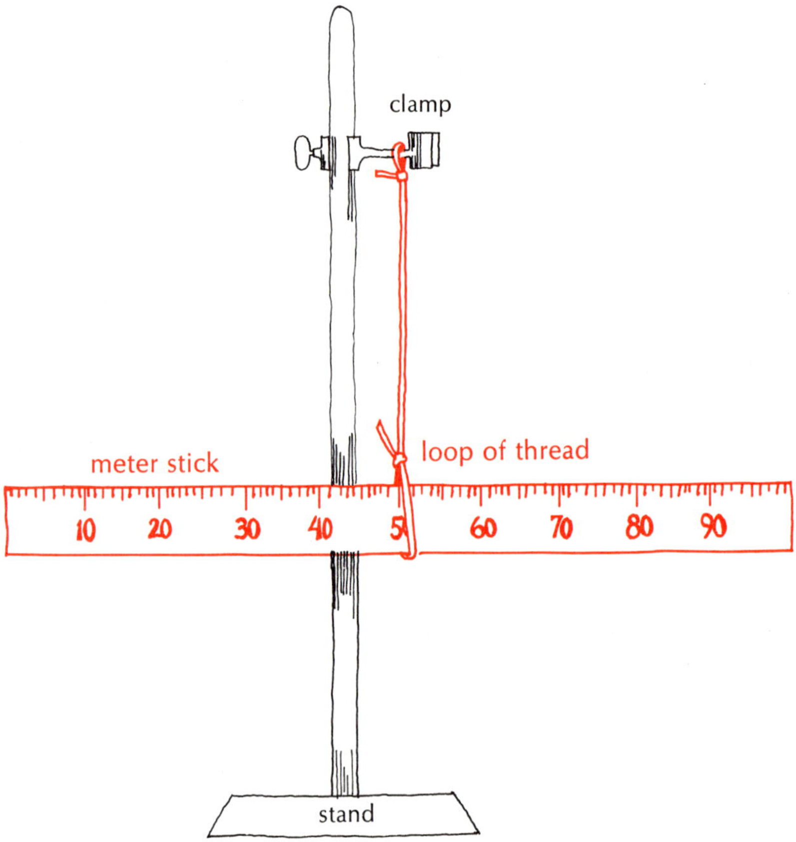

Figure 3–5
The meter stick is balanced at the center of gravity.

The Force of Gravity

If you let go of a ball held in your hand, you know that it will fall to earth. Why does it fall? The earth must attract the ball. Indeed, the earth attracts all bodies on the earth. This earth-pull is called the force of **gravity** (GRAV-uh-tee). The word *gravity* comes from a Latin word meaning "heavy."

In the 17th century Isaac Newton, an English scientist, discovered many things about this force. For example, he discovered that the greater the mass of a body the greater is the earth's force of gravity that acts upon it. The gravitational force that acts on a body is called its **weight.**

Mass and Weight

Mass and weight are quite different. The mass of an object depends only upon the amount of matter in the object. The mass is therefore constant, no matter where the object is located. On the other hand, the weight of an object varies, depending upon where it is relative to the surface of the earth.

Since the concepts of mass and weight are different, the units of mass and weight are also different. The unit of weight or force in the metric system is called the **newton** (NEW-tun) in honor of Isaac Newton. (The abbreviation for newton is **N**.) As you already know, the unit of mass is the kilogram.

The Effect of a Force on a Lever

You know that you can open a door by pushing on the doorknob. As the door opens, your hand moves through a circular path. Similarly, a lever rotates if a force is applied at a distance from the fulcrum. The direction of rotation may be clockwise or it may be counterclockwise. As shown in Figure 3–6, if the force is applied downward to the right of the fulcrum, the rotation is clockwise. On the other hand, if the force is applied downward to the left of the fulcrum, the rotation is counterclockwise.

The turning effect of a force is called **torque** (TORK). Torque depends upon two factors, the amount of force applied, and the distance from force to fulcrum. How do these two factors affect

Figure 3–6
A downward force applied to the right of the fulcrum produces a clockwise rotation. What are two ways of producing a counterclockwise rotation?

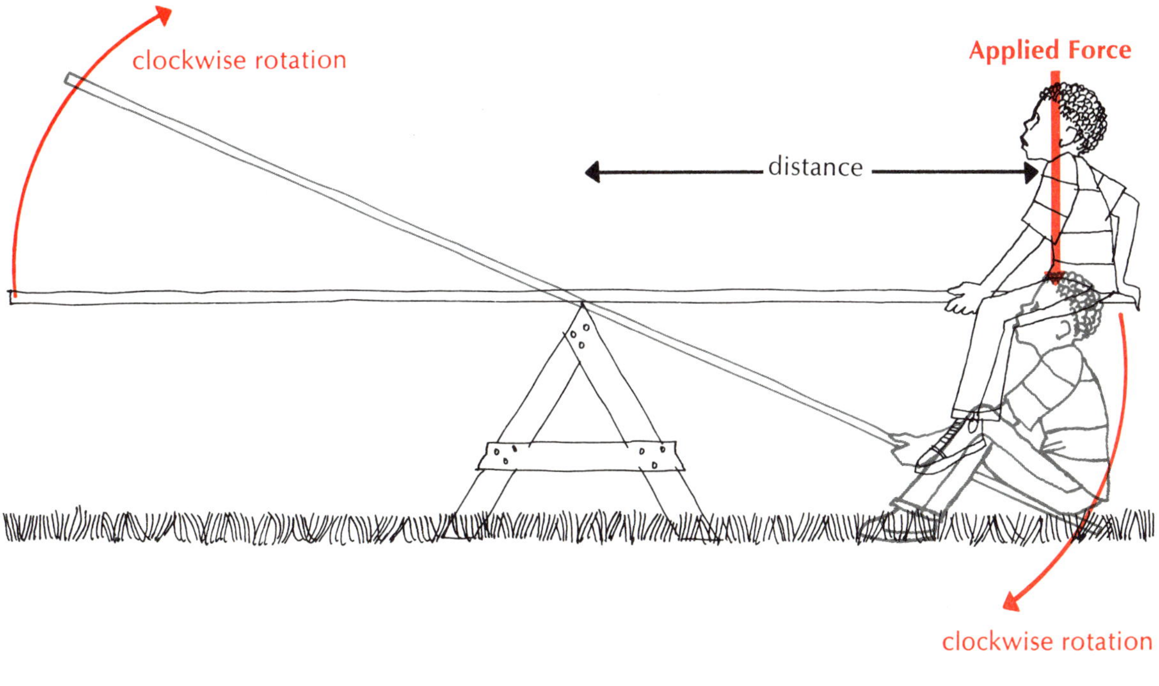

the torque? Will the torque become larger or smaller if you increase the force? What happens to the torque if the distance is changed? Would a door be easier to open if the doorknob were moved closer to the hinges?

experiment 3–3 A Lever with Two Equal Forces

Materials

meter stick; loop of thread; stand; clamp; 2 sets of 5 washers each; spring scale

Procedure

Set up the balanced meter stick as in Experiment 3–2. Hang sets of five washers by loops of thread, one set on each side of the center of gravity position. Adjust the thread loops until the lever and suspended forces are in balance or, as it is usually called, in **equilibrium** (ee-kwi-LIB-ree-um).

Draw a large neat diagram of the lever in your notebook. Record the data as shown in Figure 3–7.

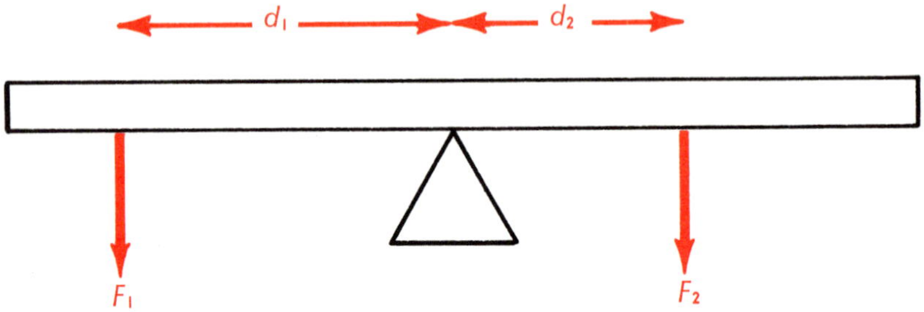

Figure 3–7
Record your forces and lengths on a diagram like this.

Weigh each set of washers with a spring scale. Record these weights in newtons as F_1 and F_2. Also record the lengths of the two lever arms d_1 and d_2. These lengths should be recorded in meters. Remember that 100 cm = 1 m, so 1 cm = 0.01 m.

Since a torque involves force and distance, the unit of torque also involves force and distance. It is called the **newton-meter,** abbreviated to **N · m.**

Example: If a force of 5 N is applied at a point 40 cm (0.40 m) from the fulcrum,

$$\text{torque} = 5 \text{ N} \times 0.40 \text{ m}$$
$$= 2.0 \text{ N} \cdot \text{m}$$

Computations

1. What is the clockwise torque on your lever? (Multiply the value of F_2 by the value of d_2.)

2. What is the counterclockwise torque? ($F_1 \times d_1$)

Questions

1. How do these torques compare with each other?

2. What conclusion can you draw concerning clockwise and counterclockwise torques when the lever is balanced at its center of gravity?

experiment 3–4 A Lever with Two Unequal Forces

Materials

meter stick; 2 loops of thread; stand; clamp; set of 5 washers; set of 2 washers; spring scale

Procedure

Using the same arrangement as in Experiment 3–3, hang a set of five washers on one side of the fulcrum and a set of two washers on the other side. Move the loops of thread back and forth until balance is restored. Draw a large scale diagram of the lever and enter your data as in Experiment 3–3.

Computations

1. Compute (a) the clockwise torque, and (b) the counterclockwise torque.

2. How do the computed torques compare with each other?

3. Consider the lever in Figure 3–8. (a) What is the clockwise torque? (b) What is the counterclockwise torque?

4. What is the force F_2 shown in Figure 3–9?

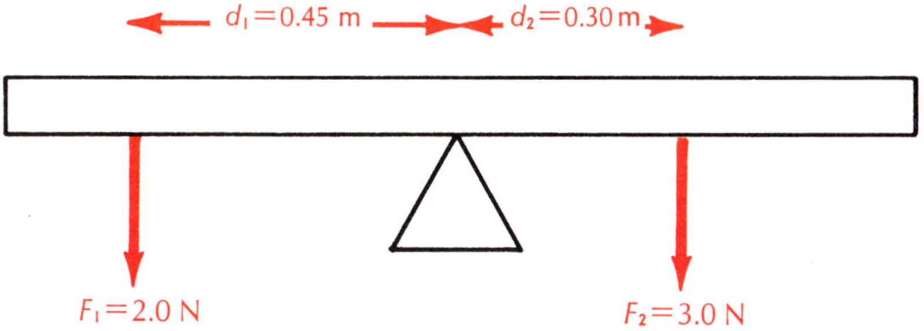

Figure 3–8
Diagram for Computation 3.

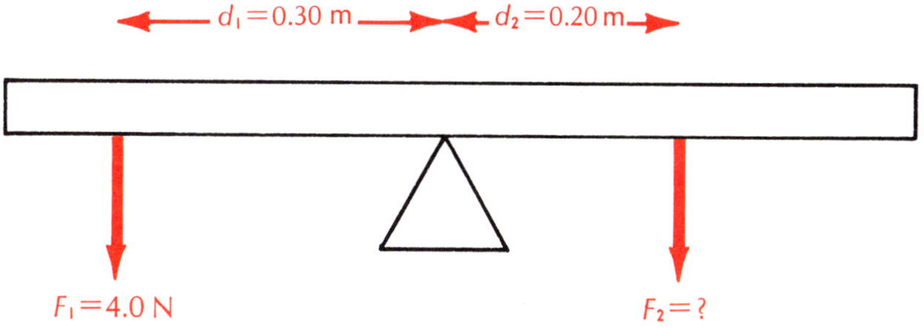

Figure 3–9
Diagram for Computation 4.

experiment 3–5

A Lever with Three Forces

Materials

meter stick; 3 loops of thread; stand; clamp; 3 sets of 5 washers; set of 2 washers; spring scale

Procedure

Use the same basic arrangement as in Experiment 3–4. Tie together two sets of five washers to make a set of ten. Hang this set of ten on one side of the fulcrum. On the opposite side of the fulcrum hang a set of five washers from one point on the meter stick and also a set of two washers from a different point. Again adjust the loops of thread until balance is restored. Enter the data upon a large diagram of the lever.

Computations

1. Compute the torque produced by the five washers and the torque produced by the two washers. Then find their sum.

2. Compute the torque produced by the largest force, the set of ten washers.

3. How does the torque produced by the largest force compare with the sum of the torques produced by the two smaller forces?

Questions

1. A lever that does not rotate is said to be in **rotational** (roh-TAY-shun-ul) **equilibrium.** What must be true if rotational equilibrium is to be maintained in a lever?

experiment 3–6

To Find the Mass of a Meter Stick by Torques

Laboratory Test

Materials

meter stick; loop of thread; stand; clamp; set of 5 washers; spring scale; equal-arm balance

Procedure

Record the number of your meter stick. Find the center of gravity of the meter stick and record this position in your notebook. Now hang the stick by a loop of thread at the 25.0-cm mark. Using only one set of five washers, balance the stick. Record all the data on a lever diagram.

Computations

1. Using what you learned from the last three experiments, compute the weight of the stick in newtons.

2. From the value computed above, calculate the mass of the stick. (Remember that a mass of one kilogram has a weight of 9.8 newtons. See Perspective 3–6.)

3. Now find the mass of the stick by means of an equal-arm balance. Compute the percentage difference between the two mass values for the stick. (You learned about percentage difference in Experiment 2–5.)

experiment 3–7 To Find the Weight of a Rock by Torques

Laboratory Test

Materials

meter stick (the same stick as used in Experiment 3–6); loop of thread; stand; clamp; rock

Procedure

Using the meter stick whose weight you computed in the previous laboratory test, find the weight of the rock given to you by your teacher.

WORK

What Is Work?

The effect of applying a force depends upon the circumstances. For instance, a force can stretch a wire or a force can rotate a lever about a fulcrum. In both cases the force is said to do work. Work is done when a force causes a body to move. Or, stated another way, **work** is done when a force is exerted through a distance. That is,

$$\text{work done} = \text{force} \times \text{distance}$$

In symbols,

$$W = F \times d$$

However, the distance must be measured in the direction in which the force is applied. Thus, if a force (F) pushes an object a distance (d) as shown in Figure 3–10, the work done is $F \times d$.

It is important to observe that the word *work,* when used in a scientific sense, does not necessarily mean the same as physical labor. For example, you may push hard on a car stuck in a snowdrift. But if you fail to move the car you have done no work even though the pushing has made you tired.

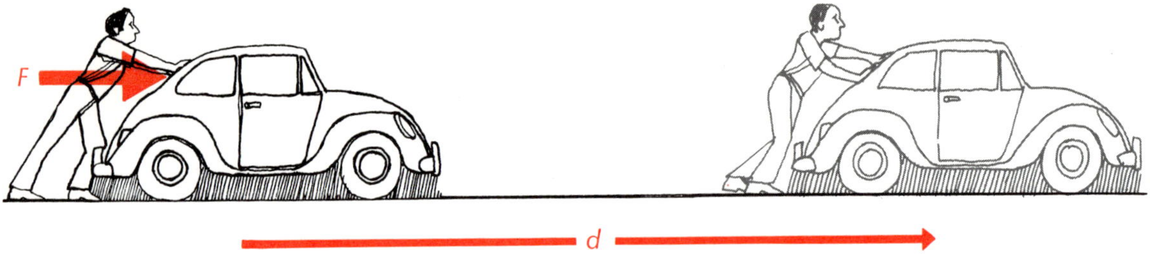

Figure 3–10
Work involves both force and distance.

Figure 3–11
Pushing this box took one newton-meter of work.

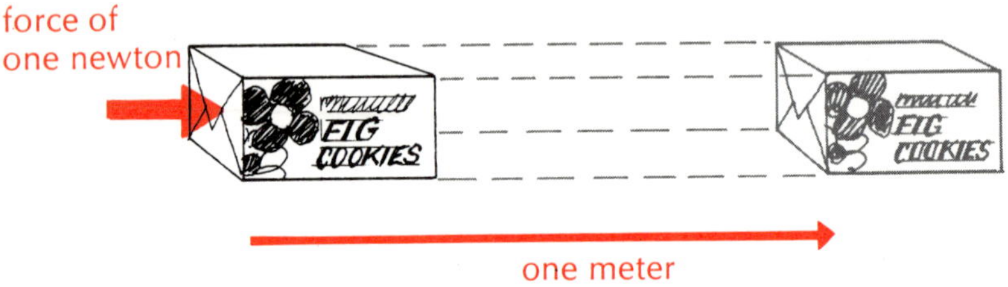

The unit of work must include the unit of force (a newton) and the unit of distance (a meter). It is called a **newton-meter,** abbreviated **N·m.** Thus, a newton-meter of work is the amount of work done if a force of one newton is applied through a distance of one meter. This is shown in Figure 3–11. Another name for a newton-meter is a **joule** (JOOL). You will investigate the concept of work in Experiments 3–8 and 3–9.

experiment 3–8 The Work Done in Lifting a Textbook

Materials loop of thread; 20-newton spring scale

Procedure Place a loop of thread between the pages of your textbook and pull the thread back against the binding. Hook a spring scale to the loop and lift the book free of the bench. Read and record the weight of the book in newtons.

Computations **1.** If the book is lifted vertically a distance of 2 m, how much work is done?

weight on the horizontal axis. The origin should represent zero values of friction and weight.

Number of Books	Weight of Books in Newtons	Force of Friction in Newtons
1	? N	? N
2	? N	? N
3	? N	? N
4	? N	? N

r Experiment

1. What happened to the force needed to pull the books as the number of books increased?

2. Can the graph be used to predict the force needed to pull more than four books? Explain.

3. How does the weight of a body affect the force needed to move the body horizontally across the surface of another body?

ment

The Force of Friction and the Area of Contact between Surfaces

loop of thread; spring scale; cardboard (30 cm by 25 cm); centimeter ruler; scissors; double-coated masking tape

re

In Experiment 3–10 the force pressing the two surfaces together was changed, while the area of contact between the surfaces remained unchanged. Now you will keep the force constant but vary the area of contact.

In this experiment you will pull pieces of cardboard of different sizes along the table. Each piece of cardboard will be loaded with a book to make sure there is enough friction to measure. Attach a loop of thread to a book as in Experiment

Figure 3–12
Pull slowly and steadily.

experiment 3–9

The Work Done in Pulling a Textbook along a Horizontal Surface

Materials

loop of thread; spring scale

Procedure

Place the textbook with its attached loop of thread on the table. Hook a spring scale in the loop and pull the book along the table at a constant speed. Keep the scale close to the table and pull parallel to the table. This arrangement is shown in Figure 3–12.

Record the reading of the spring scale when the book is moving at constant speed.

Computations

1. Compute the amount of work done if the book is pulled along the table a distance of 2 m.

2. Compare the force required (a) to lift the book, and (b) to pull the book horizontally along the table. (c) Are these forces the same or different? Why?

3. Compare the work done in (a) lifting the book through a vertical distance of 2 m and (b) pulling the book through a horizontal distance of 2 m.

FRICTION

What Is Friction?

You have already explored three different kinds of forces — a push or pull, elasticity, and the force of gravity. As you have shown, a force is needed to move a book along the top of a table. When you apply such a force, you must be overcoming a force in the opposite direction, just as when you lift a book, you overcome its weight. The force that opposes the book's motion is called **friction** (FRIK-shun). This is shown in Figure 3–13.

What causes friction? Friction is a force between two surfaces in contact. It is caused by several factors, the most important being the interlocking of the two surfaces. No matter how smooth a surface appears to be, there are slight irregularities in it, tiny ridges and hollows. As shown in Figure 3–14, ridges on each surface are meshed into hollows on the other. Thus if a force tries to pull one surface over another, it will be opposed by a force resulting from the interlocking effect.

When you lift an object, you do work against the weight of the object. The applied force is then equal to the weight of the object. The work done equals the applied force multiplied by the distance through which the object is raised. However, if you pull the object along the surface of a table, you no longer do work against the weight. This is because the direction of the weight is at right angles to the direction of the motion. Instead, you do work against the force of friction because the direction of friction is along the same line as the applied force. The three forces (weight, friction, and applied force) are shown in Figure 3–13.

You will investigate the force of friction in the next four experiments.

experiment 3–10

The Force of Friction and the Weight of a Body

Materials spring scale; loop of thread; 4 textbooks

Procedure Four textbooks will be needed for this experiment, so you should work in groups of four.

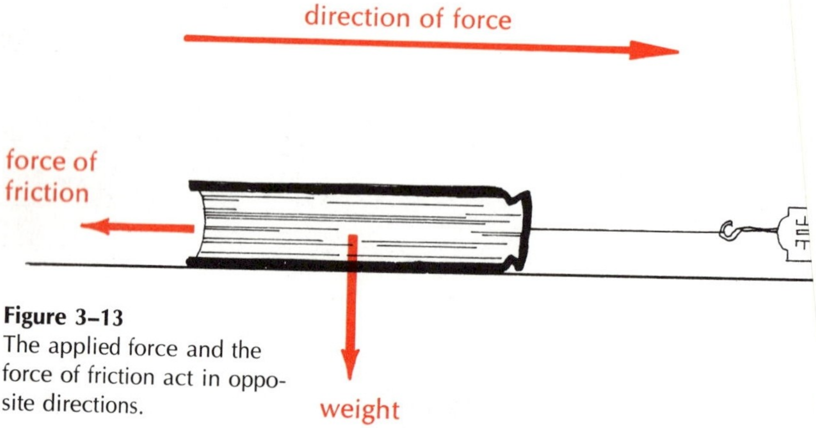

Figure 3–13
The applied force and the force of friction act in opposite directions.

Figure 3–14
Every surface is slightly rough. This roughness is the chief cause of friction.

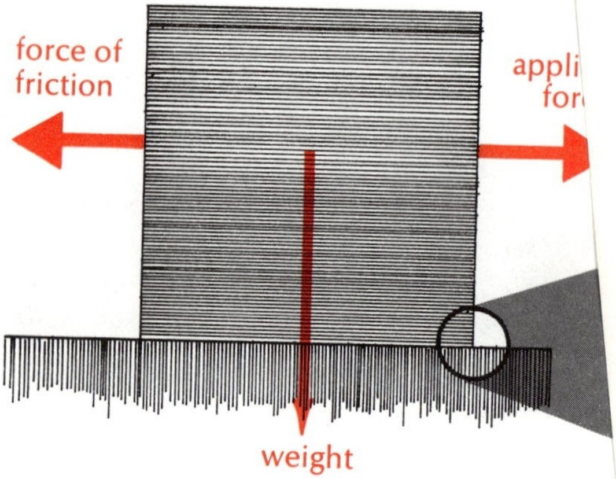

Find the weight of a book by means of
record the weight. Then, using a loop of th
along the table at uniform speed and record
as registered on the spring scale. Now place
top of the first, drag the two books along t
record the force of friction.

Continue this procedure until a stack of fo
dragged along the table. (Assume that all t
same weight.) Record the data in a table like
3–15.

Construct a full-page graph of the data,
Figure 3–16. Plot the force of friction on the

Figure 3–1
Data table
3–10.

Questic

expe
3–1

Mate

Proc

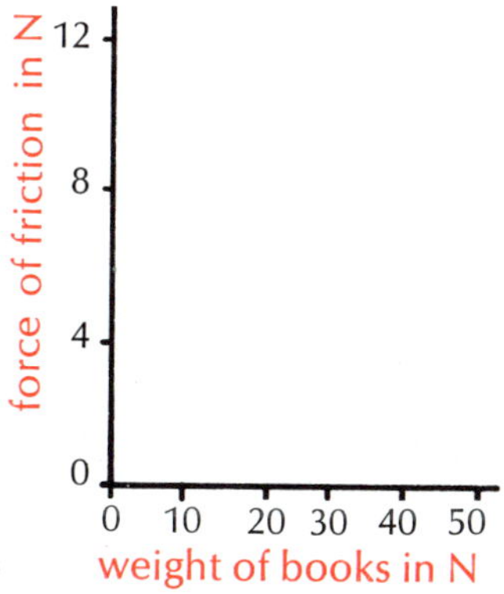

Figure 3–16
Graph axes for Experiment 3–10.

3–10. Place the book on a piece of thick cardboard about 30 cm long and 25 cm wide. (It may be necessary to place a piece of double-coated masking tape between the book and cardboard to prevent the book from sliding over the cardboard surface.)

Pull the book and cardboard across the table at a steady speed by means of a spring scale. Record the scale reading. This reading is the force of friction between cardboard and table. Measure the length and width of the cardboard, compute the area, and record the value of the area.

Now cut the cardboard in half. Using one of the pieces, repeat the procedure described in the last paragraph.

Again cut one of the smaller pieces in half and again repeat the procedure with one of the smallest pieces.

Record your data as in Figure 3–17.

Figure 3–17
Data table for Experiment 3–11.

Size of Cardboard	Area of Contact	Force of Friction
Large piece ? cm X ? cm	? cm^2	? N
Smaller piece ? cm X ? cm	? cm^2	? N
Smallest piece ? cm X ? cm	? cm^2	? N

Questions

1. Compare the forces of friction in the three trials.

2. On the basis of your data, what effect does area of contact have on the force of friction?

experiment 3–12

The Force of Friction and the Nature of the Surfaces

Materials

loop of thread; spring scale; waxed paper; bath towel

Procedure

In this experiment, both the area of contact and the force pressing the two surfaces together will be kept constant, but the nature of the surfaces in contact will be changed.

First pull a book along the table by means of a loop of thread and a spring scale. Record the scale reading.

Now place a piece of waxed paper, about 30 cm wide and 125 cm long, on the table. Hold the waxed paper with one hand and pull the book along the length of the waxed paper with the other hand. Record the scale reading.

Finally, place a bath towel on the table. Hold the towel in place with one hand and find the force needed to pull the book along the length of the towel.

Record the data as in Figure 3–18.

Questions

1. On the basis of your data, how is the force of friction affected by the nature of the surface?

Figure 3–18
Data table for Experiment 3–12.

Surface	Smoothness of Surface	Force of Friction
tabletop	medium	? N
waxed paper	smooth	? N
bath towel	rough	? N

experiment 3–13

Rolling Friction and Sliding Friction

Materials

loop of thread; spring scale; 2 6-inch lengths of $\frac{1}{4}$ inch wooden dowels, or 2 round pencils

Procedure

Experiments 3–9 through 3–12 involved sliding friction, the sliding of a book along a surface. You will now investigate rolling friction and compare it with sliding friction.

Pull a book along the table by a loop of thread and a spring scale. Record the scale reading.

Then place the wooden dowels or round pencils under the book. Pull the book along the bench on the dowels or pencils and record the reading of the spring scale.

Questions

1. What conclusions can you draw from this experiment?

THE INCLINED PLANE AND EFFICIENCY

The Inclined Plane

Suppose a heavy box weighing 1500 N (about 330 lb) has to be lifted onto the tailgate of a truck 1.5 m above the ground. Only a man with extraordinary strength could exert such a force. But even an ordinary man could do the job by using a type of machine called an **inclined plane** — that is, by arranging a wooden plank with one end on the ground and the other on the truck as shown in Figure 3–19. The man would then apply a force (F) parallel to the plank and either push the box or pull on a rope tied around the box. How much force is needed to push the box up the plank?

Regardless of the length of the plank or how the force is applied, the box has to be lifted a vertical distance of 1.5 m. Therefore

$$\begin{aligned}\text{work that must be done} &= \text{weight of box} \times \text{vertical distance}\\ &= 1500\ \text{N} \times 1.5\ \text{m}\\ &= 2250\ \text{N}\cdot\text{m}\end{aligned}$$

However, as shown in Perspective 3–11, the force needed to push the box depends on the length of the plank. You will understand why after you have done Experiments 3–14 and 3–15.

Input Work and Output Work

In operating an inclined plane (or any machine for that matter), two different quantities of work are involved. First there is the work put into the machine. This is called **input work.** And there is the work accomplished by the machine. This is called **output work.** Thus, for an inclined plane,

input work = force applied × length of plane
output work = weight of load × height of plane

Notice that force and weight are both expressed in newtons, and length and height in meters. Therefore work comes out in newton-meters, which is correct.

In the following experiment you will set up an inclined plane and compare the input work and the output work.

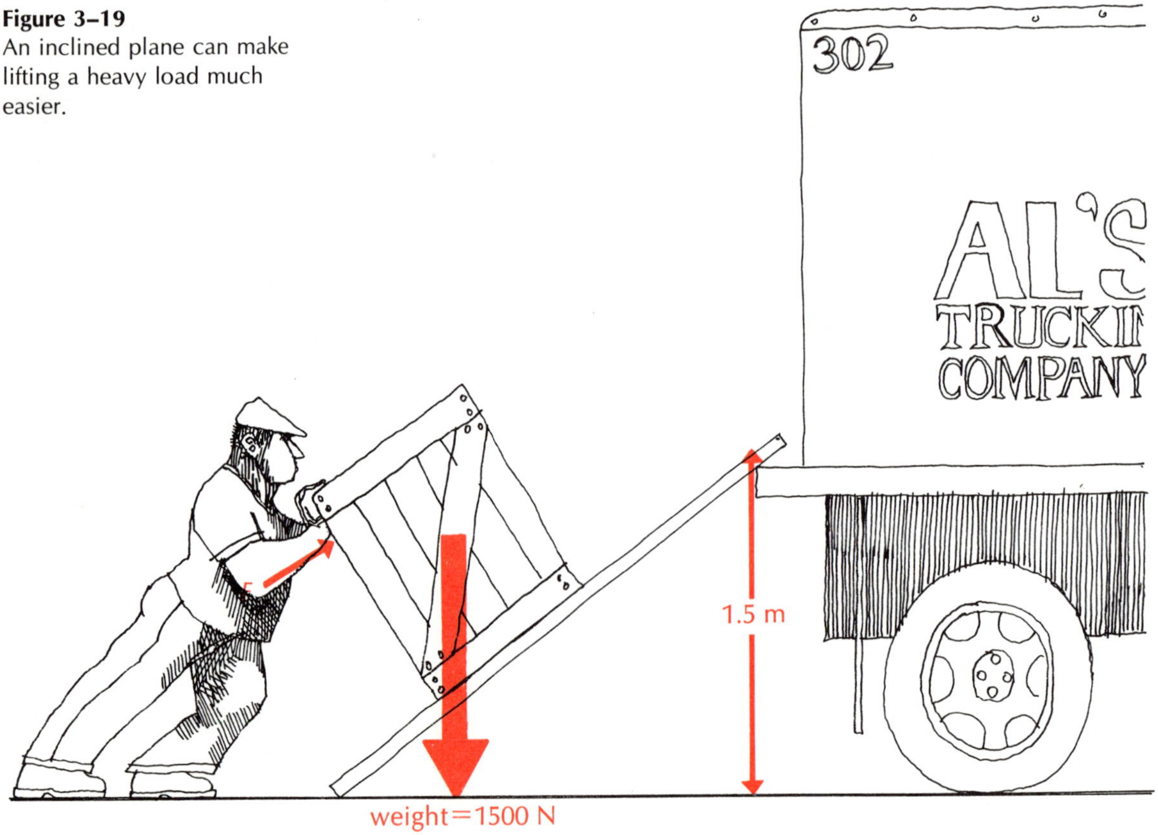

Figure 3–19
An inclined plane can make lifting a heavy load much easier.

experiment 3–14

Input Work and Output Work on an Inclined Plane

Materials

board (1 m by 15 cm by 2 cm); loop of thread; spring scale; stand; clamp; meter stick

Procedure

Place the board on the laboratory table. Weigh a textbook by means of the spring scale and record the weight in newtons. Attach a loop of thread to the book as in Experiment 3–5, and place the book at one end of the board. Using the spring scale, pull the book along the board at uniform speed. Record the scale reading. This reading represents the force of friction between the book and the board when the board is horizontal.

Now raise one end of the board to a height of 15 cm (0.15 m) and support it in this position by a stand and clamp. Then pull the book up the incline and record the reading of the spring scale.

Repeat the above procedure by supporting the end of the board at heights of 30 cm (0.30 m) and 50 cm (0.50 m). Your data should be tabulated as in Figure 3–20.

For each of the four trials, compute the input work and the output work. In each case assume that the book is moved along the whole length of the plane. Make a table of these computed values as in Figure 3–21.

Figure 3–20
Data table for Experiment 3–14.

Weight of book (W) = ? N
Length of board (L) = ? m

Height of Board (H)	Spring Scale Reading (F)
0 m	? N
0.15 m	? N
0.30 m	? N
0.50 m	? N

Figure 3–21
Computations for Experiment 3–14.

Height of Board (H)	Input work (FXL)	Output work (WXH)
0 m	? N·m	? N·m
0.15 m	? N·m	? N·m
0.30 m	? N·m	? N·m
0.50 m	? N·m	? N·m

Questions

1. In each of the four trials, how does the applied force compare with the weight of the book?

2. In each of the four trials, how does the length of the incline compare with the height the book is raised?

3. In each of the four trials, how does the input work compare with the output work?

Work and Efficiency

In all machines there are frictional forces that must be overcome. Consider again the box being pulled up the plank. Work is done in lifting the box, but work is done against friction also. Work done in lifting the box is called useful work, but work done against friction is wasted work.

If a machine were free of friction (which is never the case) the output work would equal the input work. Or, expressed in terms of the machine under frictionless conditions, the work put out by a machine equals the work put into it. This statement is known as the **principle of work.** However, because some work is always done to overcome friction, the principle of work is modified to read

$$\text{input work} = \text{useful output work} + \text{work done against friction}$$

In some machines the force of friction is small, so the work wasted in overcoming friction is also small. In such cases, as in an equal-arm balance, the useful output work is almost equal to the input work. Therefore the ratio of output work to input work

$\left(\frac{\text{output work}}{\text{input work}}\right)$ is only slightly less than one. This ratio is called **efficiency** (ih-FISH-un-see). It is usually expressed as a percentage.

$$\text{efficiency} = \frac{\text{output work}}{\text{input work}} \times 100\%$$

The higher the value of this ratio, the more efficient is the machine. In the next experiment you will find the efficiency of the inclined plane used in Experiment 3–14.

experiment 3–15 The Efficiency of an Inclined Plane

Materials none

Procedure Using the data obtained in Experiment 3–14, compute the work done against friction and also the efficiency. Record your results as shown in Figure 3–22.

Questions

1. What happens to the work done against friction as the height of the board is increased? Explain.

2. Based on your computations, how much work would be done against friction if the book were pulled the length of the board when the board was in a vertical position? Explain.

3. How does the efficiency change as the slope of the incline increases? Why is this so?

Figure 3–22
Computations for Experiment 3–15.

Height of Board	Work done against friction (input work − output work)	Efficiency $\left(\frac{\text{output work}}{\text{input work}} \times 100\%\right)$
0 m	? N·m	? %
0.15 m	? N·m	? %
0.30 m	? N·m	? %
0.50 m	? N·m	? %

experiment 3–16 To Increase the Efficiency of an Inclined Plane

Materials board (1 m by 15 cm by 2 cm); spring scale; thread; stand; clamp; meter stick

Procedure Use the same arrangement as in Experiment 3–14, plus anything else you think might help. Find one or two ways to make the setup more efficient. Make measurements like those in Experiment 3–14.

Computations

1. What is the efficiency of your system at each height?

2. How do these efficiency values compare with those you computed in Experiment 3–15?

Questions

1. If you thought of any methods that could not be tested with the equipment available, what were they?

PERSPECTIVES

perspective 3–1 Force and Friction

Isaac Newton was the first scientist to give clear meaning to the word *force*. In one of his laws of motion he stated, "Every object remains at rest or continues in a state of uniform motion unless an unbalanced force acts upon it." This means that a force (a push or a pull) is needed to start a body moving. It also means that a force is needed to stop a body if it is moving.

Let us apply this idea to the case of a man pushing a heavy box along a rough floor as shown in Figure 3–23. The man pushes on the box in one direction, and the floor pushes on the box in the opposite direction. The push exerted by the floor on the box is the force of friction. When the box is moving at a constant speed, the backward force of friction is exactly equal to the forward push exerted by the man. Thus, the two opposite forces are balanced. If the man stops pushing, the box comes to rest immediately. Therefore there is no longer

Figure 3–23
A man pushing a box along a floor is exerting a force against friction.

either a force of friction or a forward force. Expressed another way, a force of friction is exerted only when the box is in motion.

Suppose the box is moving at constant speed and the man increases the push on the box. What happens? The forward force is now greater than the force of friction. As a result an unbalanced force acts on the box. What is the effect of the unbalanced force? The unbalanced force increases the speed at which the box is moved. Therefore, in doing experiments that involve measurement of friction it is important to pull the object at constant speed, for only then will the force of friction be equal to the pull exerted on the body.

As a body starts from rest its speed increases. Thus, at the beginning of its motion the two opposing forces are not balanced; the applied force stays greater than the force of friction until the object moves at constant speed.

perspective 3–2

Elasticity

Elasticity is a property that enables a substance to recover its original size and shape after being stretched. Every material has its own particular elasticity. The greater the elasticity of the material, the greater is its restoring force. For example, a steel wire is difficult to stretch, but its elasticity is high. High elasticity means it has a strong tendency to return to its original shape and position. On the other hand, a rubber band is easily stretched but, after being stretched a number of times, it does not return to its original shape. Thus, the elasticity of rubber is low.

Steel beams are used to support large structures such as skyscrapers and bridges. Since these beams are subjected to enormous forces, it is important for the architect and engineer to know about the elasticity of steel. If the steel were unable to recover its shape after being distorted by an applied force, the structure would lose its main support and it would no longer be safe.

perspective 3–3

Levers

A crowbar is an example of a lever. So is a seesaw (Figure 3–6). Something like a crowbar was undoubtedly used by early man when prying rocks out of the ground to clear the land. A very heavy rock can be moved by applying a relatively small force.

A lever increases the force applied (the effort) if the length from fulcrum to force (called the effort arm) is greater than the length from fulcrum to load (called the load arm). See Figure 3–24. The longer the effort arm compared with the load arm, the greater is the output force compared with the applied or input force. Archimedes (ar-ki-MEE-dees), a Greek philosopher who lived in the first century B.C., was so impressed with the power of the lever that he is said to have exclaimed, "Give me a place to stand and a fulcrum on which to rest a lever and I will move the earth."

The equal-arm balance is a special kind of lever. As the name suggests, the two arms are equal in length. As a result, the output force is the same as the input force.

perspective 3–4

Gravity is a Peculiar Force

You cannot push or pull an object unless you touch it, but gravity can. Newton argued that the force of gravity is not limited to bodies on earth — its effect extends out into space.

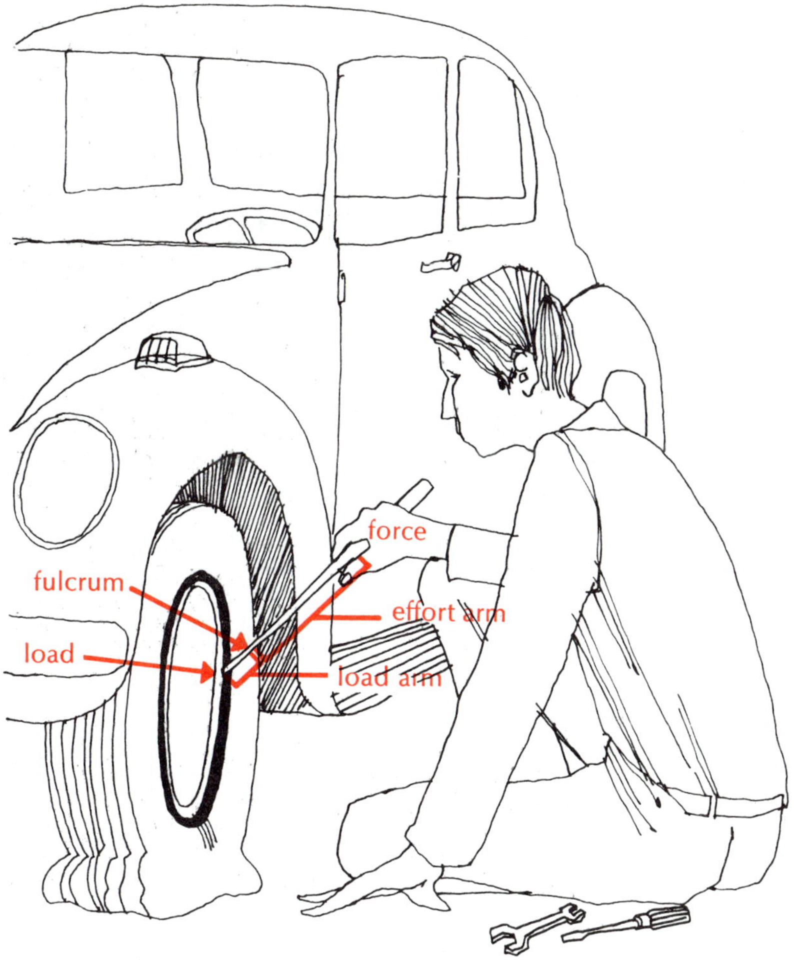

Figure 3–24
A lever is used to pry a hub-cap off a wheel. The effort arm is longer than the load arm, so this lever increases the force.

In Newton's time this was a new idea, and many thinkers rejected it. Today it is accepted without question. Indeed, the gravitational force exerted by the earth upon bodies in space (upon the moon, for example) can readily be computed.

All objects on earth are attracted to the center of the earth, which is about 6400 km (about 4000 miles) below the surface. The farther a body is from the center of the earth, the less is the earth's pull and therefore the lighter the body becomes. If an object weighs 40 N on earth, it would weigh only 10 N if the distance from the earth's center were doubled; that is, if the object were taken 6400 km above the surface of the earth. This is shown in Figure 3–25.

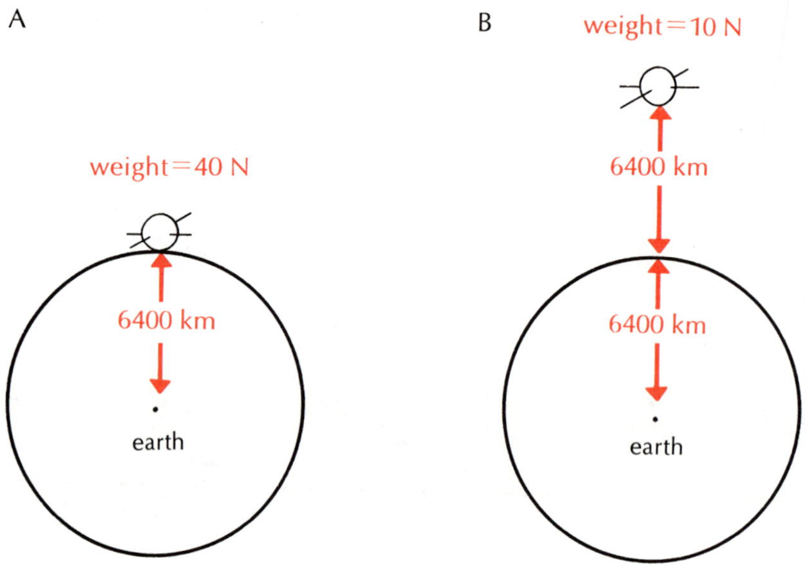

Figure 3–25
A. On the surface of the earth, 6400 km above the center of the earth, this object weighs 40 N. **B.** If the object is taken 6400 km above the surface of the earth, twice as far from the center of the earth, it weighs 10 N.

Although the force of gravity can be measured, no one really understands this force. Even now, 300 years after Newton's time, scientists do not know what gravity is. Its cause remains a mystery.

perspective 3–5 Center of Gravity

As stated in Experiment 3–2, there is a point at which a meter stick or bar can be balanced on a fulcrum, as shown in Figure 3–26.

The weight (W) of the stick presses down on the fulcrum. Therefore the fulcrum must exert an equal upward force (F) to balance the weight. Thus the meter stick acts as though all its weight is concentrated at the point of balance, usually called the center of gravity.

perspective 3–6 Mass and Weight

You recall that the mass of a body can be measured by an equal-arm balance as shown in Figure 3–27. When the masses on the two pans are balanced, they are the same. That is, the

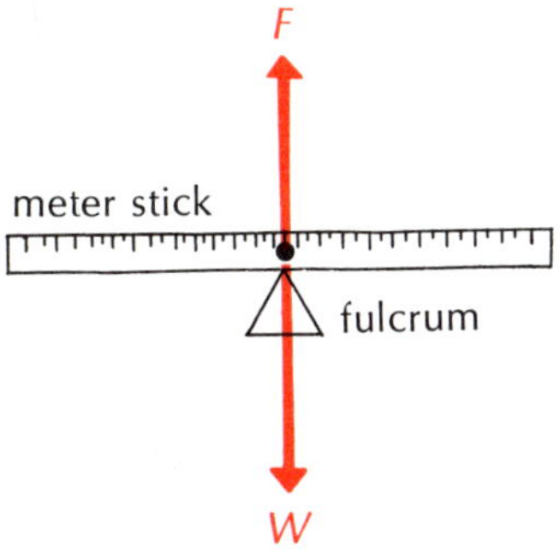

Figure 3–26
This meter stick is balanced at its center of gravity. The fulcrum exerts an upward force equal to the weight of the meter stick.

unknown mass on the left pan equals the sum of the known masses on the right pan. The earth's pull on equal masses is the same. Therefore the weight of the mass on the left pan (W_A) is the same as the weight of the masses on the right pan (W_B). If the balance and masses were carried by a rocket far above the earth's surface, they would still be balanced. However, the new weights (W_X and W_Y) would be less than the weights on earth, W_A and W_B. Thus the weight of a body is a variable force that acts upon its mass.

The best way to find the weight of a body is to use a spring as in Experiment 3–1. The washers have mass and they also have weight. The stretching of the spring is proportional to the earth's pull on the washers; that is, the greater the weight, the greater is the amount of stretching of the spring. The spring and suspended masses could be taken by rocket high above the earth's surface. The greater the altitude, the less would be the earth's pull. Therefore the greater the altitude, the less would be the stretching of the spring.

On the earth's surface it is found that a mass of 1 kg is acted upon by a force of 9.8 N. Expressed another way, 1 kg of mass weighs 9.8 N. Therefore 1 g of mass weighs 0.0098 N. Thus, to find the weight of a given mass, you multiply the number of kilograms by 9.8 N/kg.

Example: What is the weight of a mass of 200 g?

$$200 \text{ g} = 0.200 \text{ kg}$$
$$0.200 \text{ kg} \times 9.8 \text{ N/kg} = 1.96 \text{ N}$$

Figure 3–27
If masses balance on earth, they would still balance far above the earth's surface, although their new weights would be less.

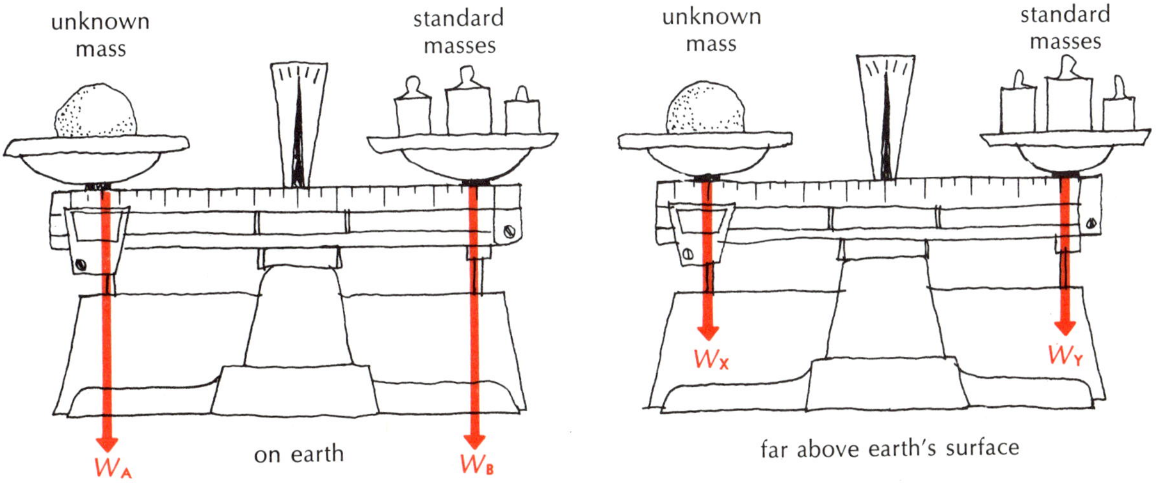

perspective 3–7

Reducing and Using Friction

The amount of friction between two surfaces depends upon the roughness of the surfaces. The rougher the surfaces, the greater is the force of friction. Friction can be reduced by lubricating the surfaces with oil or even with soap. As you probably know, the pistons of an automobile engine move rapidly up and down the cylinders when the engine is running. The cylinder walls are continuously lubricated with oil. Without oil, the engine would overheat and the cylinder walls would become scratched and worn.

Still another way to reduce friction is to change from the sliding kind of friction to the rolling kind. For instance, there is a surprisingly small amount of friction as the wheels of a heavy locomotive roll along steel railroad tracks. Ball bearings or roller bearings (Figure 3–28) are used in the axles of automobiles. As a result, rolling friction replaces sliding friction. The effect is to reduce the forces of friction a great deal.

In machinery, friction is a hindering or harmful force. In many instances, however, friction is a useful force. You have probably experienced the skidding of a car on an icy road in winter. Skidding is due to lack of friction between tires and road. Friction can be increased by using snow tires or chains. Friction is also a useful force in walking. Indeed, if there were no friction between your feet and the earth, walking would be impossible. Even an ice skater depends upon the small amount of friction between his skates and the ice. Without this force he would not be able to move.

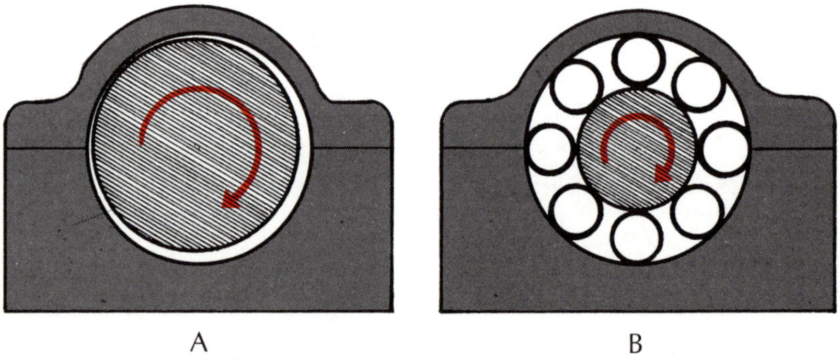

Figure 3–28
A. There is great friction when an axle turns like this. **B.** Ball bearings replace sliding friction with rolling friction.

perspective 3–8

Energy

Before a machine can do work it must be supplied with energy. **Energy** is the ability to do work. When you wind a watch, you tighten its coil spring. In doing so, you do work upon the spring. This work is stored in the spring as energy. When the spring is tightened, it can move the mechanism of the watch. That is, it can do work.

Consider the operation of a pile driver as shown in Figure 3–29. First, a heavy ram is raised above the pile. Because of its position the ram now possesses energy. The higher it is lifted, the greater is its energy. This kind of energy, the energy of position, is called **potential energy.** If the ram is released, it begins to fall. In falling, it loses potential energy. However, as it falls it picks up speed; the greater the fall, the greater the speed. Thus, while the ram loses energy of position, it gains energy due to speed. Energy due to speed is called **kinetic** (kih-NET-ik) **energy.** When the ram strikes the pile it comes to rest and therefore loses its kinetic energy. But at the same time it drives the pile into the earth. In this operation potential energy is changed (transformed) to kinetic energy and the kinetic energy is used to do work.

Figure 3–29
A pile driver changes potential energy to kinetic energy. The heavy ram is striking the pile and driving it into the earth. In the foreground are more piles ready to be driven.

perspective 3–9

Energy Transformation in a Pendulum

The transformation of potential energy to kinetic energy is a common occurrence, as when an automobile, with its engine off and its brakes released, runs down a hill. Another example of energy transformation is a swinging pendulum. The bob of the pendulum could be a set of washers. The bob is attached to a hook by a long thread.

At the start, the bob of the pendulum is at rest in position A, as shown in Figure 3–30. The bob is then pulled aside to position B and released. To raise the bob from position A to position B, work has to be done. Therefore in position B (its highest position) the maximum amount of potential energy is stored in the bob. When the bob is released, it begins the downward swing and, at the same instant, the transformation of potential to kinetic energy also begins. For example, in position C the pendulum has both potential and kinetic energy. As the pendulum passes through its lowest position at the bottom of its swing (in position A) its potential energy is at a minimum.

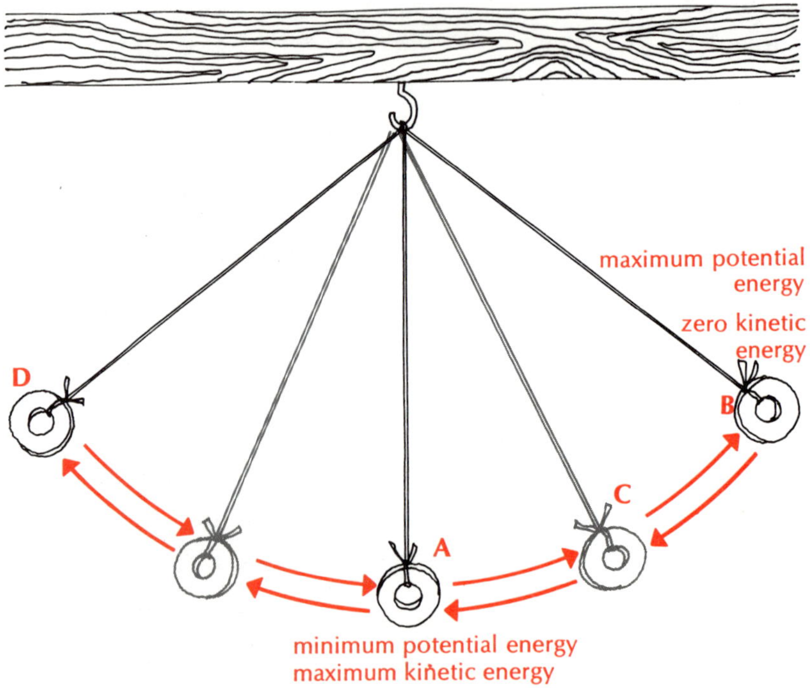

Figure 3–30
A pendulum is an example of energy transformation.

On the other hand, it moves through this position with its greatest speed so that its kinetic energy is at a maximum.

On the upward swing the reverse is true. The bob loses kinetic energy and gains potential energy. When it reaches position D (the top of its swing) the bob comes to rest for an instant. At rest, all the energy of the bob is potential. Thus the pendulum bob swings back and forth, transforming energy. Eventually the bob is brought to rest by frictional forces, mainly by air resistance.

When the bob is in position A, is its potential energy zero? What would happen if the thread broke?

perspective 3–10

Energy in Fuels and Food

Figure 3–31
What energy transformation takes place in the cylinders of an automobile engine?

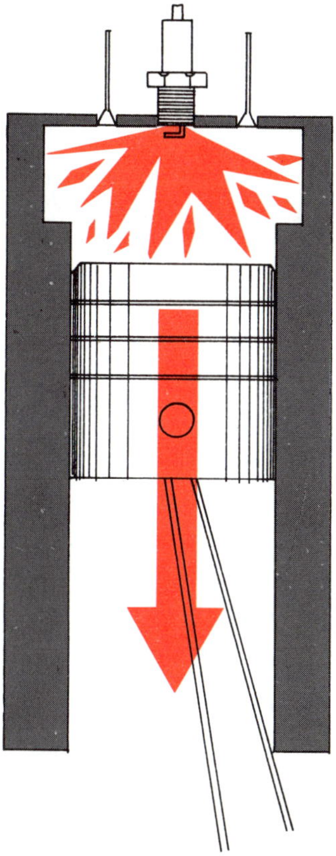

The potential energy of a pendulum bob is energy of position. Certain chemicals such as gasoline can do work too, so we say that potential energy is stored in gasoline. This is called energy of condition. So, by definition, **potential energy** is energy due to position or condition.

The energy of gasoline can be released in a chemical reaction in which the gasoline combines with oxygen of the air. As you know, the energy released in this way is used to power automobiles. In this particular energy transformation gasoline is mixed with air and exploded by sparks in the cylinders of the automobile. See Figure 3–31. Hot gases are released in the chemical reaction and these hot gases drive the pistons back and forth. That is, the potential energy of gasoline is transformed first into heat energy and then into kinetic energy of moving pistons. Then, by suitable machinery the back-and-forth thrust of the pistons is used to turn the wheels of the car.

The most important example of energy transformation concerns the heat and light from the sun. Without the sun's energy the earth would be absolutely cold and life on earth would not exist. The sun warms the earth and makes it possible for plants to grow. In the presence of sunlight, green plants can combine water vapor and carbon dioxide, present in soil and air, to form oxygen, sugar, and starch (Figure 3–32). As you know, oxygen, sugar, and starch are essential for all animals. When sugar and starch are formed, energy supplied by the sun is absorbed and stored as potential chemical energy.

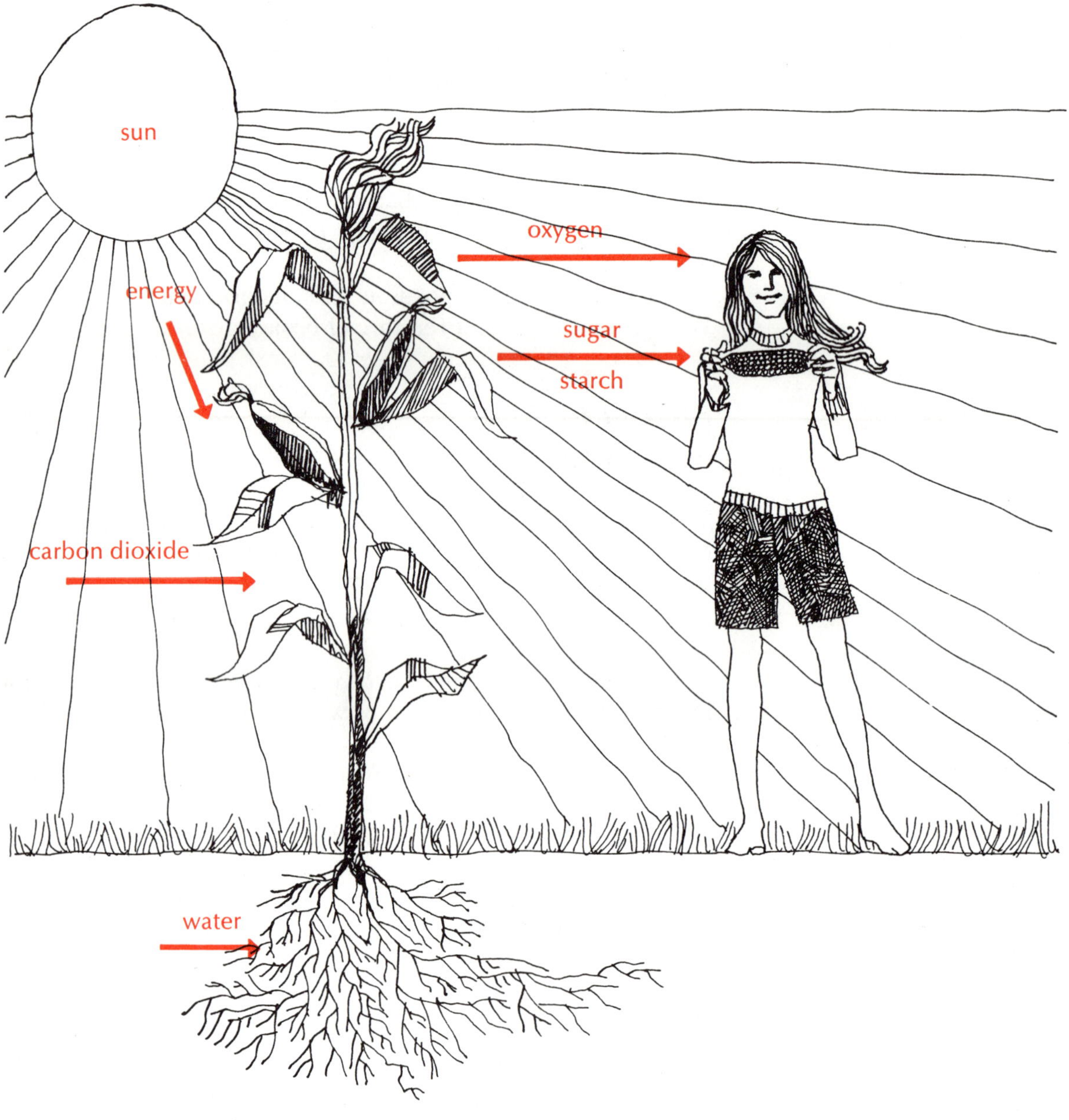

Figure 3–32
Green plants, such as this corn plant, absorb energy from the sun. They store it as chemical potential energy in sugar and starch. When you eat and digest the food, the energy is released for your body to use.

When you eat and digest the food, the potential energy is transformed into other forms of energy. Some of it is changed to heat energy. It keeps you warm and maintains your body at the right temperature. Some of it supplies the energy you need to go about your daily tasks. Thus, without the sun's heat there would be no plants and without plants animal life could not exist. Plant and animal life in the sea is the source of oil (petroleum). Gasoline, which powers our automobiles, is obtained from petroleum.

Coal is the remains of plants that lived millions of years ago. While the plants were alive, they absorbed energy from the sun, transformed it into chemical potential energy, and stored it. When the plants died they were covered by mud and sand. Under the pressure of the mud and sand the plant materials slowly decomposed and changed to coal. In other words, the energy stored in coal is energy absorbed from the sun millions of years ago. When coal is burned, this energy is released. Thus the heat of a coal fire is energy from the sun, locked up ages ago and finally released after a series of energy transformations.

perspective 3–11 The Inclined Plane

The inclined plane is a machine. A machine is a device that makes it easier to do a job. A machine does not save work but rather it allows work to be done more conveniently, usually by reducing the applied force.

Let us compute the force needed to push a heavy box up a wooden plank as described in Figure 3–33. Suppose the plank is 6.0 m in length. Work equals force times distance and therefore

$$\text{Force } (F) = \frac{\text{work done}}{\text{distance moved}}$$

The work to be done to raise the box onto the tailgate is 2250 N · m. Hence,

$$F = \frac{2250 \text{ N·m}}{6.0 \text{ m}} = 375 \text{ N}$$

The newton unit can be changed to the familiar weight unit (the pound) by using the relationship between these two units.

$$1 \text{ N} = 0.24 \text{ lb}$$

Hence,

$$375 \text{ N} = 375 \text{ N} \times 0.24 \text{ lb/N} = 84 \text{ lb}$$

Thus, by using this inclined plane, a man can push a box weighing 1500 N (330 lb) up the plane by exerting a force of only 375 N (84 lb), as in Figure 3–33. Since an ordinary man can exert a force of 84 lb, the use of an inclined plane enables the man to do the job.

Now suppose the length of the plank is increased from 6.0 m to 8.0 m. Will the push exerted on the box be increased or decreased?

$$F = \frac{2250 \text{ N}\cdot\text{m}}{8.0 \text{ m}} = 280 \text{ N or } 68 \text{ lb}$$

Thus, the longer the incline the smaller is the force needed to push the box. But, whatever the length of the plane, the work to be done is the same, 2250 N·m. If the force is decreased from 375 N to 280 N, we think of this as a gain in the force factor of the machine. Similarly, if the length of the plane is increased from 6.0 m to 8.0 m, we think of this as a loss in the distance factor of the machine. This means that if the force needed to lift a load is decreased, the reduction of

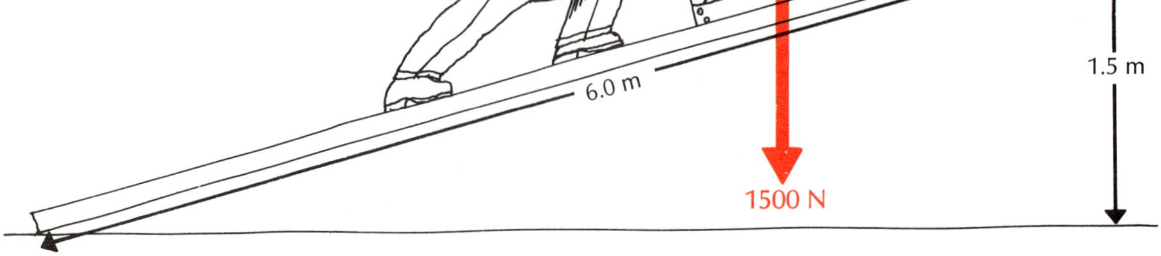

Figure 3–33
An inclined plane lets a person lift a load using a smaller applied force.

force must be "paid for" by exerting the push through a greater distance. In general terms, what is gained in force is lost in distance. This statement is a rule that applies to all machines.

There is of course a good deal of friction between the box and the plank. In our computations we disregarded the force of friction. However, because of friction, the force exerted would be greater than 375 N on the shorter plank and greater than 280 N on the longer plank.

perspective 3–12

Engineering Feats in the Ancient World

One of the greatest engineering feats of all time was the construction of the pyramids in Egypt. (See Figure 3–34.) The largest of them, the Great Pyramid near Cairo, was built about 5000 years ago. The Great Pyramid was both a tomb for the king, Cheops, and a monument to his memory. Its size is staggering. Its base is square and covers 13 acres. Its height is 480 feet, about as high as a 50-story building. The pyramid is constructed of more than two million blocks of hard rock, weighing from two to four tons each, cut to exact size and shape. How were these massive blocks lifted into place up

Figure 3–34
The construction of the pyramids in Egypt was one of the greatest engineering feats of all time. No one knows just how it was done.

Figure 3–35
Perhaps ramps of sand were used to raise the stones.

the sides of the pyramid? No one knows the answer to this question. Early historical records do not mention the method of construction, so the Great Pyramid remains an engineering puzzle. However, there is evidence that ramps were used, as shown in Figure 3–35. Most archeologists believe that sand was carried in to make a ramp and that the heavy rocks were dragged up the incline by manpower. As the pyramid became higher, the ramps were made steeper. Finally, when the construction was completed, the sand was carted away and the pyramid uncovered.

A ramp is, of course, an inclined plane. Even 5000 years ago, construction workers knew how to make simple machines that would reduce the force needed to lift loads weighing many tons. Of course they had to contend also with large frictional forces. And one wonders if they had learned that friction could be reduced by rolling the blocks of stone over the rounded trunks of trees. Who knows?

Some engineers interested in ancient structures do not accept the theory that ramps were the only means of raising the heavy blocks of stone. They argue that the ramps would have been at least a mile long. Moreover, the vast quantities of earth needed to construct the ramps would cause many additional engineering problems. The work of construction would have been greatly simplified by the use of a "load arm," a kind of lever that was used in ancient times. Such a piece of equipment is shown in Figure 3–36. It consists of a lever

Figure 3–36
Or maybe large levers, called load arms, were used.

made of heavy timber about 20 feet in length. The lever is on a sturdy fulcrum about six feet high. To use this device in a lifting operation, the lever must first be moved so that its short arm is above a heavy stone. A sling is then put under the stone and attached to the short arm. Rocks are then piled in another sling at the end of the long arm. When enough rocks have been added, the heavy stone is raised as shown in Figure 3-36. Sand would then be piled under the heavy stone and the lifting operation repeated at the greater height. A single lever would raise a heavy stone by only one or two feet at each lift. But if many levers were in operation, many stones could be lifted at the same time. Moreover, if load arms were used, long ramps of earth would be unnecessary and fewer workers would be needed.

QUESTIONS

A

1. Which of the following words represent forces?
a elasticity **b** friction **c** gravity **d** push **e** torque

2. (a) What is the center of gravity of an object? (b) Why is the center of gravity of a lever the same as the point of balance?

3. (a) What is an unbalanced force? (b) What is the unbalanced force when a match box is pushed at an increasing speed?

4. Write the two pairs of words that are associated with each other: mass, weight, newton, kilogram.

5. (a) What is a counterclockwise torque? (b) How is it measured? (c) What are the units of measurement?

6. (a) Define the term *work*. (b) What is the unit of work in the metric system, and what does its name mean?

7. Define the terms (a) energy, (b) kinetic energy, (c) potential energy.

8. Gasoline contains chemical potential energy. Explain what is meant by this term.

9. (a) What is a machine? (b) Why is an inclined plane a machine?

10. To solve the definition puzzle, read the definitions. Then think of a word that fits the first definition. Check to see that the word has the same number of letters as the number of spaces in the puzzle. Also, check to see that the word has the same letter in the given position as the one in the puzzle. Copy the puzzle in your notebook, and print each word, one letter per space.

Definitions:

1. A simple machine
2. A force that opposes motion
3. A point of support for a lever
4. A force that acts upon your body
5. A force that acts upon a stretched object in a direction opposite to an applied pull
6. A mechanical device that makes it easier to do work
7. A unit of work
8. A unit of force
9. A push or a pull
10. A force that attracts all bodies on the earth

Puzzle:

```
 1.             _ e _ _ _
 2.               f _ _ _ _ _ _ _
 3.               f _ _ _ _ _ _
 4.           _ _ i _ _ _
 5. _ _ _ _ _ _ c _ _ _
 6.       _ _ _ _ i _ _
 7.       _ _ _ _ e
 8.     _ _ _ _ _ n
 9.         _ _ _ c _
10. _ _ _ _ _ _ y
```

B **11.** Is the elasticity of putty high or low? Give a reason for your answer.

12. A wheelbarrow is an example of a lever. Where is the position of (a) the fulcrum, (b) the effort in this lever?

13. Does the weight of a rock change as it falls down a mine shaft? Explain.

14. (a) In the figure below, which forces exert (i) a clockwise torque, (ii) a counterclockwise torque? (b) Is this a balanced lever? Give a reason for your answer.

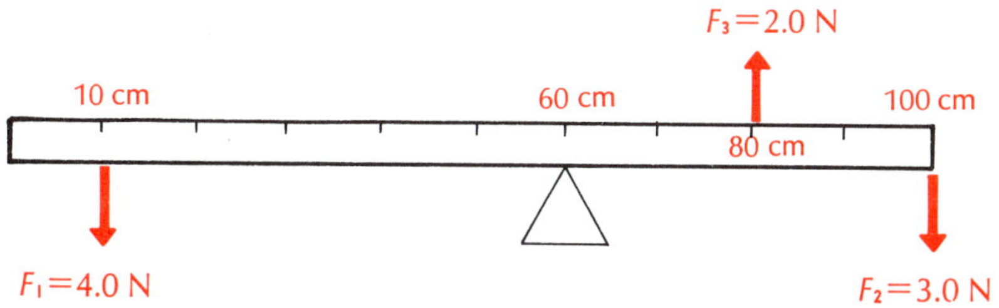

15. Friction between two metal surfaces can be reduced by oiling them. Why?

16. If a baseball is thrown upward, it reaches a certain height and then returns to earth. Describe the energy transformations that occur in this operation.

17. An inclined plane is a machine. Would the plane be a machine if it were not inclined? Explain.

18. A rule that applies to all machines is, "What is gained in force is lost in distance." The rule can also be stated as, "What is gained in force is lost in speed." Explain how the second statement of the rule applies to (a) a lever, (b) an inclined plane.

19. A brick whose dimensions are 10 cm × 12 cm × 20 cm is pushed along a table. (a) Compute the areas of the three different faces. (b) Suppose the force of friction between the table and the largest surface of the brick is five newtons. What is the force of friction between the brick and each of the other two surfaces?

Each of Questions 20–24 contains three words that have one thing in common. In your notebook write the factor that is common to each set of words.

20. elasticity, friction, gravity

21. meter, joule, newton

22. ramp, seesaw, crowbar

23. potential, kinetic, heat

24. equilibrium, torque, rotation

In Questions 25–34, assume that all washers have the same mass and that each lever is uniform and is marked off in equal spaces. Select one of the following answers (**a, b,** or **c**) *for each situation and record it in your notebook.*

a The lever is unbalanced, and the clockwise torque is greater than the counterclockwise torque.

b The lever is unbalanced, and the counterclockwise torque is greater than the clockwise torque.

c The lever is in equilibrium.

25.

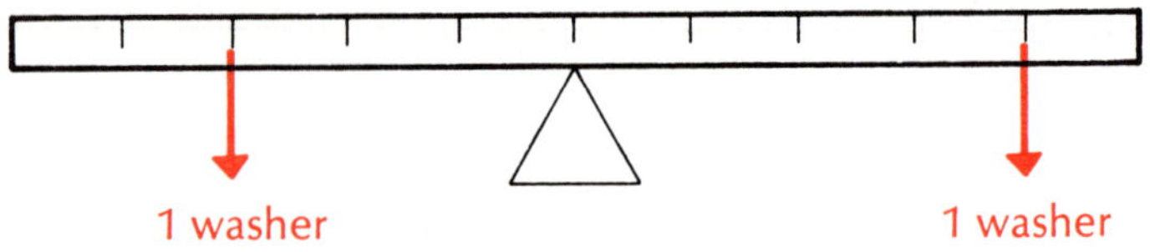

26.

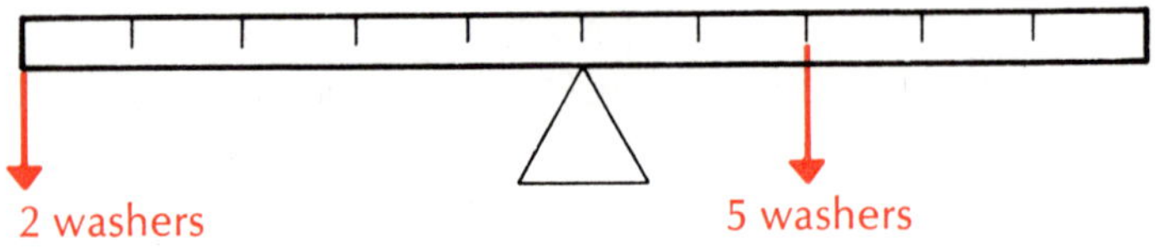

27.

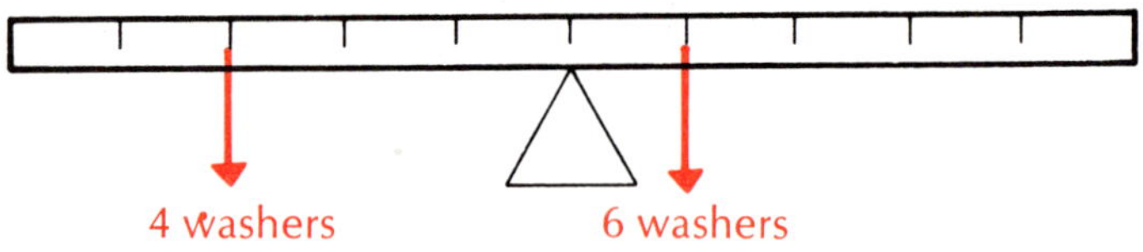

28.

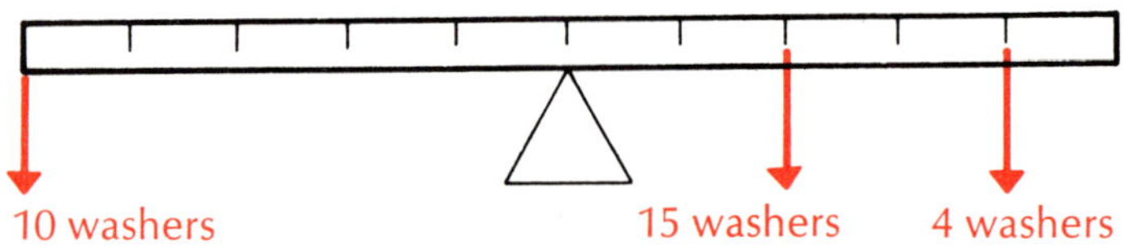

29.

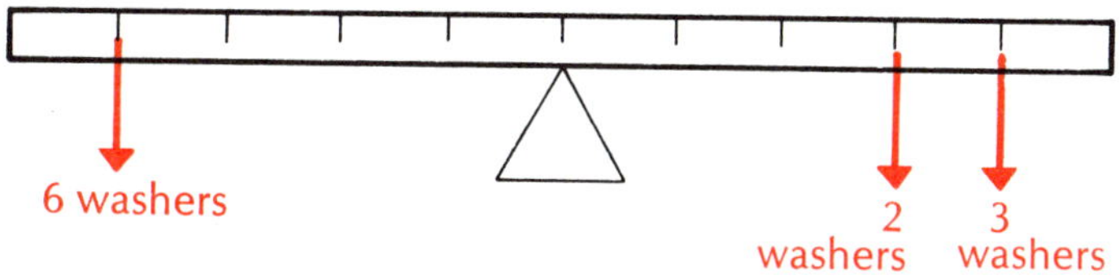

30.

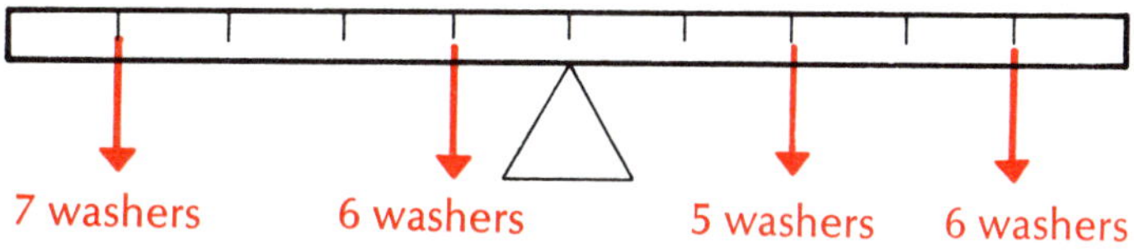

For the following, assume that the lever has a mass of 10 washers.

31.

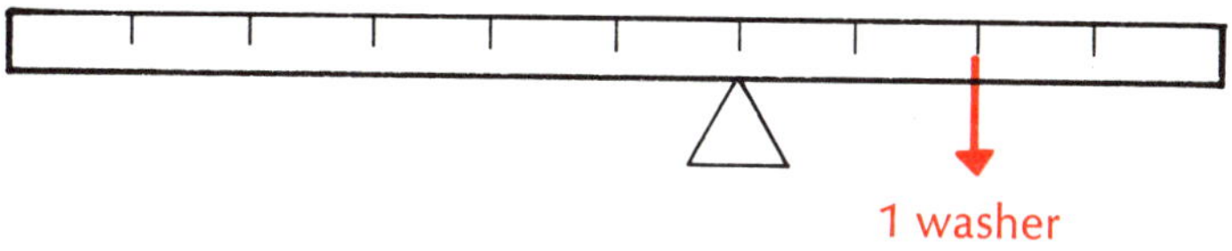

32.

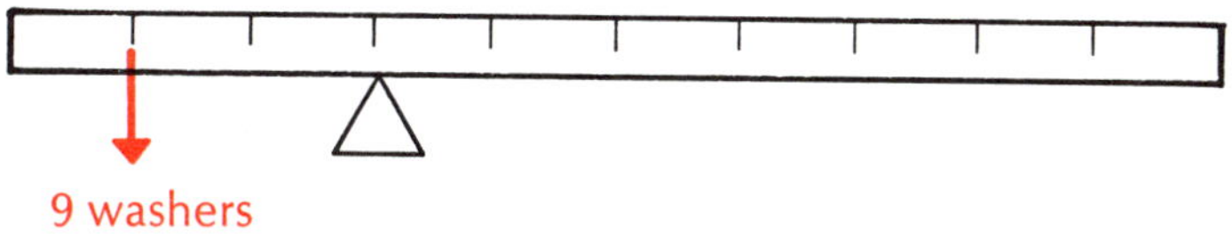

33.

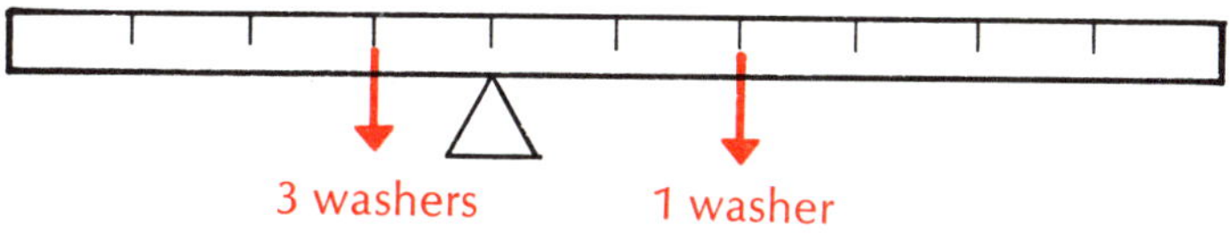

34.

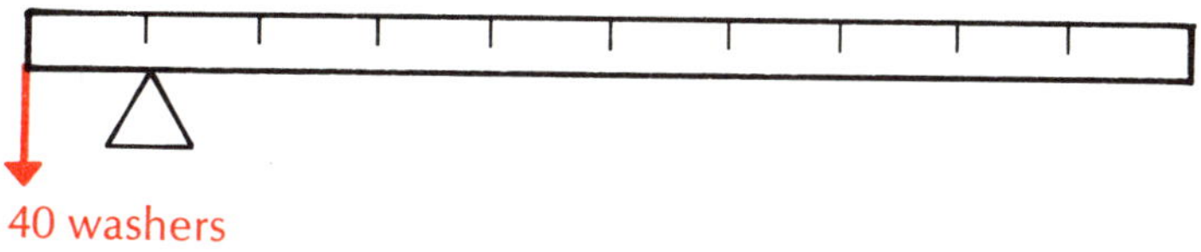

In Questions 35–38, which relationship of friction applies to each situation? The three relationships of friction are:

a the relationship between the force of friction and the weight of a body

b the relationship between the force of friction and the area of contact between surfaces

c the relationship between the force of friction and the nature of the surface

35. The outside surfaces of high-speed airplanes are very smooth and highly polished. Engineers say that the shiny surface improves the speed of the airplanes.

36. A squeaky hinge is oiled to get rid of the noise.

37. Two or more locomotives are needed to pull freight trains containing many cars.

38. An ice skater may have his skates sharpened to help him skate better.

PROBLEMS

1. A spring suspended vertically from a stand is stretched 2.3 cm when a mass of 150 g is suspended from it. It is stretched 6.9 cm by a mass of 450 g. (a) By what amount would the spring be stretched by a mass of 300 g? (b) A mass suspended from the spring stretches the spring by 13.8 cm. What is the mass?

2. When a book is suspended at the zero mark of a 0.200-kg meter stick, the stick balances at its 20.0-cm mark. Compute the weight of the book in newtons.

3. The center of gravity of a 7.0-m tapering pole, weighing 200 N, is 1.7 m from the thick end. The pole is lying on the ground. (a) What force is needed to lift the thin end of the pole? (b) What force is needed to lift the thick end of the pole?

4. A man balances on his shoulder a pole from which a 175-N object and a 300-N object are suspended. Disregarding the weight of the pole, what force must the man's shoulder exert to support the load?

5. In Problem 4, if the "weightless" pole is 2.0 m long and if the objects are suspended at the ends of the pole, how far is the 175-N object from the man's shoulder?

6. Compute the work done by raising a 2-kg mass vertically through 10 m.

7. Compute the work done by exerting a force of 2 N horizontally to push a 10-kg mass horizontally through a distance of 20 m.

8. A force of 110 N is required to push a 50-kg mass up an inclined plane which is 2 m high and 10 m long. Compute (a) the work input, (b) the work output, (c) the work done against friction.

9. A shovel weighing 40 N has a handle 1.60 m long, and the center of gravity of the shovel is 1.40 m from the end of the handle. (a) If a man grasps the handle with his left hand, 1.00 m from the handle end, what force must he exert with his right hand at the end of the handle to keep the shovel in a horizontal position? (b) If the man grasps the end of the handle with his right hand, what force must he exert with his left hand 1.00 m from the handle end to keep the shovel in a horizontal position?

10. A boy catches a fish and wishes to weigh it at once. He uses his baseball bat which weighs 17 N, as a lever and a small twig as a "ruler." The balance point of the bat is ten twig lengths from the handle end. When the fish is hung from the handle end, the balance point is four twig lengths from the handle end. How much does the fish weigh?

11. Al and Ben each weigh 800 N. They each have to carry two sacks, weighing 200 N each, up a ladder 10 m high. Al climbs up the ladder carrying both his sacks. Ben climbs up twice, carrying one sack each time. (a) How much force does each man exert when he climbs the ladder with his load? (b) How much total work does each man do?

Chapter 4

TEMPERATURE AND HEAT

Fire and ice can act together. There was a fire in this house in the winter, and water from the fire hoses froze.

You have probably warmed your hands on a cold day by rubbing them together. Where does the heat come from? If you hammer a nail into a thick board, the nail becomes hot, almost too hot to touch. Again, where does the heat come from? The falling hammer has kinetic energy. When the hammer strikes the nail, the kinetic energy disappears. Is there any relation between the disappearance of kinetic energy and the appearance of heat?

To start finding out what energy has to do with heat, get a thermometer. You will put energy into the thermometer and see how hot it gets.

First, read the thermometer and record the reading. Next, firmly grasp the bulb of the thermometer with the thumb and forefinger of your left hand. Hold the thermometer with your right hand as in Figure 4–1 and briskly rub the bulb back and forth between the thumb and forefinger of your left hand for about one minute. Read and record the thermometer reading.

In this experiment three different factors can be identified. (1) Temperature was taken at the beginning and end of the experiment. (2) Work was done against friction. (3) Heat was given off by the work done.

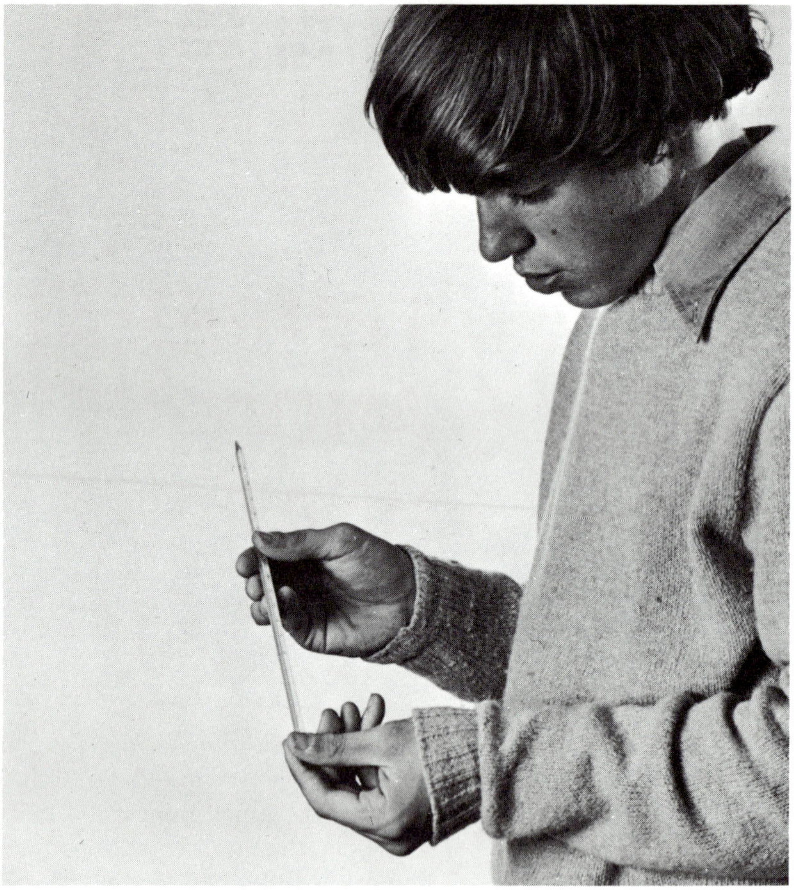

Figure 4–1
What happens when you rub a thermometer bulb?

What does the work done have to do with the heat generated? What does temperature have to do with heat? What is temperature? What is heat? The experiments you do in this chapter will help you answer these questions.

TEMPERATURE

The Thermometer

You will be measuring temperature in all of the experiments in this chapter, so it is important that you learn how to use a thermometer. A thermometer has a scale. The numbers on the scale of your thermometer may range from a high reading of 110 to

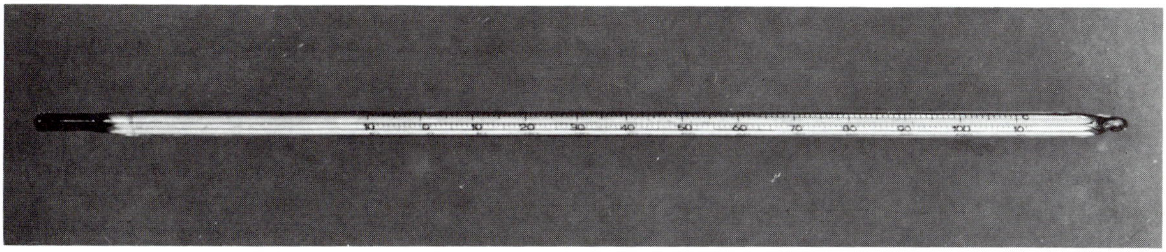

Figure 4–2
You will use a thermometer like this in your experiments.

a low reading of −10, as shown in Figure 4–2. What do these numbers mean?

Before a scale can be made for a thermometer, two fixed temperatures, called **fixed points,** must be agreed upon. The two selected fixed points are the temperatures at which water freezes and boils. The number 0 was assigned to the temperature at which water freezes, and the number 100 was assigned to the temperature at which water boils. Temperature is measured in units called degrees. The symbol for a degree is °. The temperature at which water freezes was called 0°, and the temperature of boiling water was called 100°. The scale was obtained by dividing the distance between these fixed points into 100 equal divisions, or degrees.

This particular scale was devised by a Swedish astronomer named Anders Celsius (SELL-see-us) in the 18th century. Other scales have also been devised but for most purposes the Celsius scale is the most convenient. Temperature readings on this scale are called "degrees Celsius." Thus, the boiling point of water is one hundred degrees Celsius, usually written as 100°C, where the C stands for Celsius. The letter C is also used to stand for centigrade, a word from Latin that means one hundred degrees and refers to the 100 degrees between the two fixed points. So a thermometer with the Celsius scale is sometimes called a centigrade thermometer. The boiling point of water is then sometimes called one hundred degrees centigrade.

In the next experiment you will learn how to use and read your thermometer. You will check the readings of your thermometer against the known temperatures of the two fixed points.

experiment 4–1

To Check the Fixed Points of a Thermometer

Materials

thermometer; 150-ml beaker; stand; 4-inch ring; gauze; burner; crushed ice

A. The Temperature at which Water Freezes

Procedure

The freezing temperature of water is precisely the same as the melting temperature of ice. Since it is easier to melt ice than to freeze water, you will find the temperature at which ice melts.

Half fill a beaker with finely crushed ice. Put the bulb of the thermometer in the ice and read the temperature to the nearest 0.5°C. Record the temperature at one-minute intervals for about 15 minutes. Make a table for your readings as in Figure 4–3.

Figure 4–3
Data table for Experiment 4–1.

Time	Temperature
0 min	?°C
1 min	?°C
2 min	?°C

Questions

1. How long does it take to reach a steady temperature reading?

2. What is the melting temperature of ice?

3. What is the freezing temperature of water?

4. Is there an error in the scale of your thermometer? If so, what is the error?

5. If there is an error, how would you change the scale reading to give the correct temperature?

6. Why are *two* fixed points needed in making a scale, rather than just one?

B. The Temperature at which Water Boils

Procedure

Empty the beaker of ice and half fill it with tap water. Attach a ring to a ringstand and place a burner under the ring. Adjust the height of the ring so that it is about 10 cm above the top of the burner. Place a gauze on the ring. Place the beaker of water on the gauze. The arrangement is shown in Figure 4–4.

Light the burner and heat the water. Lower the thermometer bulb into the water and stir the water. The thermometer bulb should not touch the bottom of the beaker. Record the tem-

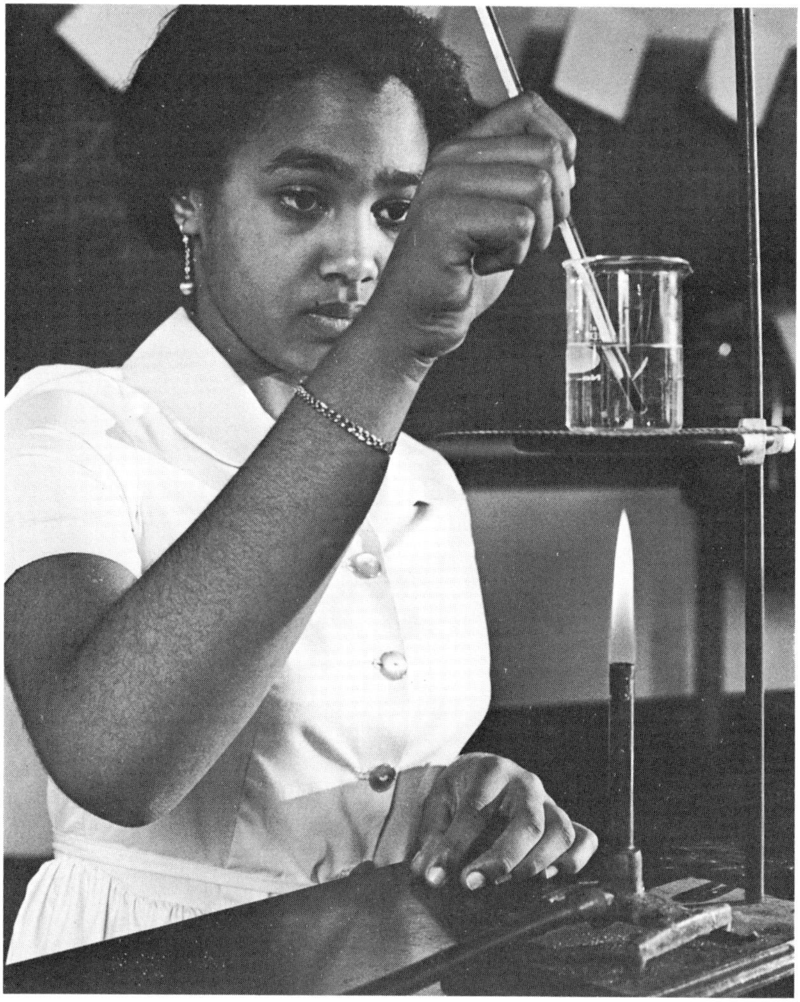

Figure 4–4
Why should the thermometer bulb not touch the bottom of the beaker?

perature of the water at intervals of one minute, and continue to heat and stir the water for at least five minutes after boiling has begun.

Questions

1. What is the reading of the thermometer at the temperature of boiling?

2. Is the scale on your thermometer accurately marked? Why or why not?

3. What happens to the heat of the flame transmitted to the water below its boiling temperature?

4. What happens to the heat of the flame transmitted to the water at its boiling temperature?

HEAT

Heat and Temperature

You have seen that a thermometer measures the degree of "hotness" of a body. Suppose you have a kettle of boiling water and also a cup of boiling water. The temperature of the water is the same in each container. Although the two samples of water are at the same temperature, there is a marked difference between them. What is this difference? The next experiment might help you to answer this question.

experiment 4–2

Cooling Hot Water

Materials

600-ml beaker; 150-ml beaker; thermometer

Procedure

Almost fill a 600-ml beaker with hot water from the hot-water tap at the laboratory sink. Pour about 50 ml of the hot water into a 150-ml beaker.

Take the temperature of both samples of water and record them in your notebook. Then, placing the thermometer first in one beaker and then in the other, read and record the temperatures of both samples of water at intervals of one minute. Continue the readings for about 15 to 20 minutes. Tabulate your readings as in Figure 4–5.

Figure 4–5
Data table for Experiment 4–2.

Time	Temperature of water in large beaker	Temperature of water in small beaker
0 min	? °C	? °C
1 min	? °C	? °C
2 min	? °C	? °C

Questions

1. What was the temperature of the hot water in each beaker at the beginning of the experiment?

2. Compare the temperatures of the water in the two beakers after (a) 5 minutes, (b) 10 minutes, (c) 15 minutes.

3. Does the amount of water seem to have an effect on how fast the water cools?

4. During the cooling of the two samples of water, you probably discovered a difference in the range of temperature. How do you account for this difference?

experiment 4–3 Mixing Equal Masses of Hot and Cold Water

Materials

100-ml graduate; 2 styrofoam cups; thermometer; hot water; cold water

Procedure

When water in a container cools as in Experiment 4–2, the heat given off is transferred to the wall of the container and also to the surrounding air. This transfer of heat is indicated by a drop in the temperature of the water. What are the important factors in measuring the amount of heat transferred? You will investigate this question in the next three experiments.

Using a 100-ml graduate, pour 100 ml of cold water into a styrofoam cup. Then pour 100 ml of hot water into the other cup. Measure and record the temperatures of both the hot and cold water to the nearest 0.5°C. Then, without delay, pour the hot water into the cold water, stir with the thermometer, and record the temperature of the mixture.

Now repeat the experiment twice, using 75 ml of hot water and 75 ml of cold water, and 50 ml of hot water and 50 ml of cold water. Make a table of your data like Figure 4–6.

Computations

1. What is the rise in the temperature of the cold water in each of the three cases?

2. What is the drop in the temperature of the hot water in each of the three cases?

3. How does the rise in temperature of the cold water compare with the drop in temperature of the hot water?

4. Find the product of the mass of the cold water and its temperature change in each of the three cases. (Remember that mass, not volume, is a measure of the quantity of matter in a body. Also remember that one milliliter of water has a mass of one gram.)

5. Find the product of the mass of the hot water and its temperature change in each of the three cases.

6. Compare the products computed in Computation 4 with the products computed in Computation 5.

Questions

1. What is the significant factor in this experiment that determines how much heat is transferred?

Volume of Cold Water	Volume of Hot water	Temperature of cold water	Temperature of hot water	Temperature of mixture
1) 100 ml	100 ml	? °C	? °C	? °C
2) 75 ml	75 ml	? °C	? °C	? °C
3) 50 ml	50 ml	? °C	? °C	? °C

Figure 4–6
Data table for Experiment 4–3.

experiment 4–4 Mixing Different Masses of Hot and Cold Water

Materials

the same as those used in Experiment 4–3

Procedure

In Experiment 4–3 you mixed equal masses of hot and cold water and discovered one of the factors that determines the amount of heat transferred. In this experiment you will mix different masses of hot and cold water. Will your data reveal still another significant factor that determines how much heat is transferred?

Repeat the procedure of Experiment 4–3, but use different volumes of water. (1) Use 100 ml of hot water and 50 ml of cold water. (2) Use 75 ml of hot water and 50 ml of cold water. (3) Use 50 ml of hot water and 100 ml of cold water.

Tabulate your data as in Experiment 4–3.

Computations

1. What is the rise in temperature of the cold water in each of the three cases?

2. What is the drop in temperature of the hot water in each of the three cases?

3. How does the rise in temperature of the cold water compare with the drop in temperature of the hot water in each of the three cases?

4. Find the product of the mass of the cold water and its temperature change in each of the three cases.

5. Find the product of the mass of the hot water and its temperature change in each of the three cases.

6. Compare the products computed in Computation 4 with the products computed in Computation 5.

Questions

1. Does this experiment point to a second factor that must be considered in computing the amount of heat transferred? If so, what is this factor?

2. From the information you gathered in Experiments 4–3 and 4–4, what is probably true concerning the internal energy lost by the hot water and the internal energy gained by the cold water?

experiment 4–5

Mixing Corn Oil and Warm Water

Materials

styrofoam cup; corn oil; 100-ml graduate; thermometer; warm water (40–45°C)

Procedure

In the last two experiments, you poured hot water into cold water. You investigated two factors that might be considered in computing the amount of heat transferred. In this experiment different substances will be used. You will use corn oil (the kind you can get in a grocery store) instead of cold water. Will the evidence you get in this experiment indicate that still another factor is involved in computing heat?

Find the mass of the cup and then ask your teacher to pour about 75 ml of corn oil into the cup. Find the mass of cup and oil, and record this value. Pour exactly 50 ml of warm water into the graduate. Measure and record the temperature of both water and oil. Then pour the water into the oil. (See Figure 4–7.) Notice that they do not mix. Stir the mixture thoroughly with the thermometer until both layers are at the same temperature. Record this value as the final temperature of the mixture.

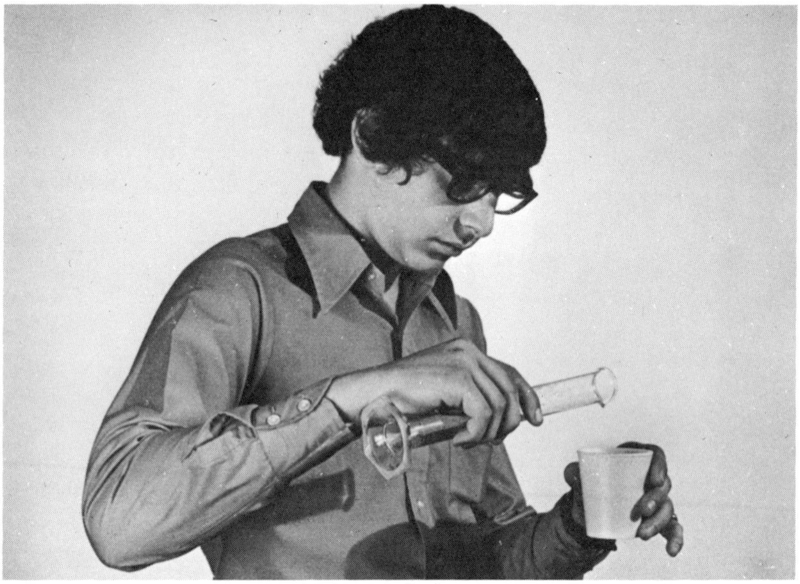

Figure 4–7
Pour the warm water into the corn oil.

Computations

1. (a) Find the change in temperature of the water. (b) Is it a rise or a drop in temperature?

2. Compute the product of the mass of the water and its change in temperature.

3. (a) Find the change in temperature of the corn oil. (b) Is it a rise or a drop in temperature?

4. Compute the product of the mass of the oil and its change in temperature.

Questions

1. Do your answers to Computations 2 and 4 measure the heat transferred from warm water to oil? Why or why not?

Unit of Heat

You recall that both the unit of mass (a kilogram) and a unit of length (a meter) are preserved in the Bureau of Standards in Washington, D.C. for purposes of comparison. However, heat is a form of energy, so a unit of heat cannot be preserved like mass or length. But it can be defined. A unit of heat is the calorie. A **calorie** is the amount of heat energy that must be supplied to one gram of water to raise its temperature through one Celsius degree (1 C°).

Notice that three essential factors (mass, temperature change, and a particular substance, water) are involved in this definition.

The calorie (abbreviated **cal**) is a very small unit. A larger unit is often needed. For instance, in measuring the heat released in chemical reactions, a unit of 1000 calories is preferred. This unit is called the *kilocalorie* (abbreviated to *kcal*).

$$1 \text{ kcal} = 1000 \text{ cal}$$
$$1 \text{ cal} = 0.001 \text{ kcal}$$

Heat Capacity

The answer to Computation 4 in Experiment 4–5 seems to suggest that the mass of corn oil times its change in temperature is not a measure of the heat gained by the oil. One calorie of heat is needed to raise the temperature of one gram of water through one C°, but less than one calorie is needed to raise the temperature of one gram of corn oil through one C°. That is, corn oil is said to have a smaller heat capacity than water. The **heat capacity** of a substance is the number of calories of heat needed to raise the temperature of one gram of that substance through one C°.

Every substance has its own particular heat capacity. For example, the heat capacity of water is one calorie per gram per Celsius degree, which is abbreviated to $1.00 \dfrac{\text{cal}}{\text{g} \times \text{C}^\circ}$. By comparison, the heat capacity of copper is 0.093 calorie per gram per Celsius degree or, using symbols, $0.093 \dfrac{\text{cal}}{\text{g} \times \text{C}^\circ}$.

Thus, to measure the amount of heat transferred, we are concerned with the three factors (1) mass, (2) temperature change, and (3) heat capacity. Therefore, the heat gained by the corn oil in Experiment 4–5 is found by multiplying its mass by its rise in temperature by its heat capacity. It is convenient to use symbols for these factors, such as m for mass, t_1 and t_2 for the initial and final temperatures, and c for heat capacity. Thus, the rise in temperature of the corn oil = $t_2 - t_1$, and the heat gained by the corn oil = $m(t_2 - t_1)c$.

The units of these factors are grams (g) for mass, C° for temperature change, and $\dfrac{\text{cal}}{\text{g} \times \text{C}^\circ}$ for heat capacity. The product of these units is $\text{g} \times \text{C}^\circ \times \dfrac{\text{cal}}{\text{g} \times \text{C}^\circ}$. The units g and C° cancel, leaving an answer expressed in calories. And this is what you expect.

Example: Find the calories of heat needed to raise the temperature of 60 g of alcohol from 21°C to 37°C. In this example, $m = 60$ g, $t_1 = 21°C$, $t_2 = 37°C$, and $c = 0.55 \frac{\text{cal}}{\text{g} \times \text{C}°}$

$$\text{heat gained} = 60\,\cancel{\text{g}} \times 16\,\cancel{\text{C}°} \times 0.55 \frac{\text{cal}}{\cancel{\text{g}} \times \cancel{\text{C}°}}$$
$$= 530 \text{ cal}$$

In the next experiment you will find the heat capacity of iron.

experiment 4–6 The Heat Capacity of Iron

Materials

2 styrofoam cups; thermometer; 6 iron washers; warm water (35–40°C); equal-arm balance

Procedure

Find the mass of one of the styrofoam cups. Place the iron washers in the cup and find the mass of the cup plus the washers. Set the thermometer in the cup so that the bulb touches the washers. See Figure 4–8.

Half fill the other cup with warm water. Read the thermometer and record the temperature of the washers. Also find and record the temperature of the water. Pour just enough water onto the washers to cover them. Stir gently with the thermometer. When the temperature is constant, record this value as the temperature of the mixture. Find the mass of the cup plus washers and water, and record this value.

Computations

1. Find the mass (m_i) of the iron washers.

2. Find the mass (m_w) of the water added to the washers.

3. Find the rise in temperature ($t_2 - t_1$) of the washers, where t_1 is the temperature of the washers and t_2 the temperature of the mixture.

4. Find the drop in temperature ($t_3 - t_2$) of the water, where t_3 is the temperature of the warm water.

5. Compute the heat lost by the water from $m_w \times (t_3 - t_2) \times c_w$. Remember that c_w is known to be $\frac{1 \text{ cal}}{1 \text{ g} \times 1 \text{ C}°}$.

6. Express the correct unit of the answer in Computation 5.

7. The heat gained by the washers is $m_i \times (t_2 - t_1) \times c_i$, where c_i is the heat capacity of iron. Substitute your experimental values for

Figure 4–8
The thermometer bulb must touch the washers.

m_i and $(t_2 - t_1)$ and multiply them. For example, the values might be $200 \text{ g} \times 15 \text{ C}^\circ \times c_i$.

8. In the transfer of heat, the heat gained by the iron in Computation 7 equals the heat lost by the water in Computation 5. Suppose that the heat lost by the water = 360 cal. Then, $200 \text{ g} \times 15 \text{ C}^\circ \times c_i = 360 \text{ cal}$ and, by computation, $c_i = 0.12 \frac{\text{cal}}{\text{g} \times \text{C}^\circ}$. Now compute the heat capacity of iron from your own data.

9. Ask your teacher for the correct value for the heat capacity of iron, and then compute your percentage error.

percentage error =

$$\frac{\text{difference between correct and experimental values}}{\text{correct value}} \times 100\%$$

In this experiment some of the heat of the warm water went into the styrofoam cup. However, the mass of the cup is very small, so very little heat was transferred to the cup. For this reason the heat gained by the cup is not considered in the heat equation.

experiment 4–7

The Heat Capacity of Marble

Laboratory Test

Materials

same as those used for Experiment 4–6, except use marble chips in place of iron washers

Procedure

Follow the procedure of Experiment 4–6, using about 200 g of marble chips and just enough water to cover them.

Collect all the necessary data and compute the heat capacity of marble. Then ask your teacher for the correct value of the heat capacity of marble and record it in your notebook. Finally, calculate the percentage error of your computed value.

CHANGES IN STATE

Changes in the State of Matter

You know that water can exist in three different states: solid, liquid, and gas. The solid state is ice, the liquid state is water, and the gaseous state is water vapor. All substances can exist in these three states. The state depends on the temperature. At ordinary temperatures water is a liquid but aluminum is a solid. However, if aluminum is heated to 658°C, it melts and changes to a liquid. If it is heated to 2467°C, it boils and becomes a vapor. The change of state from solid to liquid is called melting. The temperature at which melting occurs is called the melting point. Conversely, the change from liquid to solid is called freezing, and the temperature at which freezing occurs is called the freezing point. Actually the melting and freezing points of a given substance are the same.

In the next experiment you will investigate the change of state from liquid to solid of common moth crystals. This substance is a white solid at ordinary temperatures. Its chemical name is paradichlorobenzene (pair-ah-dy-klor-oh-BEN-zeen), usually written p-dichlorobenzene.

experiment 4–8

The Cooling Curve of a Substance that Changes from the Liquid to the Solid State

Materials

p-dichlorobenzene; large test tube; thermometer inserted into one-hole stopper; stand; clamp

Procedure

Pour crystals of p-dichlorobenzene into the test tube to a depth of about 4 cm. To do this, first fold a piece of paper in half. Press down on the fold and then unfold the paper. Pour the crystals onto the paper. Bend the paper at the fold and pour the crystals into the tube. This method of pouring solid from paper to tube is shown in Figure 4–9.

Hold the tube in a horizontal position and insert the thermometer in the tube. Now hold the tube upright. In the upright position the bulb of the thermometer must be completely covered by the p-dichlorobenzene when the stopper is snugly fitted into the tube. If the bulb is not covered or centered, remove the stopper from the tube and adjust the thermometer (that is, force it slightly in or out of the stopper) until it is in the correct position.

Your teacher has a large beaker of water on a hot plate. Write your initials on your tube and place it in the hot water until all the p-dichlorobenzene has melted. Then remove the tube from the beaker and clamp it to the stand as shown in Figure 4–10.

Read and record the temperature of the p-dichlorobenzene every minute until the thermometer registers about 45°C. Estimate each temperature reading to the nearest 0.5°C. In recording your time and temperature readings, set up a table like the one in Figure 4–11.

Finally, replace the test tube in the beaker of hot water until the solid is remelted. Then remove the stopper and thermometer from the liquid. Return both the tube of p-dichlorobenzene and the stopper and thermometer to your teacher.

Prepare a full-page graph, plotting temperatures (from 45°C to 60°C) along the vertical axis and the corresponding time values in minutes along the horizontal axis, as shown in Figure 4–12.

Questions

1. Judging from the different slopes of the graph line, how many different sections make up your graph?

2. Explain what physical process happened in each of the sections.

3. (a) What is the freezing point of liquid p-dichlorobenzene? (b) Why did you choose this temperature?

4. (a) What is the melting point of solid p-dichlorobenzene? (b) Why did you choose this temperature?

5. In what way would you expect the cooling curve for water to be (a) the same as for p-dichlorobenzene, and (b) different from it?

Figure 4–9 (right)
How to pour a solid into a test tube.

Figure 4–10 (below)
When your solid has melted, clamp the tube to a stand.

Figure 4–11 (below right)
Data table for Experiment 4–8.

Time	Temperature
0 min	? °C
1 min	? °C
2 min	? °C

experiment 4–9

Heat Needed to Melt Ice

Materials

150-ml beaker; stand; ring; gauze; burner; thermometer and stopper; clamp; crushed ice

Procedure

Usually when heat is put into a body, its temperature rises. But the experiment with p-dichlorobenzene suggests that this is not always the case. What happens when heat is put into ice?

Pour crushed ice into the beaker until it is about three-quarters full. Attach the ring to the stand at the correct height for heating. Put the gauze on the ring and set the beaker on the gauze. Put a clamp around the stopper and thermometer used in Experiment 4–8, and attach the clamp to the stand. Lower the clamp until the bulb of the thermometer is completely covered by the ice. The tip of the bulb should not touch the bottom of the beaker. The arrangement is shown in Figure 4–13.

Read and record the temperature of the ice. Light the burner and heat the ice, using only a very small flame. Record the temperature of the ice every 30 seconds until all the ice is melted.

Questions

1. What was the temperature of the ice at the beginning?
2. What was the temperature of the ice after (a) about one quarter of it was melted, (b) about one half of it was melted, and (c) about three quarters of it was melted?
3. What happened to the heat put into the ice during the melting?

Heat of Fusion

Figure 4–12
Graph axes for Experiment 4–8.

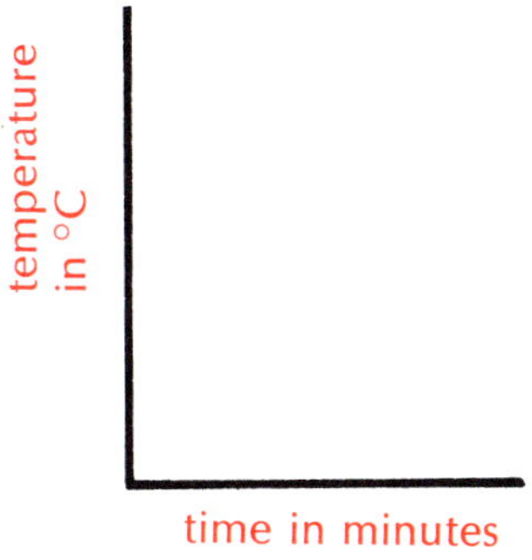

Scientists of the 17th and 18th centuries could not account for the apparent disappearance of heat during the melting of ice. Not only did heat disappear when ice melted, but heat mysteriously appeared when water froze. Today, the heat of melting ice is no longer a mystery. The heat put into the ice is transformed into potential energy. Scientists today use the term "heat of fusion." The word *fusion* (FYOO-zhun) means melting. The **heat of fusion** of ice is the number of calories of heat that must be supplied to one gram of ice at its melting point to change it to water at the same temperature.

In the next experiment you will find the heat of fusion of ice.

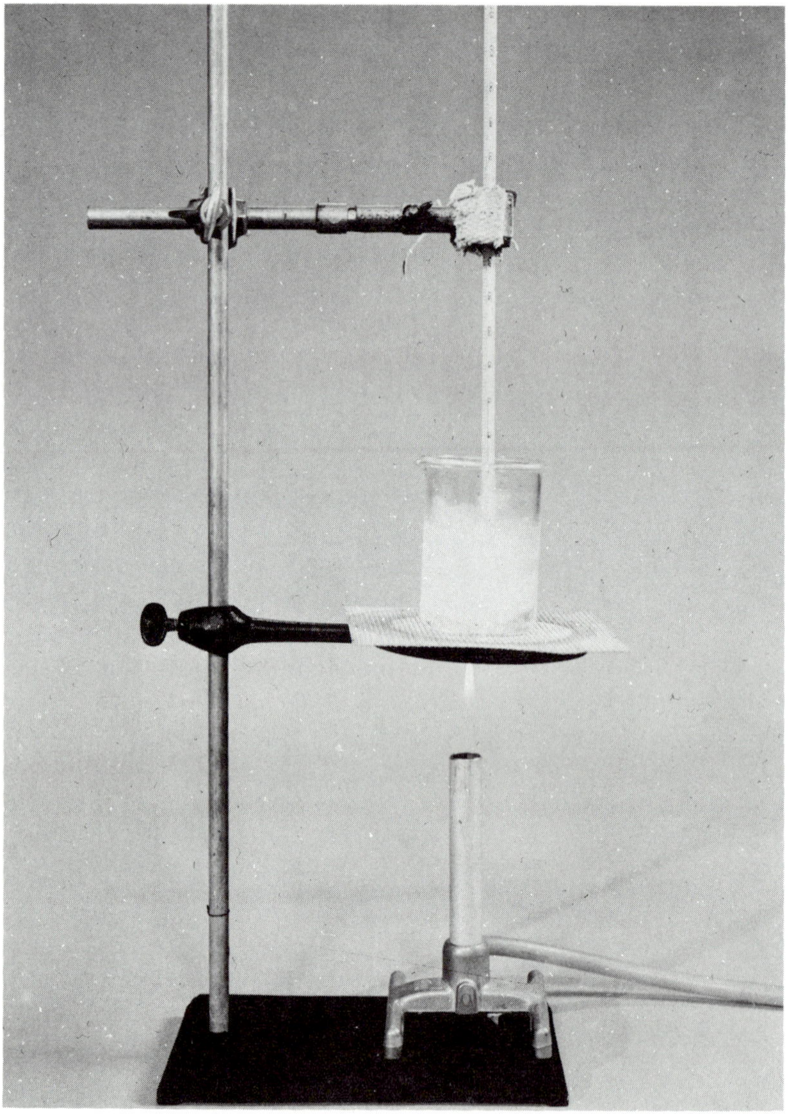

Figure 4–13
Why should you use only a small flame?

experiment 4–10

The Heat of Fusion of Ice

Materials

styrofoam cup; equal-arm balance; cotton or paper towel; thermometer; warm water (about 40°C); ice (pieces about the size of marbles)

Procedure

Find the mass of the styrofoam cup, and record it. Half fill the cup with warm water. Find the mass of the cup and water, and record it.

Figure 4–14
If you did not dry the surface of the ice, how would your results be affected?

Place some pieces of ice on the towel. The volume of ice should be about one-third the volume of the warm water.

Take the temperature of the warm water and record it. Dry the surface of the ice with the towel and then slide the dried pieces of ice from the towel into the cup. (See Figure 4–14.) Stir the mixture of ice and water with the thermometer and record the lowest temperature reached. If you followed the instructions correctly, all the ice will be melted at this lowest temperature. But if the temperature falls to 0°C and fragments of ice still remain in the cup, you have taken too much ice. In this case, repeat the experiment, using less ice.

Find the mass of the cup plus water. The mass of ice added is the difference between this mass value and the initial mass of the cup plus warm water.

Computations

1. Find the mass of the warm water.

2. Find the mass of ice melted.

3. Find the temperature change of the warm water.

4. Find the temperature change of the melted ice. Assume that the initial temperature of the melted ice (cold water) is 0°C.

5. The heat lost by the warm water equals the mass of warm water times its temperature change times the heat capacity of water. Compute this value.

6. The heat gained by the melted ice (the cold water) equals the mass of cold water times the temperature change of cold water times the heat capacity of water. Compute this value.

7. (a) What is the difference between the number of calories in Computations 5 and 6? (b) What does this difference in calories represent?

8. Compute the heat fusion of ice.

9. Ask your teacher for the correct value for the heat of fusion of ice. Then compute the percentage error of your answer.

PERSPECTIVES

perspective 4–1

Heat

A. What Is Heat?

The early scientists were familiar with certain facts about heat. They stated a theory that to them answered the question, "What is heat?" They knew, for example, that if hot water is poured into cold water, the hot water becomes cooler and the cold water becomes hotter. They also knew that a hot object does not lose mass as it cools, even though it loses heat. They argued that heat must be something that *flows*, and that it is therefore a fluid. Moreover, they said it must be a weightless fluid. This mysterious fluid was called caloric (ka-LOR-ik).

According to the caloric theory, the fluid caloric is present in all substances in varying amounts. This theory certainly explained the exchange of heat when hot and cold water were mixed. According to this theory some of the caloric in the hot water flowed into the cold water. However, the caloric theory did not explain the heat of friction; that is, the heat released when two bodies were rubbed together. This problem was first investigated by an American-born scientist named Benjamin Thompson (1753–1814), better known as Count Rumford.

Benjamin Thompson was born in Woburn, Massachusetts. He was a highly intelligent student and, even at the age of 14, showed unusual ability in mathematics and science. When he was 22 years old the Revolutionary War broke out. Thompson sided with the royalists. As the war proceeded, he was distrusted by supporters of the American cause, and he decided to leave the country. He emigrated to England, where he continued his scientific studies.

He later emigrated to Bavaria, a part of Germany. Here he worked for the Bavarian government, supervising the boring of cannon for the army. While engaged in this project, he began to doubt the truth of the caloric theory.

In boring the solid brass to make a gun barrel, a tool was rotated by machinery that, in turn, was operated by a horse. (See Figure 4–15.) The brass shavings were very hot. Heat was released as long as the tool was rotated. There appeared to be no end to the amount of heat that could be released by the solid brass. Where did the heat come from? According to the caloric theory, heat was squeezed out of the brass. If this were so, the heat left in the shavings after they had cooled should be less than the heat in the same mass of the original solid brass. Rumford tested this idea experimentally by finding the heat capacity (see Experiment 4–6) of the solid brass and the brass shavings. But he could find no difference. Thus the caloric theory was not supported by experimental evidence, and Rumford rejected it. But still the question remained, "Where did the heat come from?"

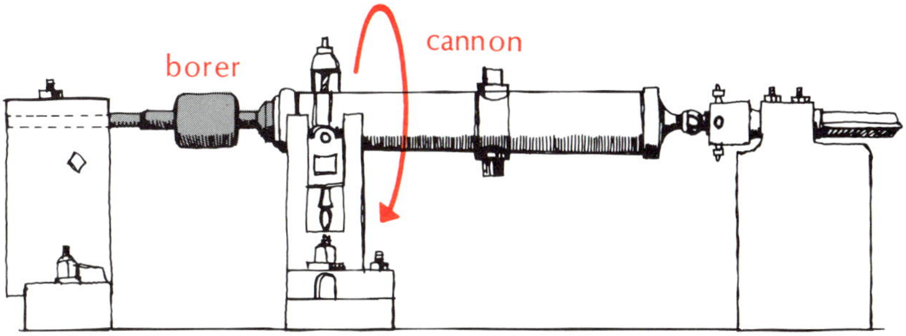

Figure 4–15
Count Rumford studied heat while he bored cannons with machinery like this.

Rumford's imagination turned to the horse. He suspected that somehow the work done by the horse was changed to heat. In other words, he figured that the mechanical energy put into the system was transformed into heat. He said that *heat is a form of energy*. This was a revolutionary idea. Today it is accepted by all scientists and it accounts for all the various kinds of physical change associated with heat.

The Emperor of Bavaria recognized the significance of Thompson's contribution to science and honored him by making him a Count. Thompson chose the title of Rumford from the name of a town in New Hampshire where he had lived.

B. Particles of Matter

Rumford stated that heat is a form of energy. What is the nature of this energy? All matter is made up of tiny particles. Heat energy is related to the motions of these particles. As you know, matter may be a solid, or a liquid, or a gas. The particles of gases have more freedom of motion than the particles of liquids or solids. The particles that make up a gas are called **molecules.** Can molecules be split into still smaller particles? The answer to this question is Yes. These smaller particles are called **atoms.** A molecule consists of two or more atoms attached together.

There are different kinds of atoms. A substance that consists of atoms of only one kind is called an **element.** Copper is an element, and so are aluminum, hydrogen, and oxygen. The molecules of a substance may be made up of two or more different kinds of atoms. Such a substance is called a **compound.** The most familiar compound is water. It is made up of atoms of the elements hydrogen and oxygen. The chemical shorthand for a molecule of water, H_2O, shows that in every molecule of water there are two atoms of hydrogen and one atom of oxygen.

C. The Internal Energy of a Body

Atoms and molecules are always in a state of motion. In a gas there are relatively large spaces between the molecules, so they can move freely in all directions, as shown in Figure

4–16A. The molecules of liquids are also free to move. (See Figure 4–16B.) Indeed, if this were not the case, liquids would not be able to flow. But what about solids? A solid (Figure 4–16C) cannot flow, but it would be quite wrong to assume that the atoms or molecules of a solid cannot move. It is believed that they vibrate back and forth somewhat like the motion of the bob of a pendulum. That is, molecules and atoms of solids, liquids, and gases possess kinetic energy.

Telephone wires become longer on hot summer days. (See Figure 4–17.) If a bar of copper is heated by a burner, it expands. Why do the wires become longer? Why does the copper expand? Atoms attract each other with an electrical force. Heat supplies energy to force the atoms a little farther apart. Atoms separated in this way possess potential energy, energy of position or condition. Thus both the kinetic energy and the potential energy of the atoms are increased by the heat. The total of the kinetic and potential energies of the atoms or molecules of a substance is called the **internal energy** of that substance.

You know that heat flows from a hot object to a cold one. When this happens the hot object loses some of its internal energy and the cold object gains internal energy. Thus heat means a *change* in internal energy. Heat transferred to an object is stored in that object as internal energy, not as heat. If heat enters the object, the internal energy of the object is increased. If heat leaves the object, the internal energy of the object is decreased. In other words, once an object is in a steady state (that is, if it is neither receiving nor emitting heat), it does not contain heat. Instead it contains internal energy.

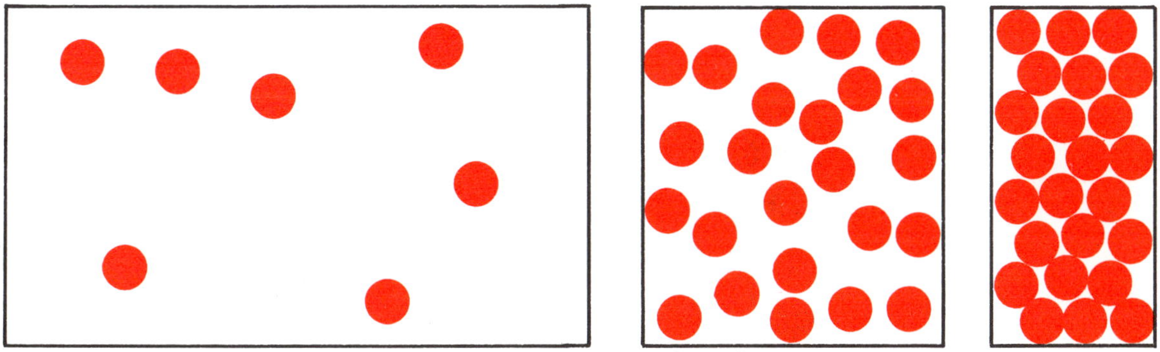

A gas **B** liquid **C** solid

Figure 4–16
A. In a gas, the molecules have room to move freely. **B.** In a liquid, too, the molecules have some room to move. **C.** In a solid, the molecules can vibrate.

Figure 4–17
In winter, wires are straight. In summer, they stretch and sag. How do workers putting up wires allow for these changes?

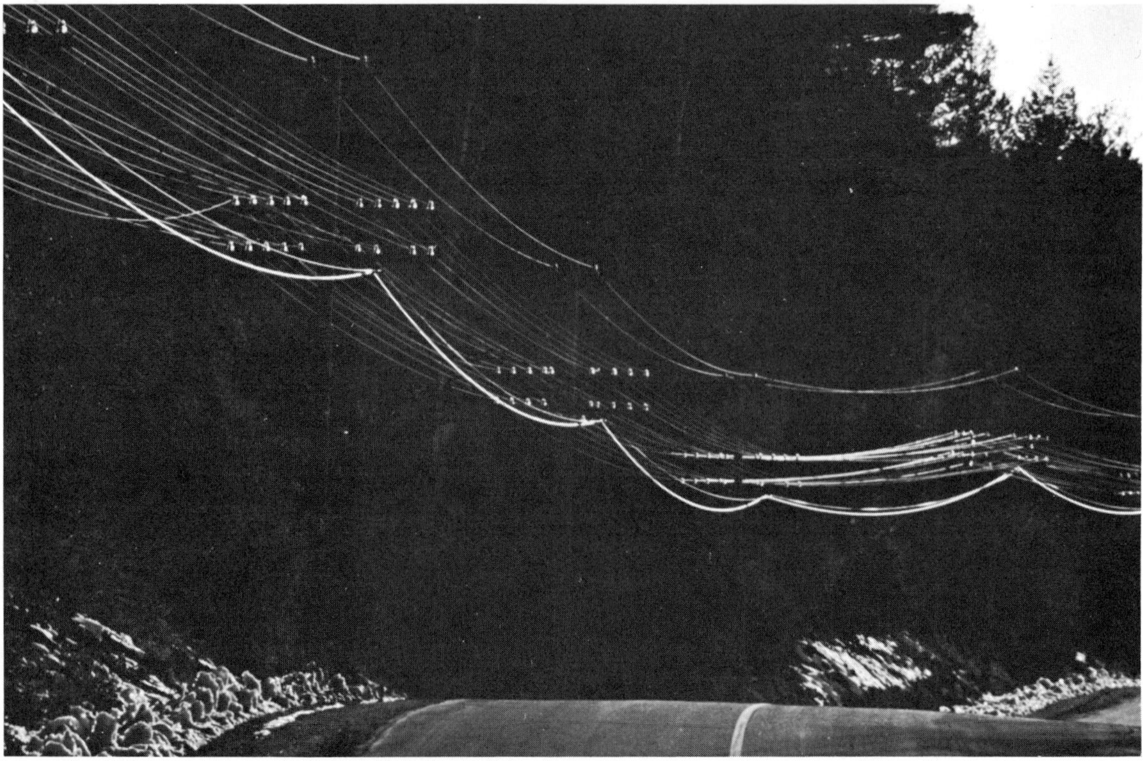

perspective 4–2

Temperature

A. What Is Temperature?

The word *hot,* as used in ordinary speech, really means high temperature, and the word *cold* means low temperature. Hence temperature refers to the hotness or coldness of a body. If heat is put into a body, its temperature usually rises. But temperature is not the same as heat. Temperature is not a form of energy. Then what is temperature?

The **temperature** of a substance is related to the *speed* of its atoms or molecules; the higher the temperature, the faster they move. How can temperature be measured? Because you cannot measure the speeds of molecules, you cannot measure temperature directly. However, you can measure it indirectly. For instance, a body expands if its temperature is raised. Since you can measure the amount of expansion, you can use the change in length or volume as an indirect way of measuring temperature. Indeed, this is how it is done in most thermometers. A common type of thermometer contains mercury. Another contains alcohol. Since alcohol is a colorless liquid, a red dye is usually added to it so that the temperature can be more easily read.

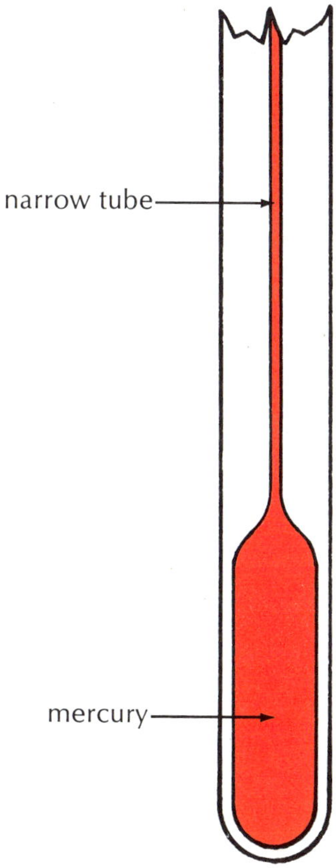

Figure 4–18
The narrower the tube, the more the mercury rises when it expands.

B. A Mercury Thermometer

A mercury thermometer consists of a glass bulb and stem. First the bulb and the stem are filled with mercury. Then the stem is sealed by melting the glass. The operation of the thermometer depends upon the fact that mercury, like other substances, expands (occupies more space) when heated, and contracts (occupies less space) when cooled. However, because there is only a small quantity of mercury in the bulb, the amount of expansion (ik-SPAN-shun) or contraction is very small. If the hollow stem were an ordinary piece of glass tubing, the expansion would hardly be noticed. However, the expansion can be magnified by using a tube that is very narrow inside. In a very narrow tube even a small expansion of the mercury can be observed as a measurable rise in the stem. (See Figure 4–18.)

To read the rise or fall of the mercury level, a scale must be marked on the stem. The first step in constructing a scale is to select two fixed points of widely separated temperatures. The usual reference temperatures are the freezing point and the boiling point of water. To mark the lower fixed point, the bulb and part of the stem are immersed in crushed ice, as shown in Figure 4–19. When this happens, the mercury contracts and the mercury level in the stem falls. When the contraction stops, the lowest level reached by the mercury is marked zero. To mark the upper fixed point, the bulb of the thermometer is placed in boiling water as shown in Figure 4–20. The mercury expands and the mercury level rises in the stem. When the expansion stops, the highest level reached by the mercury is marked 100. The space between these two fixed points is then divided into 100 equal divisions called degrees.

Figure 4–19 (below)
The lower fixed point of a Celsius thermometer is the freezing point of water, 0°C.

Figure 4–20 (right)
The upper fixed point is the boiling point of water, 100°C.

perspective 4–3

Heat Insulation

When hot and cold water are mixed as in Experiment 4–3, the fast-moving molecules of the hot water are slowed down and the slower molecules of the cold water are speeded up. Thus there is an exchange of kinetic energy. It seems reasonable to assume that the kinetic energy lost by the hot water equals the kinetic energy gained by the cold water.

However, this assumption is not quite true. Some heat from the hot water escapes into the air, and some escapes through the wall of the container. The heat that escapes cannot be transferred to the cold water. To reduce this heat loss to a minimum, an insulated container (one that minimizes the transfer of heat) should be used. Styrofoam is a good heat insulating material because it keeps heat from passing through. Whenever possible, styrofoam cups should be used instead of glass beakers in experiments that involve heat transfer.

A thermos bottle is a familiar example of a heat-insulated container. You know that hot coffee stays hot much longer in a thermos bottle than in an ordinary bottle. However, a thermos bottle insulates in a different way from a styrofoam cup.

A thermos bottle has double walls as shown in Figure 4–21. Air has been removed from the space between the walls and the inside wall has been silvered. The polished silver wall reflects heat back into the bottle and so helps to prevent its escape. However, the inner wall is hot. If this were a single wall exposed to air, the air would carry heat away from the hot liquid in the bottle. But, since it is a double wall with no air between, this loss of heat is prevented.

perspective 4–4

Heat Capacity

By definition, one calorie is the amount of heat needed to raise the temperature of one gram of water one Celsius degree. One calorie per gram per Celsius degree $\left(1.00 \frac{\text{cal}}{\text{g} \times \text{C}^\circ}\right)$ is called the heat capacity of water.

The heat capacity of a substance is a characteristic property of that substance. A few examples of the heat capacities of various substances are given in Figure 4–22.

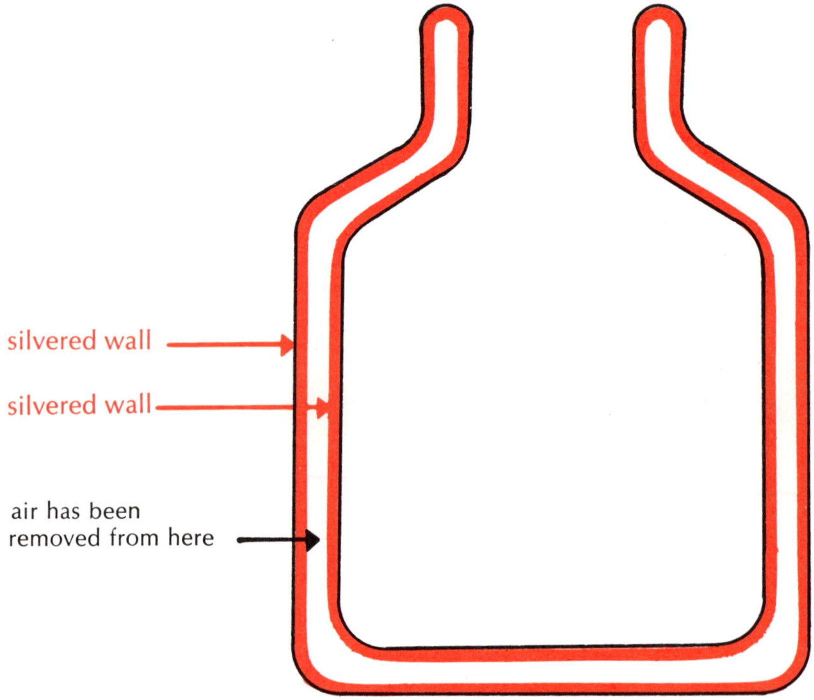

Figure 4–21
A thermos bottle insulates because the inside walls are silvered and because air has been removed from between them. Can you figure out how a styrofoam cup insulates?

Figure 4–22
The heat capacities of some common substances.

Substance	*Heat Capacity in* $\frac{cal}{g \times C^\circ}$
water	1.00
alcohol	0.55
ice	0.50
water vapor (steam)	0.50
aluminum	0.22
glass	0.16
copper	0.093
silver	0.056
mercury	0.033

Water has the highest heat capacity of all common substances. More heat is needed to raise the temperature of a given mass of water through 1 C° than to raise the temperature of the same mass of any other common substance through 1 C°. It should now be apparent that the amount of heat lost or gained by a substance depends not only upon its mass and temperature change, but also upon its heat capacity.

perspective 4–5

Change of State: Solid to Liquid and Liquid to Solid

A. The Heat of Fusion of Ice

The temperature of ice on a pond on a cold winter day is well below 0°C, and so is the temperature of ice cubes taken from the freezer section of a refrigerator. Suppose some ice cubes at −10°C are crushed, then placed in a beaker and gently warmed. What happens to the temperature of the ice? If you put a thermometer in the ice, you would see that the temperature of the ice rises until its melting point is reached. The temperature then remains steady until all the ice is melted. At this point the temperature of the water begins to rise. The temperature-heat relationship can be shown as a graph, as in Figure 4–23.

Why does the temperature of the ice remain constant during melting? What happens to the heat put into melting the ice? Ice is a solid because its molecules are bound together by strong electrical forces. The only way the molecules can move is by back-and-forth vibrations. As the temperature of the ice is raised, the rate of vibration increases until, at 0°C, molecules are able to break away from the crystal structure of the solid ice. When this happens the molecules are free to move around; we say that the ice has begun to melt. Further heating breaks

Figure 4–23
This graph shows the temperature changes as ice melts. Why is the ice line steeper than the water line?

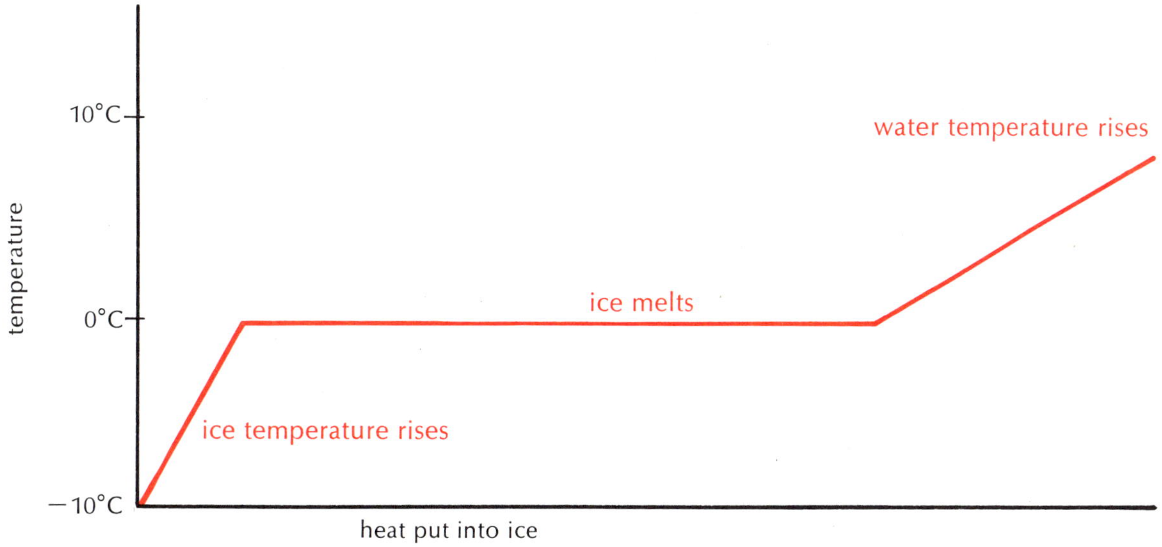

more of the crystal bonds until eventually all the ice melts. Continued heating then speeds up the moving molecules and the temperature of the water rises.

Since the temperature of melting ice is the same as the temperature of melted ice, we conclude that the speeds of the molecules in ice and melted ice are the same. Therefore the kinetic energy of ice molecules at 0°C must be the same as the kinetic energy of water molecules at the same temperature. Hence the energy put into the ice molecules is changed to potential energy. In other words, water molecules contain more potential energy than ice molecules at the same temperature. The number of calories of heat needed to melt one gram of ice is called the **heat of fusion** of ice. Every substance has its own particular heat of fusion value. That is, heat of fusion is a characteristic property of a substance. For example, the heat of fusion of aluminum is 94 cal/g, of copper 49 cal/g, of lead 5.4 cal/g, and of mercury 2.8 cal/g.

B. The Freezing of Water

Suppose water in a container is left outdoors on a cold night when the temperature drops from 10°C to −10°C. As the water cools, the rate of vibration of the molecules decreases until at 0°C the molecules begin to stick together. That is, freezing begins. As the cooling effect continues, freezing continues, but the temperature stays at 0°C until all the water is frozen. Then the temperature falls. These temperature and heat changes are shown graphically in Figure 4–24.

During the freezing process the molecules lose potential energy and this potential energy is transformed into heat. In other words, when water freezes, heat is given out.

The energy changes during melting of ice and freezing of water are represented in Figure 4–25.

C. The Effects of Freezing and Melting in Nature

When a lake, river, or part of a sea freezes in winter, the heat of fusion is released to the surrounding air. The total amount of heat released in this way is large enough to affect the local climate. These regions are warmed up, and extremely low temperatures are prevented.

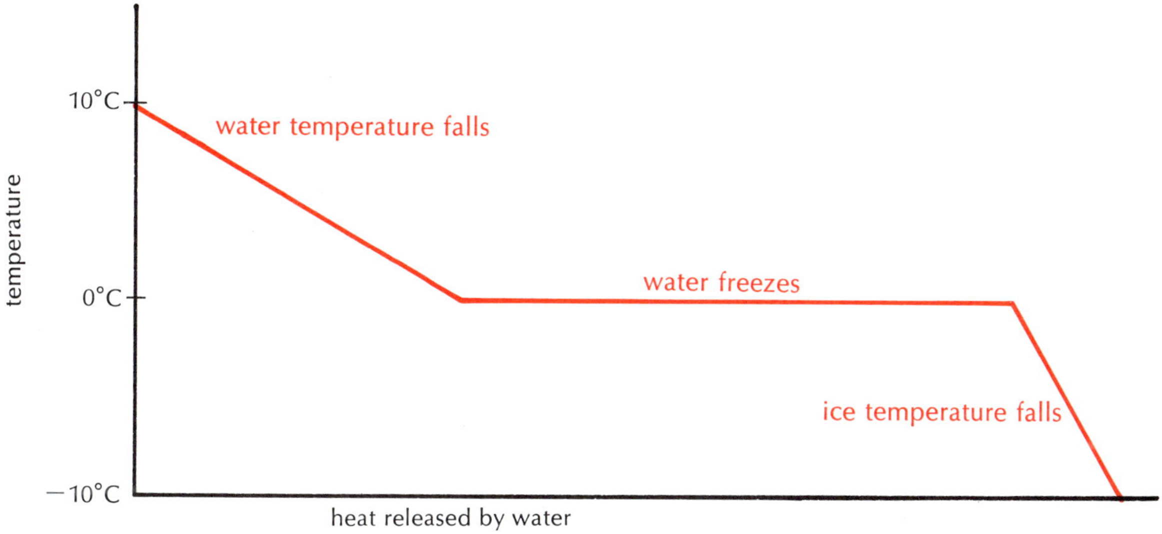

Figure 4–24
This graph shows the temperature changes as water freezes. Compare it with Figure 4–23.

On the other hand, a great deal of heat from the sun is needed to melt the ice in the spring. Because the heat of fusion of ice has a high value, the ice melts slowly and natural drainage is able to carry away the melted ice. If the heat of fusion of ice had a much lower value, snow and ice would melt rapidly under the warmth of the spring sun. The melted ice would accumulate and disastrous floods would be common.

Figure 4–25
When ice melts or water freezes, there are energy changes.

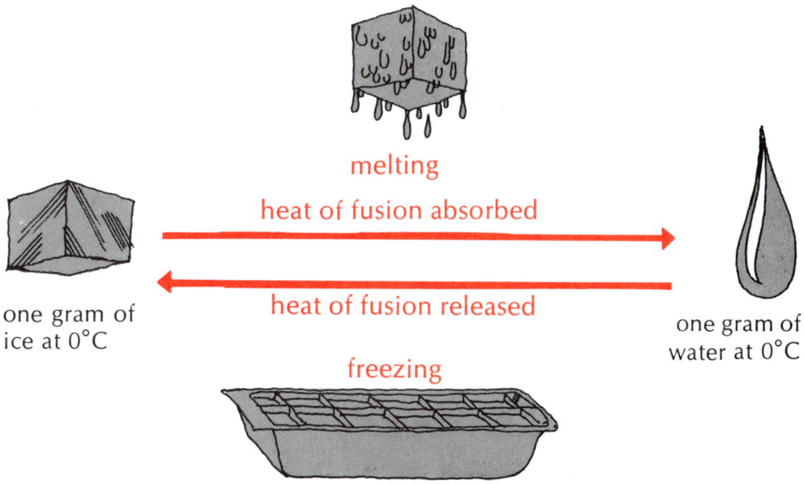

perspective 4–6

Change of State: Liquid to Gas and Gas to Liquid

A. Evaporation

Even water in a beaker does not all remain as liquid water. Some of the molecules near the surface break through the surface and escape into the air. This process, the change from liquid to vapor, is called **evaporation.** Evaporation occurs only at the surface of the water.

If you place a few drops of rubbing alcohol on the back of your hand, you notice that it feels cool. There is a similar cooling effect if you use water instead of alcohol. The cooling effect of evaporation can be observed as a drop in temperature by wrapping some cotton around the bulb of a thermometer, dipping the bulb in water and then withdrawing it from the water. As the water evaporates, the thermometer registers a marked drop in temperature. Why does evaporation have a cooling effect?

Figure 4–26
Faster-moving molecules are able to escape from a liquid.

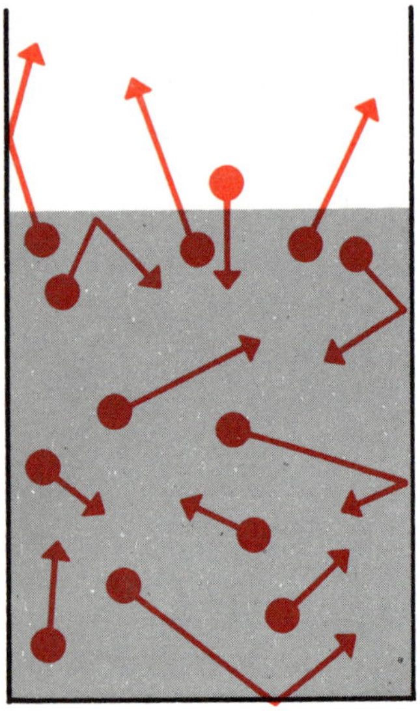

As you know, water molecules are always on the move. Some move faster than others. The faster-moving molecules break through the water surface and escape into the air more readily than the slower ones. (See Figure 4–26.) Since the faster molecules escape, the average speed of the remaining molecules decreases, and therefore the temperature of the water is lowered. Or, expressed another way, the average kinetic energy of the water molecules is decreased, and therefore evaporation cools the water that remains in the container.

The rate of evaporation depends upon the temperature of the water. The higher the temperature, the greater are the speeds of the molecules and therefore the greater is the rate of evaporation, at least until the boiling point is reached.

The reverse of evaporation (that is, the change from the gaseous state to the liquid state) is called **condensation.** These changes between the three states of matter are shown in Figure 4–27.

B. Boiling

When water is heated, the temperature rises and the rate of evaporation increases. Eventually the water becomes so hot

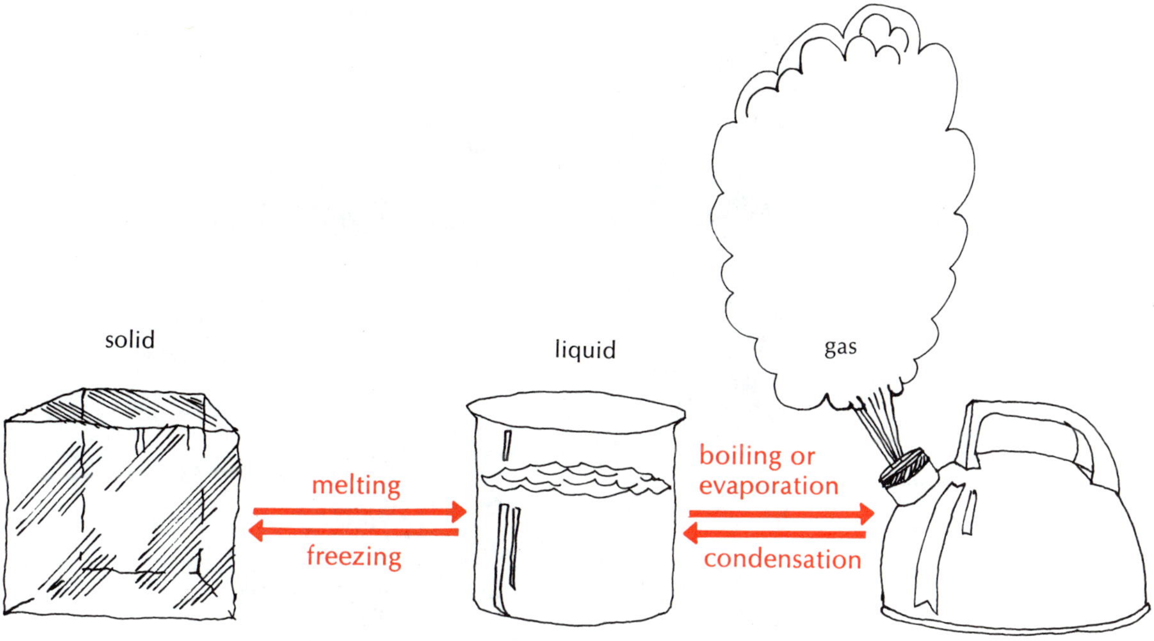

Figure 4–27
The changes between the three states of matter. Actually, steam is invisible. The cloud is tiny drops of liquid water condensed by the cool air.

that bubbles of vapor are formed within the water. The bubbles rise to the surface and "explode." When this occurs the water is said to be **boiling.** The temperature at which it occurs is the **boiling point** of the liquid.

Every substance has its own particular melting point and boiling point. These characteristic temperatures for a few common substances are given in Figure 4–28.

If water in a beaker is heated, its temperature rises steadily until it boils. However, the temperature of boiling water does not rise above its boiling point, no matter how strongly it is heated. What happens to the heat put into boiling water?

Figure 4–28
The melting points and boiling points of some common substances.

Substance	*Melting Point in °C*	*Boiling Point in °C*
water	0	100
aluminum	658	2467
copper	1080	2595
iron	1530	3000
mercury	−39	357
oxygen	−219	−184
hydrogen	−259	−252
sulfur dioxide	−73	−10

Figure 4–29
Boiling affects the whole volume of a liquid, not just the surface.

When water boils, bubbles of vapor are formed throughout the water, and these bubbles rise to the surface and burst. (See Figure 4–29.) Thus the heat put into boiling water is used to overcome the force of attraction between water molecules and push them farther apart. The temperature of the vapor is the same as the temperature of boiling water. Hence, when water is boiling, the average speed of the molecules in the liquid is the same as the average speed of the molecules in the vapor. Therefore the kinetic energy of the molecules does not change when boiling water becomes water vapor. However, the potential energy does change. The heat does work in pushing molecules apart, and this energy is stored in the molecules of vapor as potential energy. Thus heat put into boiling water does not raise its temperature. Instead the heat increases the potential energy of the molecules.

Conversely, if water vapor at 100°C is condensed to water at 100°C (that is, to water at its boiling point), the molecules become packed closer together and lose potential energy. Thus, heat is given out during condensation. The amount of heat given out during condensation is precisely the same as the amount of heat put into the water when it was boiling.

The change from liquid to vapor is called **vaporization** (vay-per-ih-ZAY-shun). Thus vaporization occurs in both evaporation and boiling. But evaporation and boiling are quite different. Evaporation is largely a surface effect and it occurs at all temperatures. On the other hand, boiling affects the whole mass of liquid, and it occurs at a particular temperature, namely the boiling point.

C. The Water Cycle

Figure 4–30
The water cycle is very complicated. It affects our lives in many ways, including the distribution of pollution.

All life on earth depends upon the natural processes of evaporation and condensation. The heat of the sun evaporates the waters of the earth. Thousands of millions of tons of water are changed to vapor every day. This vast amount of vapor is carried by winds, condensed to clouds, and eventually falls to earth as rain. The rain drains from the land as streams and

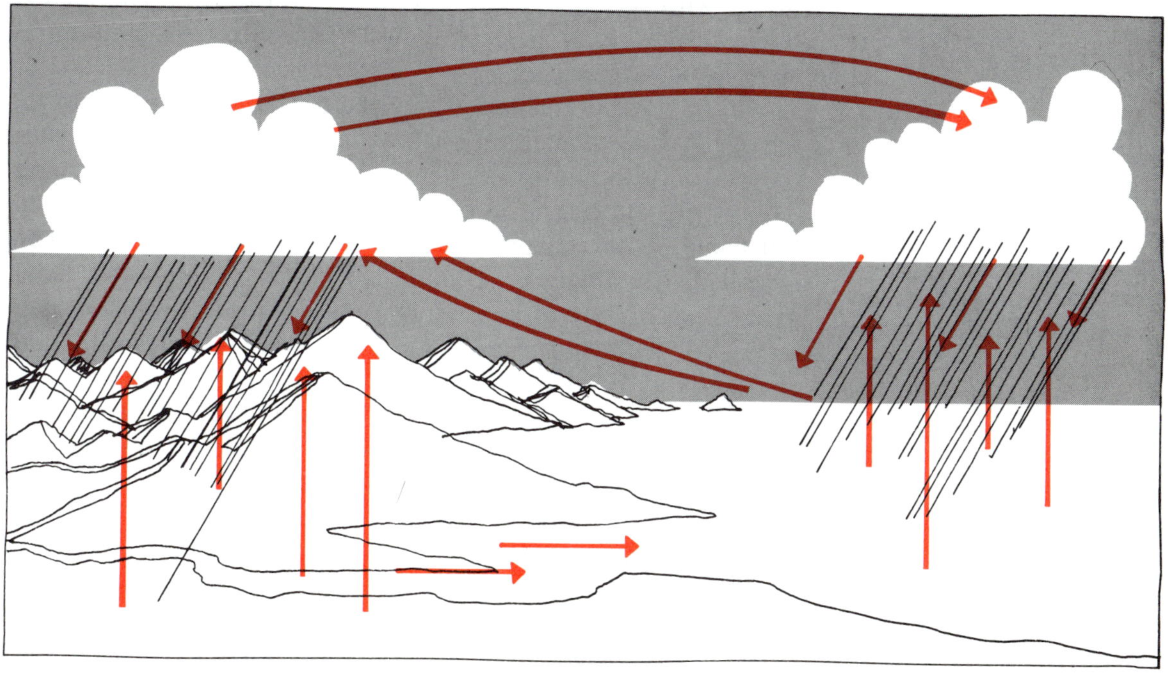

rivers. Eventually the water returns to the sea, where the cycle of evaporation and condensation begins all over again. (See Figure 4–30.) The water cycle has been in operation for billions of years. The fertility of the earth depends upon rain. Indeed, without a constantly recurring water cycle, plants could not grow and the earth could not support life.

perspective 4–7

Sources of Heat

A. Bunsen Burner

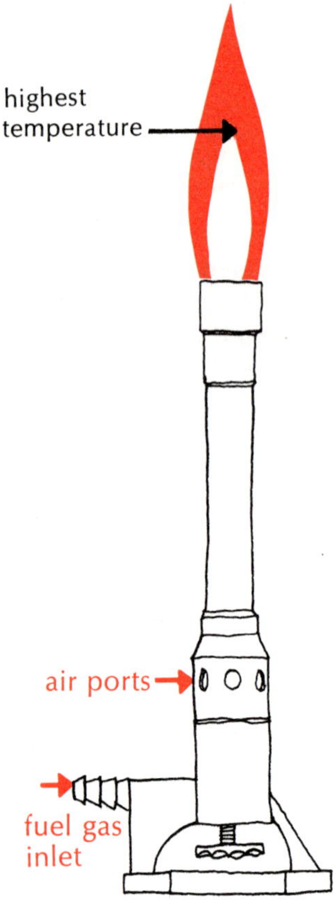

Figure 4–31
The flame of a Bunsen burner can be adjusted easily.

In most of the experiments in this chapter you have probably used a gas burner. It is difficult to realize that until the middle of the last century candles and stoves were the only sources of heat used for experimental work. In 1855 a German scientist named Robert Bunsen changed all this by designing an efficient and convenient burner that is still in use today.

A Bunsen burner burns fuel gas, which is led through an inlet tube near the base. A movable ring with openings called air ports is attached to the bottom of the barrel, as shown in Figure 4–31. Air enters the barrel through the air ports and mixes with the fuel gas. By rotating the ring, you can change the size of the air port openings and control the amount of air entering the barrel.

The mixture of air and gas burns in a flame at the top of the barrel. A fuel gas cannot burn without air. If the air ports are closed, only the air around the flame is available for burning. This is not enough to burn the gas completely and produce a hot flame. If the air ports are closed, the flame is white. This is called a **luminous** (LOO-mih-nus) flame. A luminous flame gives very little heat and deposits black soot on objects held in it. For most purposes a luminous flame is undesirable and should be avoided.

As the air ports are opened, the flame becomes less luminous and much hotter. The air ports should be opened and adjusted until the flame is non-luminous and the parts of the flame, as shown in Figure 4–31, can be clearly seen. The gas should burn relatively quietly. A noisy or roaring flame means that too much air is entering through the air ports. Such a flame is cooler than it should be.

B. Alcohol Burner

In an alcohol burner, the fuel is alcohol, a liquid. A cotton wick dips into the alcohol as shown in Figure 4–32. The liquid alcohol climbs up the wick and escapes as a vapor from the wick. Alcohol vapor is flammable. Hence, if a burning match is applied to the top of the wick, the vapor burns with a hot, almost colorless flame.

C. Propane Torch

A propane torch, shown in Figure 4–33, is like a Bunsen burner. Propane, the fuel, is a gas. It is contained in the small tank and delivered from the tank to the tip of the attached burner. Instructions for using the torch come with it.

Figure 4–32 (below)
An alcohol burner is portable and does not require gas pipes.

Figure 4–33 (right)
A propane torch produces a very hot flame.

QUESTIONS

A

1. What are fixed points on a thermometer scale?

2. What are the fixed points on a Celsius thermometer?

3. How many fixed points are necessary for a scale?

4. State the caloric theory of heat.

5. State Rumford's definition of heat.

6. List the three factors that are involved in computing the amount of heat lost or gained by a substance.

7. Compare the amount of heat required to raise 1 g of water 1 C° and 1 g of corn oil 1 C°.

8. What is the difference between melting and freezing?

9. What is the difference between the melting point and the freezing point of a substance?

10. What is meant by the internal energy of a body?

11. (a) What is an element? (b) Give an example of an element.

12. (a) What is a compound? (b) Give an example of a compound.

13. The particles of a liquid and a solid respond differently when heated. What is this difference?

14. Define a kilocalorie of heat.

15. What substance has the greatest heat capacity of all common substances?

16. The heat capacity of iron is less than the heat capacity of water. What does this statement mean?

17. What is meant by the term *heat of fusion of ice?*

18. What is evaporation?

19. Compare evaporation and condensation.

20. What happens when water boils?

21. What is meant by the term *water cycle?*

B

22. The word *thermometer* comes from two Greek words, *therm* meaning heat and *meter* meaning a measure. (a) Do you think the word *thermometer* is a good name for this instrument? (b) If your answer is No, can you suggest a better name?

23. The boiling point of alcohol is 78°C and the freezing point is −115°C. Suppose Celsius had selected these temperatures as the fixed points of his thermometer and called them 100° and 0°. Would the size of a degree have been increased or decreased?

24. A student placed some lead shot in a tall cylinder. He closed the cylinder with the palm of his hand and then turned the cylinder over 20 or 30 times. A thermometer showed that the temperature of the lead shot rose by several degrees. That is, the lead shot gained heat. Where did the heat come from?

25. A red-hot copper bar is allowed to cool to room temperature. What happens to the particles of copper during this process?

26. When a thermometer is placed in a hot liquid the mercury expands. Why can the expansion of mercury be used as a measure of temperature change?

27. Before refrigerators were invented, ice cut from a pond was packed in sawdust and stored for use in the summer. Why was sawdust used for this purpose?

28. The heat capacity of ice is 0.50 $\frac{\text{cal}}{\text{g} \times \text{C}^\circ}$. Why is the heat unit in the numerator and why are the mass and temperature units in the denominator?

29. If water and ice are mixed they eventually reach the same temperature (0°C). At this temperature ice and water molecules have the same kinetic energy but different potential energies. Explain why.

30. If you are wet when you sit on the beach on a hot day, you cool off. Why?

31. A burn from steam at 100°C is much more severe than a burn from boiling water at 100°C. Why?

For each of Questions 32–36, write the letter that best completes or answers the statement.

32. A bucket of water is heated to 60°C. A glass of water is dipped from the bucket of water at 60°C and set alongside.

a The bucket of water and the glass of water cool at the same rate.
b No cooling of the bucket, the glass, or the water occurs.
c Your hand would probably be burned if it was placed in the bucket of water when scooping out the glassful.
d The glassful of water cools at a faster rate than the bucket of water.

33. If 1 l of boiling water is poured into a metal pail containing 1000 g of ice at 0°C, the temperature of the resulting mixture is

a approximately 50°C
b below 50°C
c slightly higher than 50°C
d about 90°C

34. If 100 g of water at 10°C are poured into 100 g of cake batter at 50°C (containing eggs, flour, and shortening), the temperature of the resulting mixture is

a approximately 50°C
b approximately 30°C
c above 30°C
d below 30°C

35. According to the table of heat capacities, Figure 4–22,

a 100 ml of hot chocolate at 80°C should cool more quickly in a cup made of copper than in a cup made of styrofoam
b 100 ml of lemonade at 5°C should warm more quickly in a container made of glass than in one made of silver
c an alcohol thermometer should respond to a change in temperature more quickly than a mercury thermometer
d 100 ml of lemonade at 5°C warms up more quickly than 100 ml of whiskey at 5°C

36. According to the graph below, which is the cooling curve of a substance,

a the freezing point of the substance is approximately 20°C
b a hot water bath was used to heat the substance
c the substance is a liquid
d the substance does not have a definite freezing point, at least not between 20°C and 120°C

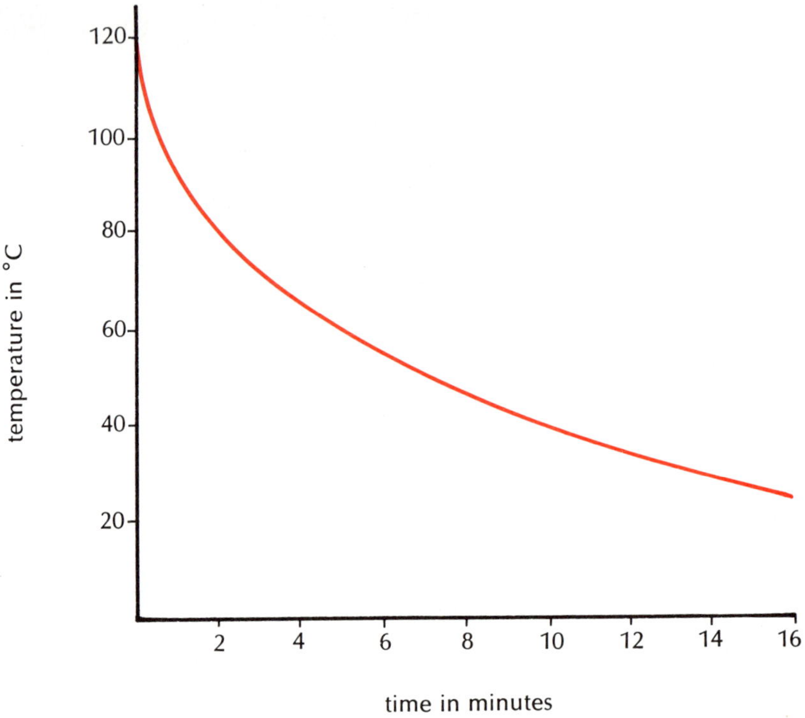

PROBLEMS

1. Compute the amount of heat needed to raise the temperature of 100 g of water from 15°C to 30°C.

2. Compute the amount of heat needed to raise the temperature of 100 g of iron from 15°C to 30°C.

$$\left(\text{Heat capacity of iron} = 0.12 \frac{\text{cal}}{\text{g} \times \text{C}^\circ}\right)$$

3. A 500-g piece of iron at 10°C absorbs 1800 cal of heat. To what temperature does the metal rise?

$$\left(\text{Heat capacity of iron} = 0.12 \frac{\text{cal}}{\text{g} \times \text{C}^\circ}\right)$$

4. 100 g of water at 70°C and 50 g of water at 25°C are mixed. What is the temperature of the mixture?

5. A sample of warm water at 80°C is added to 100 g of cold water at 10°C. The temperature of the mixture is 30°C. Compute the mass of warm water.

6. A sample of warm water at 40°C is added to 100 g of iron at 10°C. The temperature of the mixture is 30°C. Calculate the mass of warm water.

$$\left(\text{Heat capacity of iron} = 0.12 \frac{\text{cal}}{\text{g} \times \text{C}^\circ}\right)$$

7. 50 g of water at 40°C are added to 200 g of crushed rock at 20°C. The temperature of the mixture is 31°C. Compute the heat capacity of the rock.

8. What mass of ice at 0°C must be added to 200 g of hot tea at 90°C to give cold tea at 5°C? (Heat capacity of tea = heat capacity of water. Heat of fusion of ice = 80 cal/g)

9. If 20 g of ice at 0°C are added to 150 g of water at 30°C, what is the temperature of the mixture? (Heat of fusion of ice = 80 cal/g)

Chapter 5

FLUIDS

A submarine uses the properties of fluids.

Like an ordinary ship, a submarine can float on water. But it can also remain totally submerged under water. A submarine floats with about one quarter of its bulk out of water. To submerge a submarine, it must be made heavier. This is done by letting sea water into large tanks inside the submarine. To bring a submerged submarine to the surface, it must be made lighter. This is done by forcing water out of the tanks with compressed air. You can imitate the submerging and refloating of a submarine by doing the following experiment with a medicine dropper.

Almost fill an 800-ml beaker with water. Then fill a medicine dropper with water and lower it upright into the beaker. The medicine dropper sinks. Remove the dropper from the beaker and squeeze the rubber cap to force out a few drops of water. Replace the dropper in the beaker. If it sinks, force out a few more drops of water. Continue to do this until the dropper floats with the top of the rubber cap at the water surface.

Fill a 100-ml graduate with water and carefully transfer the dropper from beaker to graduate. The dropper should now float in the graduate, totally submerged except for the top of the rubber cap, as shown in Figure 5–1. Cover the top of the graduate with the palm of your hand and press down on the top of the graduate, as shown in Figure 5–2. With a little practice you can make the dropper fall to the bottom of the graduate by applying a slight pressure, and make it rise to the surface by releasing the pressure.

When you press down on the top of the graduate, what happens to the water level in the dropper? In what direction is the water pressing to make this happen? If you pressed down on a solid, would it act the same way as the water? In this chapter you will find out more about the way liquids act.

Figure 5–1
The dropper should just barely float.

Figure 5–2
If the dropper does not sink, force out a drop of water and try again.

BUOYANCY

Fluids

As you know, there are three states of matter: solids, liquids, and gases. A solid is a rigid body with a fixed volume. A liquid has a fixed volume but not a fixed shape. For example, the shape of a liter of water depends upon the shape of the vessel into which it is poured. A gas, on the other hand, has neither fixed shape nor fixed volume. A gas, unlike a liquid, will expand and completely fill a container of any shape or size.

Both liquids and gases are able to flow through pipes, or in currents such as the Gulf Stream or the trade winds. For this reason liquids and gases are called **fluids** (FLOO-idz). The Latin word *fluere* means "to flow." The most common liquid fluid is water and the most common gaseous fluid is air. An important property that applies to water and air—in fact, to all fluids—is buoyancy (BOY-an-see). You will investigate this property in the following experiments.

experiment 5–1 The Buoyant Force of Water

Materials

object denser than water; thread, about 30 cm long; 2.5-N spring scale; 250-ml beaker

Procedure

Tie the thread around the object and make a loop at the other end of the thread. Attach the loop to the hook of the spring scale and record the weight of the object in newtons.

Three-quarters fill the beaker with water. Now lower the object into the beaker as in Figure 5–3. It should not touch the bottom of the beaker. Record the weight of the object in newtons when it is submerged.

Questions

1. Does the object appear to gain or lose weight when submerged in water?

2. The upward force exerted by water upon a submerged object is called **buoyancy.** What is the buoyant force exerted upon the object by the water?

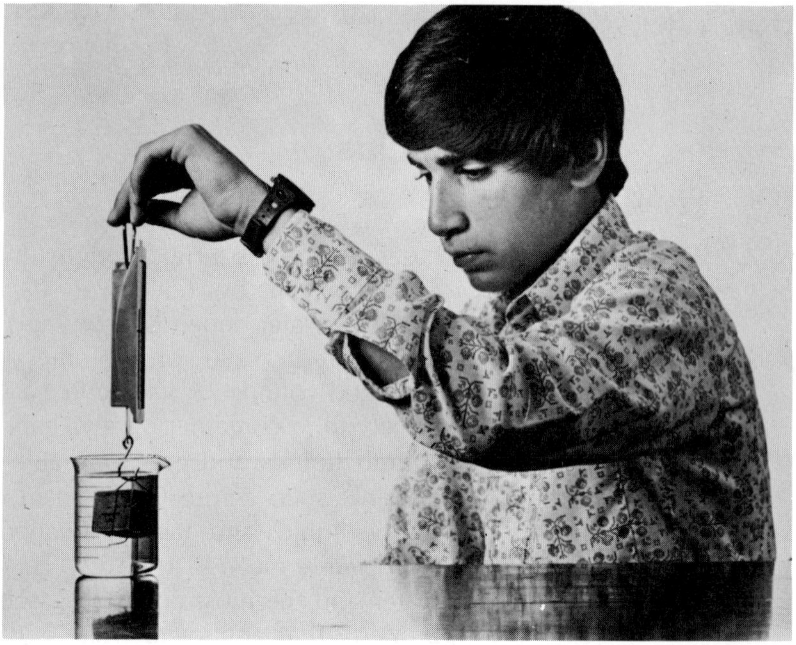

Figure 5–3
Why should the object not touch the bottom of the beaker?

Changing from Grams to Newtons

Sometimes you know the mass of an object in grams and need to know its weight in newtons. As you learned in Perspective 3–6, on the earth's surface a mass of 1 kg (1000 g) has a weight of 9.8 N. Thus, if the mass of an object is known, its weight can be computed. If its weight is known, its mass can be computed. In the next experiment you will know the mass of an amount of water, and you will be asked to compute its weight in newtons.

Example: A certain amount of water has a mass of 75 g. What is the weight of the water in newtons? A mass of 1000 g has a weight of 9.8 N, so

$$\text{weight of water} = \frac{75\ \text{g}}{1000\ \text{g}} \times 9.8\ \text{N}$$

$$= \frac{75 \times 9.8\ \text{N}}{1000} = \frac{735\ \text{N}}{1000} = 0.735\ \text{N}$$

Example: What is the mass in grams of an object whose weight is 1.5 N?

$$\text{mass of object} = \frac{1000\ \text{g} \times 1.5\ \text{N}}{9.8\ \text{N}} = 153\ \text{g}$$

Problems

1. A block of wood has a mass of 200 g. What is its weight in newtons?
2. What is the weight in newtons of 7.0 kg of salt?
3. The weight of a body is 300 N. What is its mass in grams?

experiment 5–2 The Weight of an Object Submerged in Water

Materials

overflow can or paper hot-drink cup; equal-arm balance; 150-ml beaker; thread; 6 iron washers (the same as those used in Chapter 3); spring scale

Procedure

To make an overflow can from a paper hot-drink cup, cut two slits down from the top, about 2 cm apart. The slits should be about 1 cm long. Squeeze the cup and bend the flap so that it forms a spout, as in Figure 5–4. The spout should point downward.

Pour water into the overflow can until it is full and water runs out of the spout. Using the equal-arm balance, find the mass

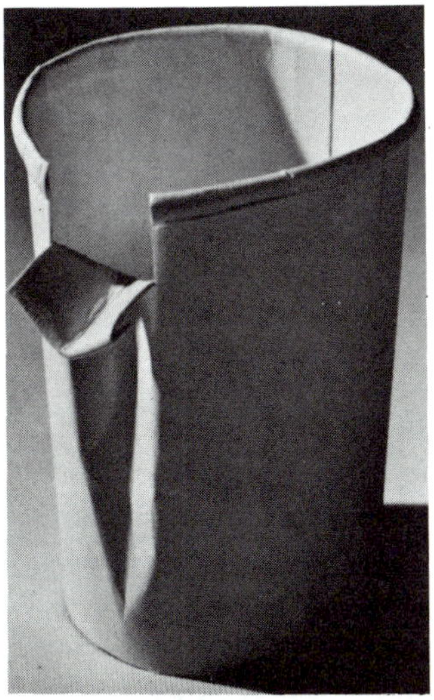

Figure 5–4 (above)
An overflow can made from a paper hot-drink cup.

Figure 5–5 (right)
How would your results be affected if the washers were not completely submerged?

of the beaker. When the last drops of water have flowed out of the overflow can, place the beaker under the spout.

Tie the thread into a loop. Pass the loop through the holes in the iron washers. Attach the ends of the thread loop to the hook of the spring scale. Read and record the weight in newtons. Still holding the spring scale, lower the washers into the can until they are completely submerged in water, as shown in Figure 5–5. Record the weight of the washers in water. The overflow water is caught in the beaker. When the overflow stops, find the mass of the beaker and overflow water.

Computations

1. Compute the mass of water displaced by the washers, that is, the mass of water collected in the beaker.

$$\text{mass of water} = \text{mass of beaker with water} - \text{mass of beaker}$$

2. Compute the weight of the displaced water in newtons.

3. Find the difference between the weight of the washers in air and their weight in water.

4. How does the weight of the water displaced by the washers compare with the change in weight when the washers are submerged in water?

Questions

1. The answer to Computation 4 illustrates a law known as **Archimedes'** (ar-ki-MEE-deez) **Principle.** State Archimedes' Principle in your own words.

experiment 5–3 The Density of the Washers

Materials

none

Procedure

In Experiment 5–2 you found the weight of the washers in air and the mass of water displaced by the washers. Since 1 ml of water has a mass of 1 g, you can easily convert the mass of water displaced into the volume of water displaced. The volume of water displaced is the same as the volume of the object. Moreover, the weight of the washers in newtons can be converted to mass in grams. With this information you can find the density of the object, because

$$\text{Density} = \frac{\text{Mass}}{\text{Volume}}$$

Make a table for your data from Experiment 5–2, as in Figure 5–6.

Figure 5–6
Data table for Experiment 5–3.

Mass of washers	=	? g
Mass of water displaced	=	? g
Volume of water displaced	=	? ml
Volume of washers	=	? ml

Computations

1. Compute the density of the washers.

2. Ask your teacher for the density of iron. Compute your percentage error. Remember that

percentage error =

$$\frac{\text{difference between correct value and experimental value}}{\text{correct value}} \times 100\%$$

experiment 5–4 To Find the Density of an Irregular Object by Buoyancy

Laboratory Test

Materials

thread; stone (about 150 to 250 g); overflow can (or the paper overflow cup from Experiment 5–2); equal-arm balance; 150-ml beaker; spring scale

Procedure

Tie the thread around the stone. Follow the procedure of Experiment 5–2. Record your data and then compute the density of the stone.

Ask the teacher for the correct value of the density and compute your percentage error.

experiment 5–5 The Buoyant Force of a Liquid Other than Water

Materials

250-ml beaker; ethylene glycol (the antifreeze used in Experiment 2–9); thread; 6 iron washers; spring scale

Procedure

Three-quarters fill the beaker with ethylene glycol. Pass a loop of thread through the six iron washers. Attach the ends of the loop to the hook of the spring scale. Record the weight of the washers in air.

Now lower the washers, still attached to the spring scale, into the ethylene glycol until they are completely submerged. Record the scale reading.

Computations

1. From the weight of the washers in air and in ethylene glycol, compute the apparent change in the weight of the washers.

2. Compare the apparent change in weight in ethylene glycol with the apparent change in weight in water as found in Experiment 5–2. How do you explain the difference in these values?

Questions

1. In general, how do you think the density of a liquid affects the apparent change in weight of an object submerged in the liquid?

experiment 5–6 The Buoyant Force on a Floating Object

Materials

overflow can (or the paper overflow cup from Experiment 5–2); wooden block; 150-ml beaker; equal-arm balance

Procedure

In all the previous experiments an object was completely submerged in water. You compared the weight of the object in air with its apparent weight in water. You will now make the same comparison for a floating object. A floating object is only partly submerged in water. Does Archimedes' Principle apply to an object that floats as well as to one that is totally submerged?

Find the masses of the block and the beaker. Fill the overflow can. Wait until the last drops of water have flowed out of the can. Then place the beaker under the spout. Gently lower the block into the overflow can until it floats freely in the water as in Figure 5–7. When the overflow stops, find the mass of the beaker plus the displaced water.

Computations

1. Compute the mass of water displaced by the floating object.

2. Compare the mass of the floating object with the mass of water that it displaced.

Questions

1. Write a statement in your own words concerning the comparison asked for in Computation 2.

2. Does Archimedes' Principle apply to a floating object? If your answer is Yes, restate the principle as it applies to both submerged and floating objects.

SOLUBILITY

Solutions

An important property of water is that it will dissolve many substances. For example, if a lump of sugar is placed in water, it soon disappears and forms a solution (suh-LOO-shun) which is sweet to taste. A water **solution** is a mixture of water and another substance dissolved in it. Some substances dissolve more easily than others. Or, some substances are more **soluble** (SOL-yoo-bul) than others. If you compare sugar, salt, and chalk, you will find that sugar is very soluble in water, salt is fairly soluble, but chalk is hardly soluble at all. An **insoluble** (in-SOL-yoo-bul) substance is one that does not dissolve in water, or dissolves only very slightly. Thus chalk is insoluble. The term *solution* also refers

Figure 5–7
Place the block in the water gently. Do not drop it in.

to substances dissolved in liquids other than water. For example, iodine, a solid, dissolves in alcohol to form a solution. However, this book will deal only with water solutions.

A given mass of water will dissolve a certain amount of solid and no more, at a given temperature. When the maximum amount of solid is dissolved in the water, the solution is said to be **saturated** (SACH-uh-ray-ted). The mass of solid that will dissolve in 100 g (100 ml) of water to form a saturated solution is called the **solubility** (sol-you-BIL-ih-tee) of the solid. In the next experiment you will find the solubility of common salt.

experiment 5–7 The Solubility of Common Salt

Materials

100-ml graduate; 125-ml Erlenmeyer flask; common salt; equal-arm balance; 10 cm by 10 cm piece of paper; rubber stopper to fit flask

Procedure

Pour 25 ml of water into the graduate and then pour the water into the Erlenmeyer flask. Find and record the mass of the piece of paper.

Common salt is also called sodium chloride (SO-dee-um KLOR-eyed). Pour about 10 g of sodium chloride onto the paper and find the mass of the paper plus salt.

Figure 5–8
Swirl the flask to make the salt dissolve faster.

Pour a small quantity of the salt into the flask and close the flask with the rubber stopper. Swirl the flask, as in Figure 5–8, to mix the contents. Do not shake the flask. If you do, some of the salt will stick to the upper part of the flask. The salt should dissolve in the water after the swirling. Now add a little more salt to the flask and repeat the swirling until the salt dissolves. Repeat this process until a little of the salt remains undissolved. The undissolved salt shows that the water has dissolved as much salt as it can. That is, it shows that the solution is saturated. Finally, find the mass of paper plus the salt that remains upon it.

Pour the rest of the salt from the paper into the flask. Close the flask with the stopper and set it aside for use in the next experiment.

Computations

1. How many grams of sodium chloride were dissolved in the 25 ml of water?

2. Compute the solubility of sodium chloride. The solubility is the number of grams of sodium chloride that would dissolve in 100 ml (100 g) of water.

experiment 5–8 To Find the Solubility of Sodium Chloride by Evaporating a Saturated Solution

Materials

100-ml graduate; evaporating dish; equal-arm balance; stand; 4-inch ring; wire gauze; burner

Procedure

Pour about 20 ml of the saturated solution of salt prepared in Experiment 5–7 into the graduate. Find the mass of the evaporating dish, pour the solution into the dish, and find the mass of the dish plus solution.

Attach the ring to the stand so that the ring is about 10 cm above the top of the burner. Cover the ring with the wire gauze, and place the dish and solution on the gauze as shown in Figure 5–9.

Light the burner and heat the dish. When the solution is nearly evaporated, turn the flame down low and heat the dish very gently. Otherwise some salt will spatter out of the dish and be lost. When the salt appears to be dry, heat the dish strongly with the full flame of the burner. The strong heating drives the last trace of water out of the salt and leaves it completely dry.

Allow the dish to cool. Then find the mass of the dish plus the salt residue.

Figure 5–9
Watch the solution carefully as it heats.

Computations

1. How many grams of sodium chloride were dissolved in the sample of solution used in the experiment?

2. Compute the solubility of sodium chloride.

3. Compare the solubility value of sodium chloride you found by the evaporation method to the value you found by the method of Experiment 5–7.

experiment 5–9 The Solubility of an Unknown Compound

Laboratory Test

Materials

100-ml graduate; evaporating dish; equal-arm balance; stand; 4-inch ring; wire gauze; burner

Procedure

Obtain about 20 ml of a saturated solution of a compound from your teacher. Following the procedure of Experiment 5–8, find the solubility of the compound.

Ask your teacher to write the correct value of the solubility under your own recorded experimental value. From these two values, compute your percentage error. Your experimental value may be higher than the correct value, or it may be lower. In either case, try to account for your error.

experiment 5–10

The Solubility of Potassium Nitrate at Different Temperatures

Materials

large test tube; 250-ml beaker; solid potassium nitrate; thermometer; equal-arm balance; burner; stand; test-tube clamp (or buret clamp); 100-ml graduate; towel; piece of paper, about 10 cm by 10 cm

Procedure

Figure 5–10
Clamp the test tube to a stand and heat it.

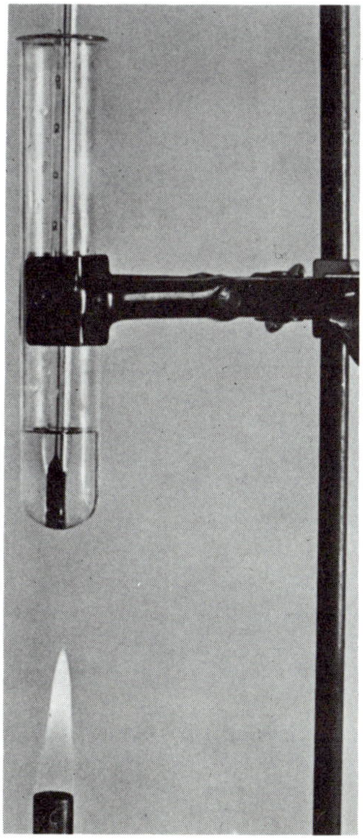

In Experiments 5–7 and 5–8, you found the solubility of common salt at room temperature. What happens to the solubility of a substance if the temperature of the solution is changed? To answer this question you will find the solubility of a substance called potassium nitrate (po-TAS-see-um NY-trayt) first at a high temperature and then at lower and lower temperatures.

Place the thermometer in the test tube, and set the tube in the beaker. Find the combined mass of the beaker, tube, and thermometer to the nearest 0.1 g.

Place the piece of paper on the left pan of the equal-arm balance. Pour about 10 g of finely powdered potassium nitrate onto the paper. Pour all of the potassium nitrate into the test tube and again find the total mass of beaker, test tube, potassium nitrate, and thermometer. Write all your data in a table, and find the mass of the potassium nitrate.

Measure out about 8 ml of water in the graduate. Pour the water into the test tube. Clamp the tube vertically to the stand so that the bottom of the tube is about 10 cm above the top of the burner. (See Figure 5–10.)

Heat the tube with a small, faintly luminous flame (not more than 2 or 3 cm in height) until all the potassium nitrate has dissolved. Stir the solution with the thermometer during the heating. When the solution is clear, remove the burner. Continue to stir the solution until a trace of solid appears in the tube. The solution is then saturated. At that instant read the thermometer. Record the temperature as the temperature of the saturated solution. Then place the tube and its contents in the 250-ml beaker and find the total mass. Compute the mass of water present in the solution, to the nearest 0.1 g.

Pour about 4 ml of water into the tube. Again heat the tube until the solid dissolves. Again allow the solution to cool, with stirring, until a trace of solid appears, and record the temperature of saturation. Finally, place the tube and contents in the beaker and find the total mass. Also find the total amount of water present in the solution.

Repeat this procedure two more times, adding about 4 ml of water each time.

Finally, rinse the thermometer with water and dry it with a towel. Then empty the test tube and wash it out.

Computations

1. Calculate the solubility of potassium nitrate at each of the four temperatures. (Note that the mass of potassium nitrate remains the same, but the mass of water increases from about 8 g in the first run to about 20 g in the last.)

Graph

Prepare a full-page graph of your results, plotting solubilities along the vertical axis and temperatures along the horizontal axis as in Figure 5–11. Such a graph is called a solubility curve. What does the solubility curve for potassium nitrate indicate?

Look up the exact solubility values for potassium nitrate in a book such as the *Handbook of Chemistry and Physics*. Plot these values on the same graph. Draw a small square around each of the points. Then

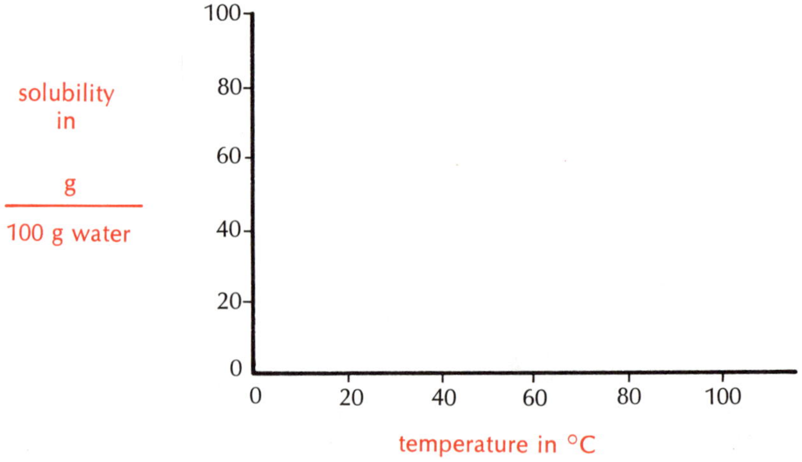

Figure 5–11
Graph axes for a solubility curve.

draw the correct solubility curve for potassium nitrate. Label each curve. How do you account for any difference between the two solubility curves?

Problems

1. A solution is prepared at 60°C by dissolving 20.0 g of potassium nitrate in 40.0 g of water. At what temperature would this solution be saturated? Use your own solubility curve in answering this question.

experiment 5–11

The Heat of Solution of Potassium Nitrate

Materials

100-ml graduate; styrofoam cup; thermometer; 10 cm by 10 cm piece of paper; equal-arm balance; powdered potassium nitrate

Procedure

In Experiment 5–10, you used heat energy from a burner to dissolve potassium nitrate. In this experiment you will use the heat energy from water. From your data you will be able to compute the heat of solution of the compound (Perspective 5–6).

Pour 100 ml of water into a styrofoam cup and find the temperature of the water to the nearest 0.5°C. Find the mass of the piece of paper to the nearest 0.1 g. Add about 7 g of potassium nitrate to the paper and find the mass of paper plus potassium nitrate.

Pour the potassium nitrate into the cup. Stir the mixture with the thermometer until all the solid has dissolved and the temperature is constant. Record this temperature.

Computations

1. Find the fall in temperature.

2. Compute the amount of heat in calories absorbed by the solution. To do this, use the equation

$$\begin{matrix}\text{heat}\\ \text{absorbed}\end{matrix} = \begin{matrix}\text{mass of water}\\ \text{in grams}\end{matrix} \times \begin{matrix}\text{drop in temperature}\\ \text{in Celsius degrees}\end{matrix} \times \begin{matrix}\text{heat capacity}\\ \text{of water}\end{matrix}$$

3. Find the mass of potassium nitrate in the solution.

4. Compute the number of calories of heat absorbed when one gram of potassium nitrate is dissolved in water. That is, divide the heat absorbed (Computation 2) by the mass of potassium nitrate (Computation 3).

5. The correct value for the heat of solution in Computation 4 is 85 cal/g. Compare your result with the correct value, and then compute your percentage error. How do you account for the difference between the correct value and your experimental value?

GASES

The Expansion of a Gas

As you already know, one of the properties of any substance, whether it is a solid, liquid, or gas, is that it expands when it is heated. However, the molecules of a gas are more widely separated than those of a liquid or a solid. Hence the attractive forces between gas molecules are much weaker than the forces between molecules of liquids or solids. As a result, gases expand much more than liquids or solids.

The expansion of various gases was first studied experimentally by a French scientist named Jacques Charles (SHARLZ) in the latter part of the 18th century. He found the rate at which gases expand. To do this, he first measured the volume of a gas at 0°C. Then he found the increase in this volume when the gas was heated through one Celsius degree. That is, he found the increase in volume when the gas was heated from 0°C to 1°C, or from 10°C to 11°C, or from 20°C to 21°C. He discovered that all gases expand at the same rate. In this respect gases behave differently from solids and liquids. Every solid and liquid has its own particular rate of expansion.

There is no simple apparatus for measuring the rate at which a gas expands as its temperature is raised. However, you can observe actual expansion or contraction of a gas in the following demonstration.

experiment 5–12

The Expansion and Contraction of Air

Demonstration

Materials

capillary tube, about 20 cm long; burner; mercury; thermometer; large test tube; stand; clamp; ruler; cold water (0°C)

Procedure

Close one end of the capillary tube by strongly heating it. After the tube has cooled, heat it gently in a flame to drive out some of the air. Then, as it cools, dip the open end of the tube below the surface of mercury in a container. As air in the tube cools, it contracts and a pellet of mercury rises in the tube. The pellet of mercury seals off a column of air in the tube. The length of

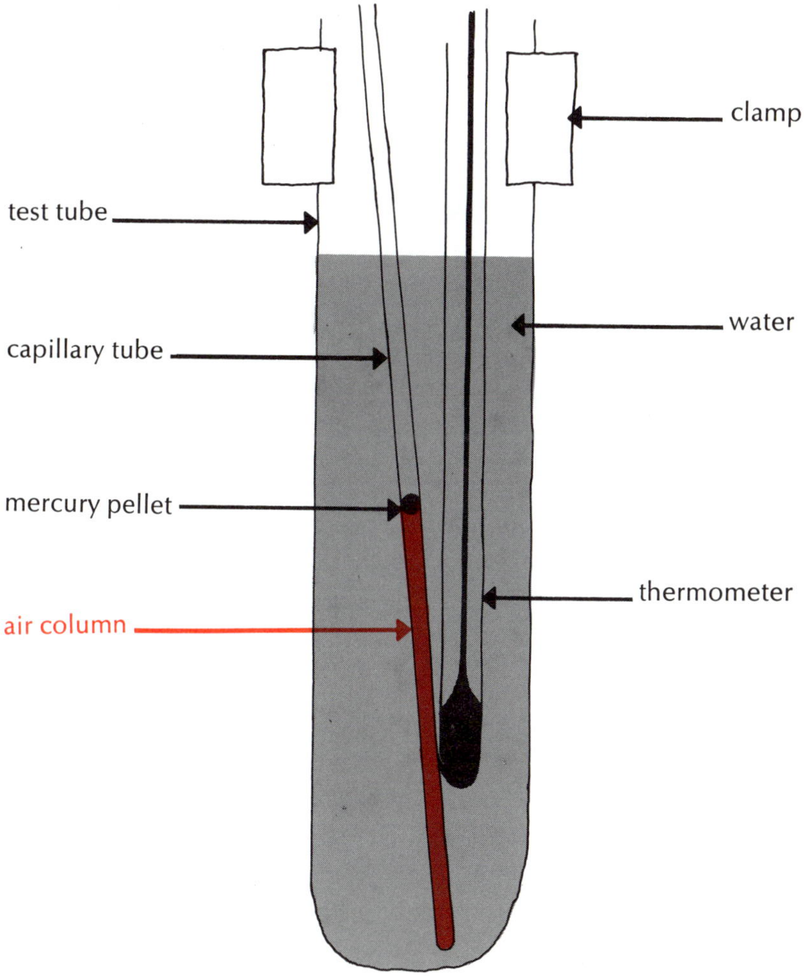

Figure 5–12
Apparatus for observing the expansion or contraction of a gas.

the air column in the tube at room temperature should be about 8 to 9 cm.

Place the capillary tube and a thermometer in the test tube. Clamp the tube to a stand. (See Figure 5–12.) Read the thermometer and read the length of the air column by means of a ruler. Record these values.

Fill the test tube with ice water, and record the temperature and the length of the air column. Heat the test tube and contents with a small flame. Take several pairs of readings (temperature and length of air column) until the temperature is about 90 to 95°C. Summarize your data as in Figure 5–13.

The air column is cylindrical in shape, so its volume equals length times area of cross-section. In these readings the length

Figure 5–13
Data table for Experiment 5–12.

Temperature of Air	Length of Air Column
? °C	? cm
? °C	? cm
? °C	? cm
? °C	? cm
? °C	? cm

changes, but the cross-section remains the same. Therefore the length of the air column can be used to represent the volume of air.

Questions

1. Write a statement concerning the relationship between the change in temperature and the change in volume of a gas. This relationship is called **Charles' Law.**

2. Is it reasonable to assume that the temperature of the trapped air is the same as the temperature of the water? Give a reason for your answer.

PERSPECTIVES

perspective 5–1

Buoyancy in Water

If you have lifted a rock out of water, you may have noticed that it seems to get heavier. On the other hand, if you wade into deep water, you become aware of a force pushing up on your body, and you seem to get lighter. If the water is deep enough you are able to lie flat on the surface and float. When this happens you appear to have no weight at all! The upward force exerted on objects submerged, or even partly submerged, in a fluid is called buoyancy.

What causes buoyancy? Molecules of water, like molecules of other liquids, are free to move in all directions. If the moving molecules strike a surface, they bounce back. When they bounce back, they exert a force on the surface. Consider what happens when a boat is partly submerged in water as in Figure 5–14. Molecules A, B, C, D, and E strike the surface of the boat and bounce back along their own paths. Molecules A and E produce side forces opposite to each other. Molecules B, C, and D produce an upward force. The total upward force of the millions of upward-moving molecules is the buoyant force. Clearly the direction of the buoyant force is opposite to the direction of the weight. That is why the boat seems to lose weight when it is in water.

Figure 5–14
Buoyant force on an object is caused by molecules bouncing off it.

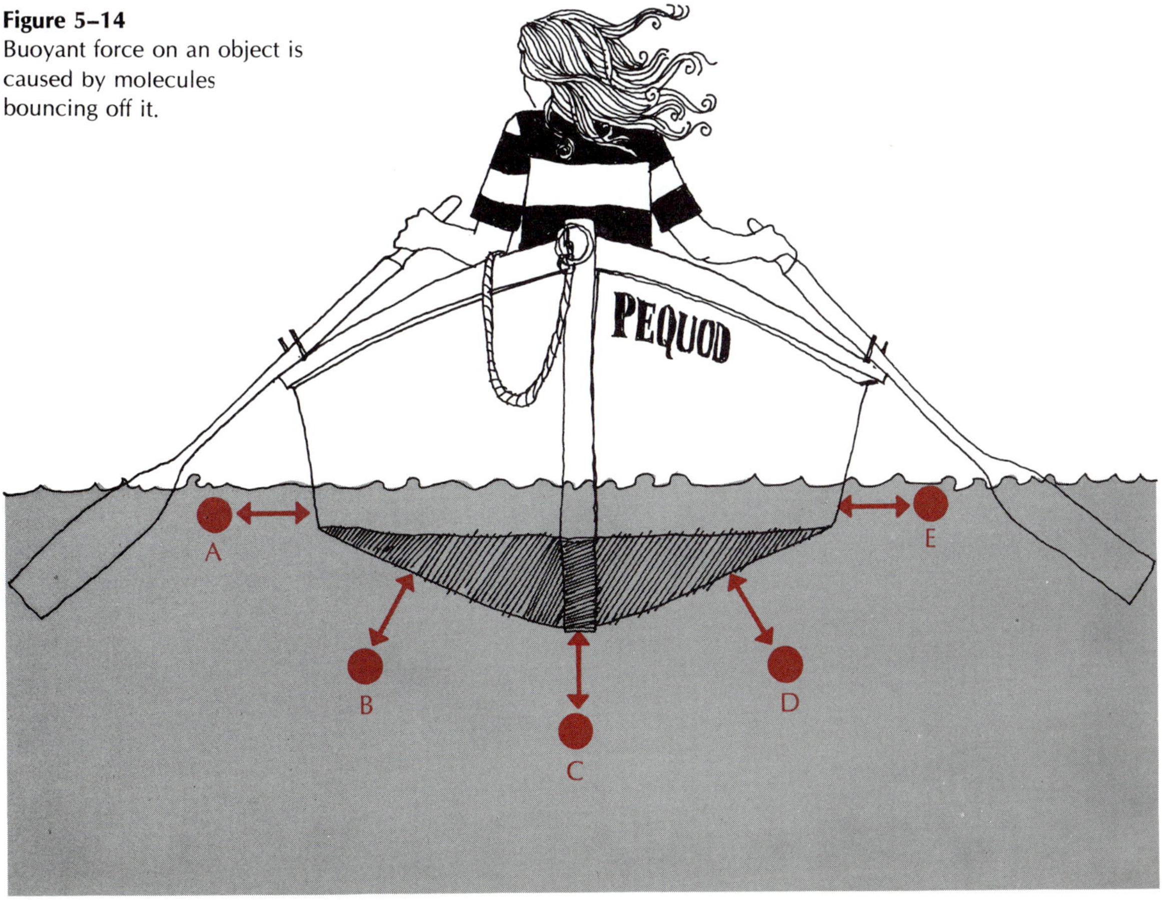

perspective 5–2

Archimedes, the Discoverer of Buoyant Force

In Experiment 5–2 you discovered a scientific law concerning the relationship between the weight of a body and the buoyant force exerted upon it. You were not the first to make this discovery. The law was discovered 2000 years ago by the Greek philosopher Archimedes (ark-ih-MEE-deez). He, you may recall, was so impressed by the power of a lever that he said he could move the earth if only there were a place in space where he could stand. He discovered the law of buoyancy in a dramatic way.

Archimedes, although a Greek, lived in the city of Syracuse in Sicily, an island off the southern tip of Italy. Hero, the king of Syracuse at that time, asked his goldsmith to make a crown from a carefully-weighed piece of pure gold. A beautiful crown, weighing exactly the same as the original piece of gold, was made. But the king suspected that the goldsmith had replaced some of the gold with silver by melting the two metals together. A solid mixture of gold and silver still looks like gold if there is not very much silver in it. The king asked Archimedes to find out if the gold in the crown had been mixed with silver.

This was a challenging problem. The mass of the gold was known and so was its density. From these values Archimedes computed the volume of the gold and then argued somewhat as follows. If the gold had not been mixed with another metal, the volume of the gold would also be the volume of the crown. However, the density of silver is about half the density of gold. So if silver had been mixed with the gold, the volume of the crown would be greater than the original volume of the gold. But how was it possible to find the volume of a solid of such irregular shape as a crown? This was the problem. In Archimedes' time there was no known method of finding the volume of an irregular object.

Archimedes pondered this problem for many days. Strangely enough, the answer came when he went to bathe in one of the public baths. He stepped into a tub filled to the brim with water. Water spilled over and, at the same time, he appeared to lose weight. He suddenly realized how he could compute the volume of the crown. History records that in his excitement he jumped out of the bathtub and ran naked down the street, shouting "Eureka!" which means "I have found it!"

Testing his idea by an experiment, Archimedes found that the volume of the crown was greater than it should have been. Clearly the goldsmith had stolen some of the gold. Thus, not only had Archimedes discovered an accurate method of finding the volume of an irregular solid object, but he was the first person to solve a crime by applying scientific principles to the case. Today, as you know, this is a common practice.

perspective 5-3

Buoyancy of Air

All bodies on earth are submerged in air. Since Archimedes' Principle applies to gases as well as liquids, all bodies are buoyed up by air. The density of air (1.29 g/l) is very low compared with the density of water (1000 g/l). Therefore its buoyant effect is not noticeable on ordinary objects. For example, an average-sized man has a volume of about three cubic feet, and the buoyant force of air on him is only 0.25 lb. That is, the man's actual weight is only a quarter of a pound more than his apparent weight in air.

A piece of cork floats in water, but it does not float in air because the buoyant force is much too small. However, buoyant force on a body can be increased by increasing the

Figure 5-15
A blimp floats in air because it is less dense than air.

volume. The volume can be increased by inflating the body with a gas lighter than air. Thus, if the weight of a body is relatively small and its volume very large, the buoyant force is large enough to make it float. A blimp, which is a large balloon, meets these conditions. A blimp is usually filled with helium gas, whose density is 0.177 g/l. If a blimp has a volume of one million cubic feet, the buoyant force on it is about 70,000 lb. A buoyant force of this size is enough to lift the fabric, framework, and gas of the blimp, and also to carry a load. (See Figure 5–15.)

perspective 5–4

Solutions

Probably the most useful property of water is its ability to dissolve many substances. Both plants and animals depend upon this property.

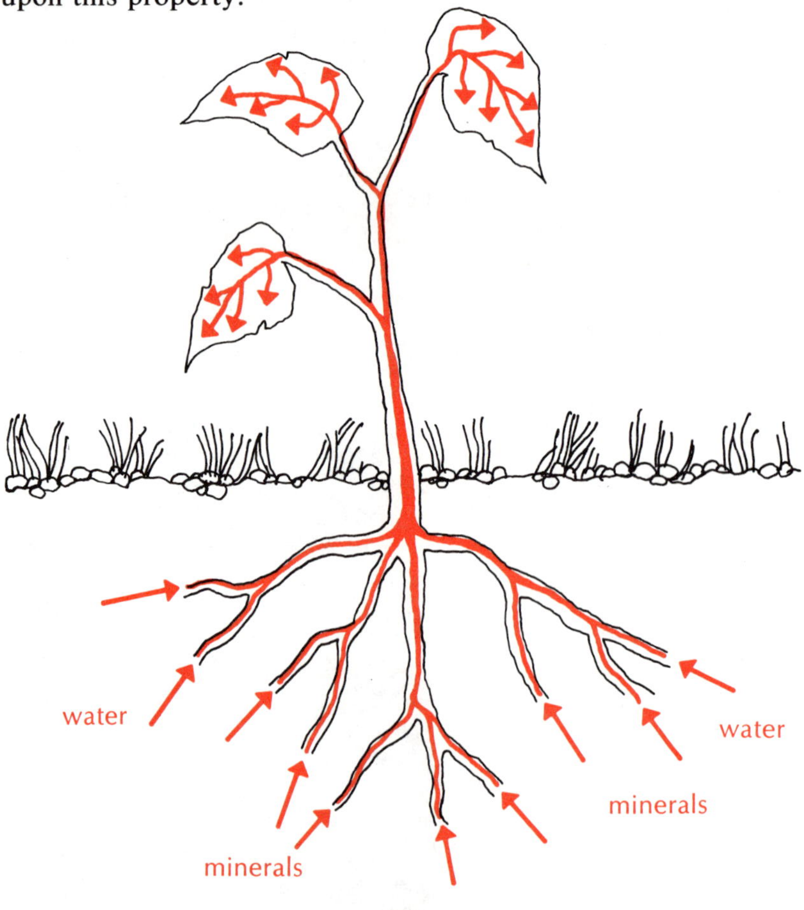

Figure 5–16
Along with energy and carbon dioxide, a plant needs water and minerals. It gets them from the soil as a solution.

Figure 5–17
Salt can be obtained by evaporating sea water.

The starches, fats, and proteins that you eat are almost insoluble in water. However, in the process of digestion, these substances are changed into soluble substances. The soluble substances are then dissolved in the blood and carried to all parts of your body.

Plants take their food from the soil. Plant food consists largely of minerals which dissolve in the water in the soil. The solution then passes through the roots and into the stem and leaves. (See Figure 5–16.) The plant grows by taking the various minerals it needs to construct its cells.

Rocks, in general, are not very soluble. However, many rocks are slightly soluble. Rivers and streams carry the soluble parts of rock to the sea. The sea has become a fairly concentrated solution of many substances. The most abundant is common salt. Seawater is about 3.5 percent dissolved solid. That is, if 100 kg of seawater are evaporated to dryness, about 3.5 kg of solid will remain. (See Figure 5–17.)

Many chemical reactions take place in water solution. A great deal of chemical knowledge has been acquired by observing reactions, even in test tubes, between dissolved substances.

perspective 5–5

Solubility

A given solid dissolves in water because the forces that hold the solid particles together are somehow weakened by the water. When this happens, the particles of solid are able to move freely among the molecules of water. However, there is a limit to this process. That is, there is a limit to the amount of solid that will dissolve. For example, if you add a small quantity of potassium nitrate to water in a test tube, it dissolves. If you add a little more potassium nitrate, it too will dissolve. If you continue to add potassium nitrate, eventually no more will dissolve. Even after vigorous shaking, some potassium nitrate remains undissolved at the bottom of the tube. The solution is then saturated. A saturated solution is one in which the maximum amount of solid is dissolved.

The ratio of the mass of the dissolved substance to the mass of water in a saturated solution is called the solubility of the substance. The solubility of a substance is the maximum mass in grams of it that will dissolve in 100 g of water.

The solubility of a substance depends upon the temperature. For most solids, the solubility increases as the temperature rises. Thus, for potassium nitrate at 20°C, 32 g will dissolve in 100 g of water. At 40°C, 86 g of potassium nitrate will dissolve in 100 g of water. If a saturated solution of potassium nitrate is heated, the solid at the bottom of the tube dissolves, so the solution is no longer saturated.

perspective 5–6

Heat of Solution

Before a solid can dissolve in water, the particles must be pulled away from the solid structure. This requires energy. The only available source of energy is the water itself. The water supplies energy to break down the structure of the solid. As a result, the water loses energy and its temperature drops. The energy needed to dissolve a solid is called its **heat of solution.** The amount of heat needed depends on the kind of solid dissolved, as well as on the amount of solid. In Experiment 5–11, you found the heat needed to dissolve one gram of potassium nitrate in water.

perspective 5–7 The Absolute Zero of Temperature

If the air in the tube in Experiment 5–12 were cooled to lower and lower temperatures, the spaces between the molecules would become smaller and smaller. Eventually there would be no space left. There would be no remaining volume of air. The temperature at which this would occur is called the **absolute zero** of temperature. The absolute zero is 273 Celsius degrees below the freezing point of water, or −273°C.

Actually, the absolute zero has never been reached experimentally. This is because gases condense to liquids before this temperature is reached. For instance, air changes to a liquid at about −200°C. Some gases liquefy at temperatures much higher than −200°C, and others at lower temperatures. Hydrogen liquefies at −253°C. Helium, the gas used in blimps, liquefies at about −269°C.

The absolute zero of temperature is the temperature at which molecules would be so closely packed that a "gas" would have neither volume nor pressure. It is a "natural" zero of temperature, not an artificial zero like the freezing point of water on the Celsius scale. For this reason, the absolute zero is a very important temperature. It is usually written as 0 K.

$$0\ \text{K} = -273°\text{C}$$

The K stands for Kelvin. Kelvin was a British scientist who first suggested that −273°C was the lowest possible temperature.

perspective 5–8 Gas Pressure

If the stopper is removed from a bottle of household ammonia, the odor of ammonia spreads and can be detected in all parts of the room. Why? Molecules of ammonia escape from solution and travel through the air. However, it takes time (perhaps several minutes) before you can smell them. This is because molecules collide with each other and so cannot travel in a straight line. In every cubic centimeter of air there are millions of molecules. A molecule of ammonia makes a vast

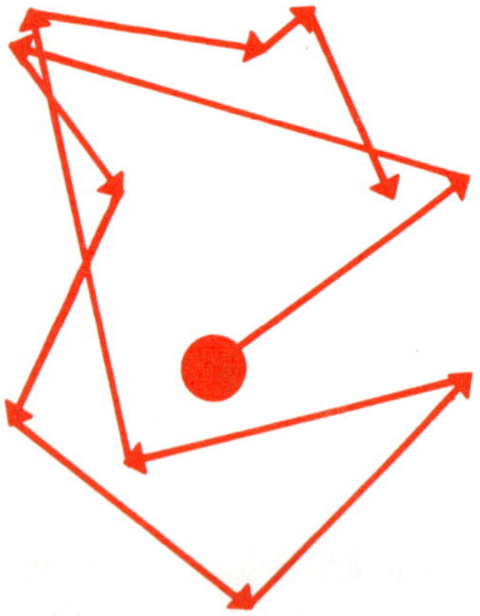

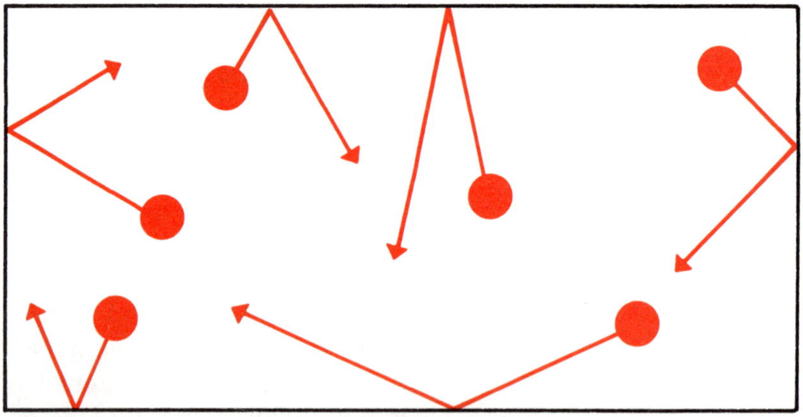

Figure 5–18 (left)
A gas molecule bounces off other molecules and travels in a zigzag path.

Figure 5–19 (right)
Molecules bouncing off the walls of a container exert a force called gas pressure.

number of collisions with molecules of air, even when it travels only a short distance. At every collision the ammonia molecule bounces away and changes direction, following a zigzag path like the one shown in Figure 5–18.

If gas molecules are in a closed container, they strike the walls of the container and bounce off, as shown in Figure 5–19. As each molecule bounces, it exerts a force upon the wall. The force exerted by a single molecule is exceedingly small. But since there are millions of molecular impacts upon each square centimeter of wall, the total effect is considerable. The force exerted by gas molecules upon each square centimeter of surface area is called **gas pressure.**

perspective 5–9

The Pressure of the Atmosphere

The atmosphere extends far above the surface of the earth. The atmosphere fades out gradually, so the distance to the top is uncertain. But it is at least 500 miles. As you would expect, the atmosphere is densest at the earth's surface. It gets thinner and thinner the higher it goes. The changing density in a column of air is shown in Figure 5–20. To reduce air resistance and to avoid rough air currents, airplanes commonly fly at a height of about six miles (about 30,000 to 35,000 feet). However, at this altitude there is not enough air to support life. In order to keep an air density about the same as that at ground level, the cabins of airplanes must be airtight. This is why airplane windows are so thick. In case of a leak, airplanes carry oxygen masks for the passengers and crew.

We on earth really live at the bottom of an ocean of air. In the early part of the 17th century, an Italian scientist named

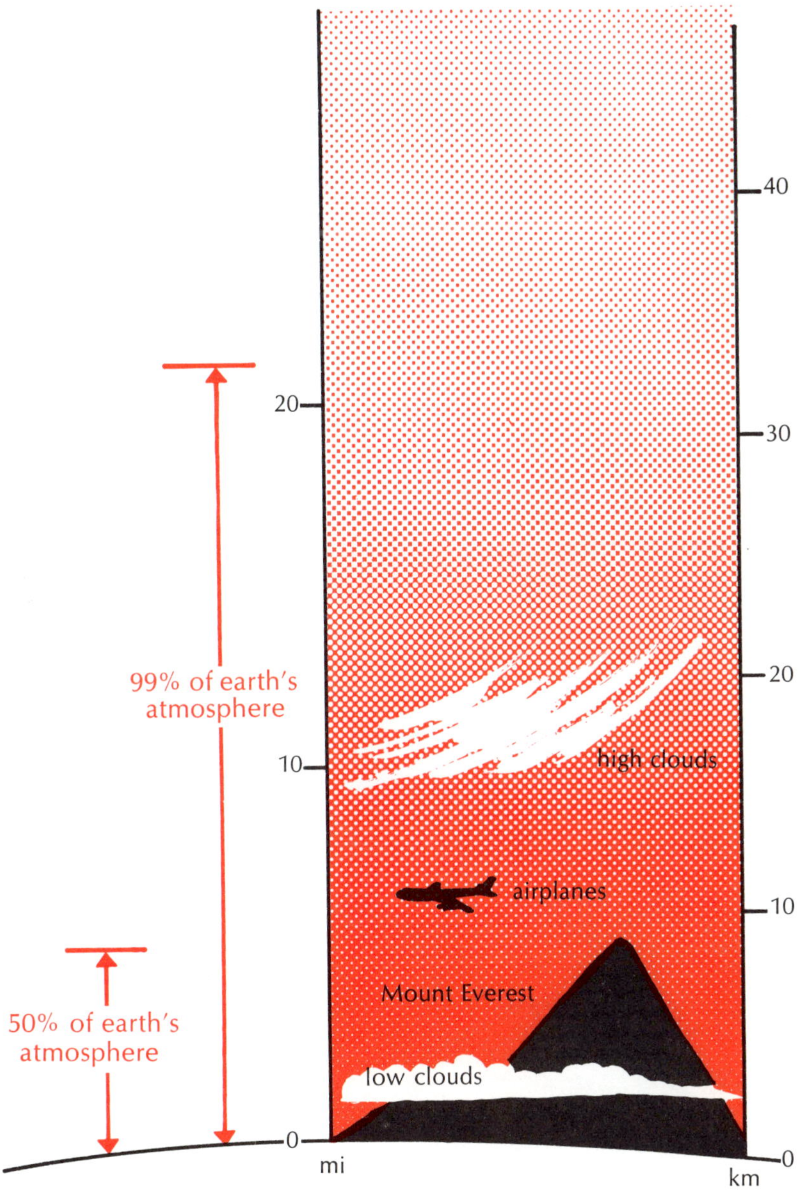

Figure 5–20
The earth's atmosphere extends up at least 500 miles, but 99% of the air is within about 20 miles of the surface. Airplanes fly higher than 50% of the air.

Torricelli (torr-ih-CHEL-ee) proposed that the atmosphere is made up of layers of air. Each of the higher layers presses down on those below. Thus, pressure of the bottom layer is what Torricelli described as atmospheric pressure.

Torricelli invented a device to measure the pressure of the atmosphere. He called it a *barometer,* from a Greek word that means "a measure of weight." Torricelli knew that mercury is nearly 14 times as dense as water and thousands of times as dense as air. So he had the idea that a column of air extending into space could be balanced by a short column of mercury, which could be seen and easily measured.

To make a barometer, he sealed one of the open ends of a glass tube about 1 m in length. If the tube were not sealed, the air would press down on the mercury. He completely filled the tube with mercury, closed the end with his fingers, inverted the tube and placed the open end in a dish of mercury. He then removed his finger from the open end and the column of mercury dropped until it was about 76 cm high. This arrangement is shown in Figure 5–21. Notice that one fluid

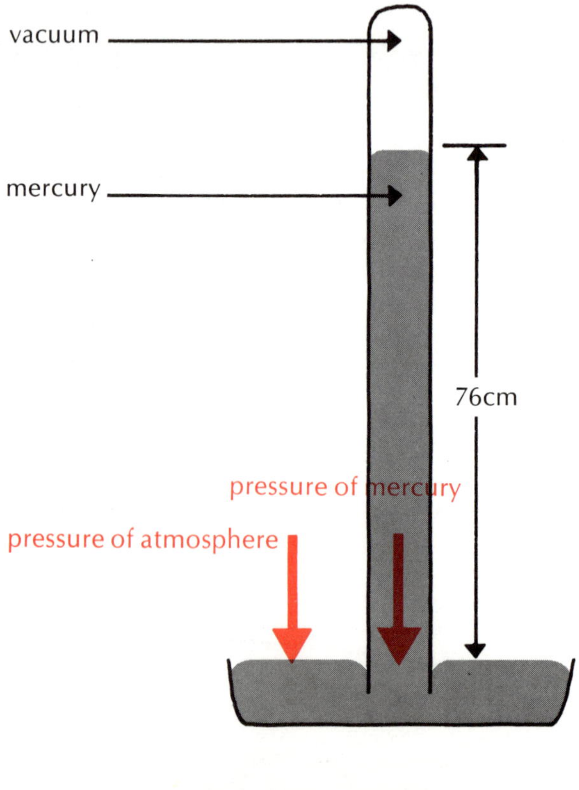

Figure 5–21
In a mercury barometer, two fluids balance each other.

is balancing another. A column of air about 500 miles high is balanced by a column of mercury about 76 cm high. Thus the height of the mercury column can be used as a measure of atmospheric pressure. The height of the mercury column varies slightly from day to day, but an average height is about 76 cm at sea level. A pressure of one atmosphere is defined as the pressure exerted by a column of mercury 76 cm high. Even today Torricelli's barometer is still the most accurate device for measuring air pressure.

When the mercury level in Torricelli's barometer tube dropped to about 76 cm, there was an empty space at the top of the tube. Torricelli claimed that this space was empty—that he had created a vacuum. But this was a controversial idea. It was against a popular belief inherited from Aristotle, a Greek philosopher who lived about 300 B.C. Aristotle had said that a vacuum was impossible or, in his own words, that "nature abhors a vacuum." So it is not surprising that in Torricelli's time most scientists would not agree that he had created a vacuum. Nevertheless, their opinions were soon changed. In 1660 an English scientist named Robert Boyle invented a pump that could suck all the air out of a flask. After Boyle had actually produced a vacuum in a flask, scientists agreed that the space above the mercury in the barometer tube was a vacuum. Boyle's invention finally put an end to the belief that a vacuum could not exist, an idea that had been accepted by men of science for nearly 2000 years.

perspective 5–10

The Effect of Pressure on the Volume of a Gas

Boyle not only made the first vacuum pump, but he was the first scientist to study the way gases behave when they are compressed. He knew that, unlike liquids and solids, gases can be easily compressed if pressure is applied to them, and he designed an apparatus to investigate this problem.

Boyle knew nothing about molecules, or the motions of gas molecules, or the cause of gas pressure. But you do, so you can investigate this problem without even doing an experiment. Consider the situation shown in Figure 5–22A. A volume of air is trapped in a cylinder by an airtight, but movable, piston. Air molecules outside the cylinder bombard the upper surface

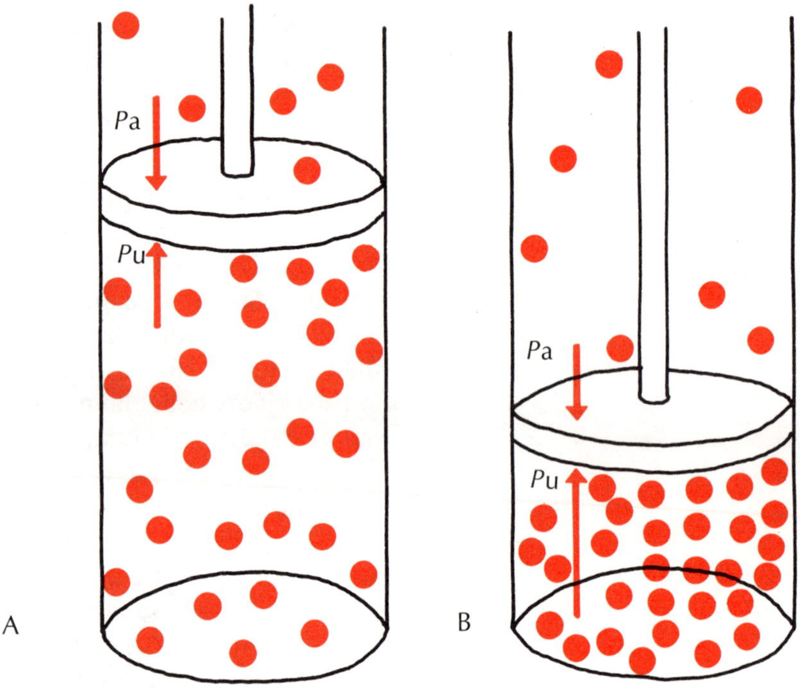

Figure 5–22
A. In a cylinder with an airtight piston, the gas pressure P_u is equal to atmospheric pressure, P_a.
B. If the volume of the gas is decreased to half the original volume, the gas pressure is doubled.

of the piston. The downward impacts make up atmospheric pressure (P_a). Inside the cylinder the upward impacts on the underside of the cylinder make up the pressure of the enclosed air (P_u). Suppose that, at the beginning, the pressure of the enclosed air is the same as the pressure of the atmosphere.

$$P_u = P_a$$

Now suppose the piston is pushed down until the volume of the enclosed air is one half of its original volume, as shown in Figure 5–22B. What is the pressure exerted by the enclosed gas? The air molecules outside the cylinder have not been disturbed, so the downward pressure remains precisely the same, namely P_a. However, the enclosed air molecules in B have been compressed into one half the space they occupied in A. Therefore, in B there must be twice as many collisions on the underside of the piston as in A. This means that the gas pressure has been doubled, so that now

$$P_u = 2P_a$$

If the volume of an enclosed gas is halved, its pressure is doubled. In general terms, if the pressure exerted upon an enclosed gas is increased, the volume of the gas is decreased by the same factor. In other words, for a gas, volume and pressure are in inverse proportion. This is just what Boyle discovered from his experimental work. His discovery is now known as **Boyle's Law.** It is usually expressed as follows: The volume of a gas varies inversely as the pressure exerted upon it.

perspective 5–11

Atmospheric Pressure and Weather

On television weather reports and in newspapers you often see weather maps like the one in Figure 5–23. On these maps there are areas marked "high" and "low" and lines marked "warm front" and "cold front." What are these areas and lines? How do highs, lows, and fronts affect the weather you get?

Air over a warm place like the Gulf of Mexico becomes warm. As it warms, it expands and becomes less dense. Therefore it rises, as in Figure 5–24. This forms an area of low

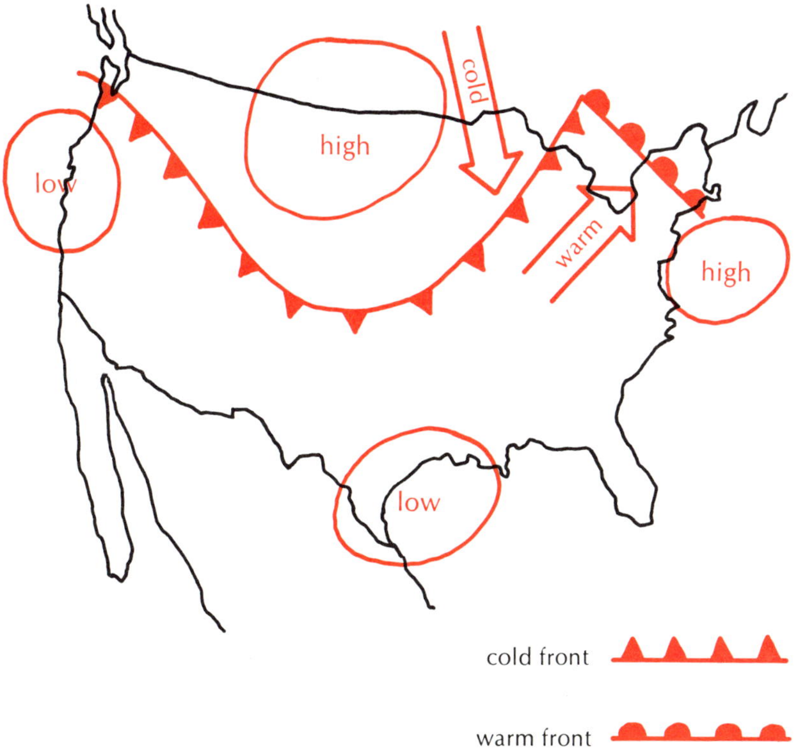

Figure 5–23
A typical weather map of the United States.

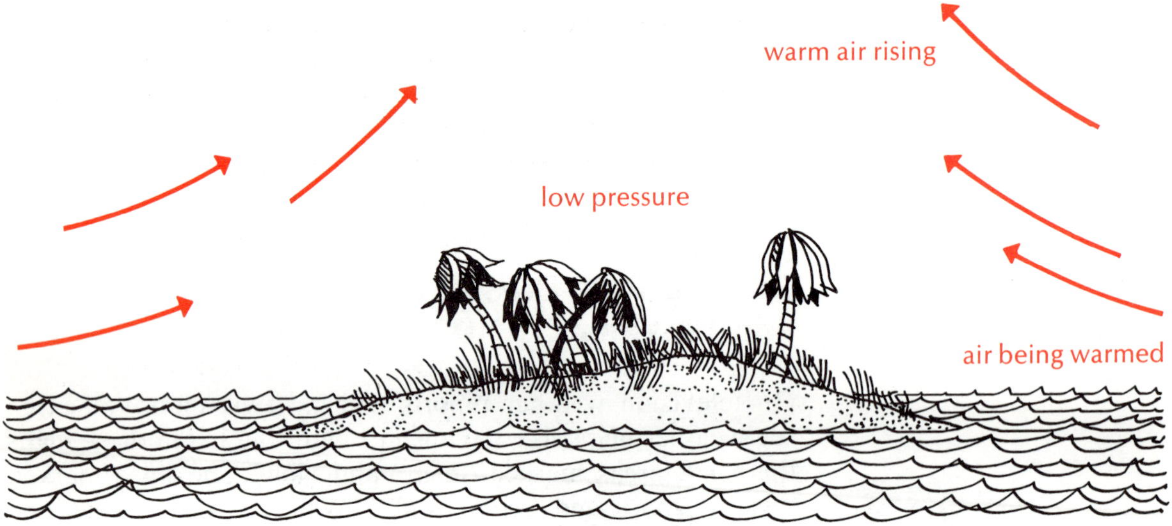

Figure 5–24
Air over a warm place may expand and rise, forming an area of low air pressure.

Figure 5–25
Air over a cold place may become denser and sink, forming an area of high air pressure.

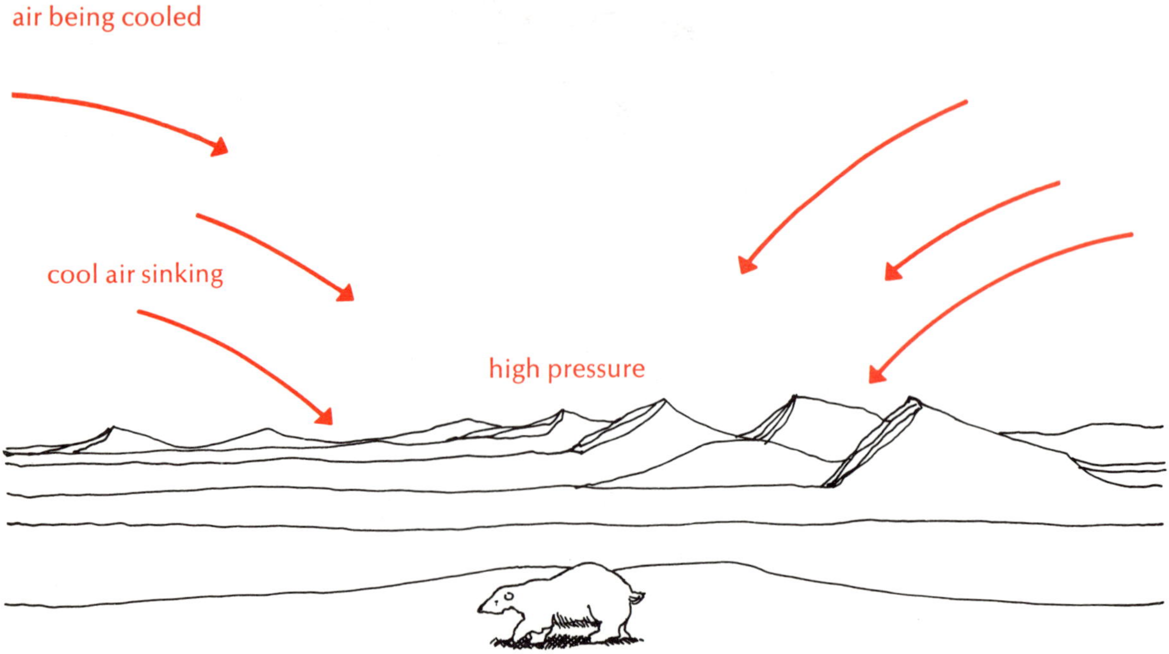

air pressure, called a **low.** On the other hand, air over a cold place like the arctic regions of northern Canada becomes cold. As it cools, the cold air becomes denser and sinks, as in Figure 5–25. An area of high air pressure, a **high,** is formed.

Once a high or a low is formed, it may start to move. In the Northern Hemisphere, most highs move toward the southeast and most lows move toward the northeast.

Remember that a low is generally warm and a high is generally cool. When a high moves into an area of lower pressure, a **cold front** is formed, as in Figure 5–26. The warm air is pushed upward and cooled off. When it cools, it can no longer hold as much water vapor as it did when it was warm. The water vapor condenses into droplets, forming clouds. Further cooling produces rain, perhaps a thunderstorm.

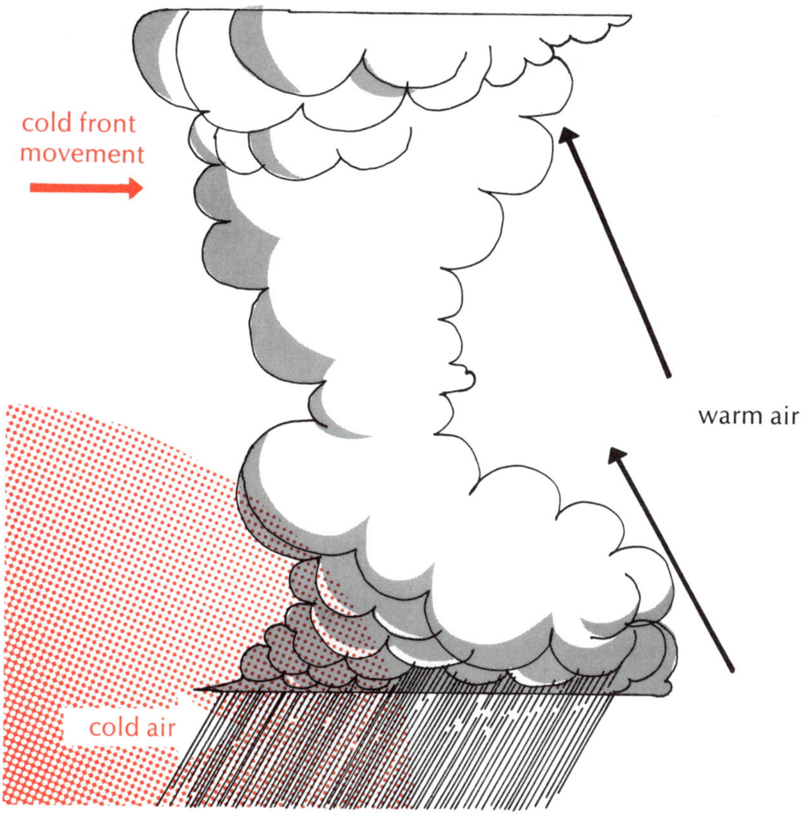

Figure 5–26
A cold front may bring a thunderstorm.

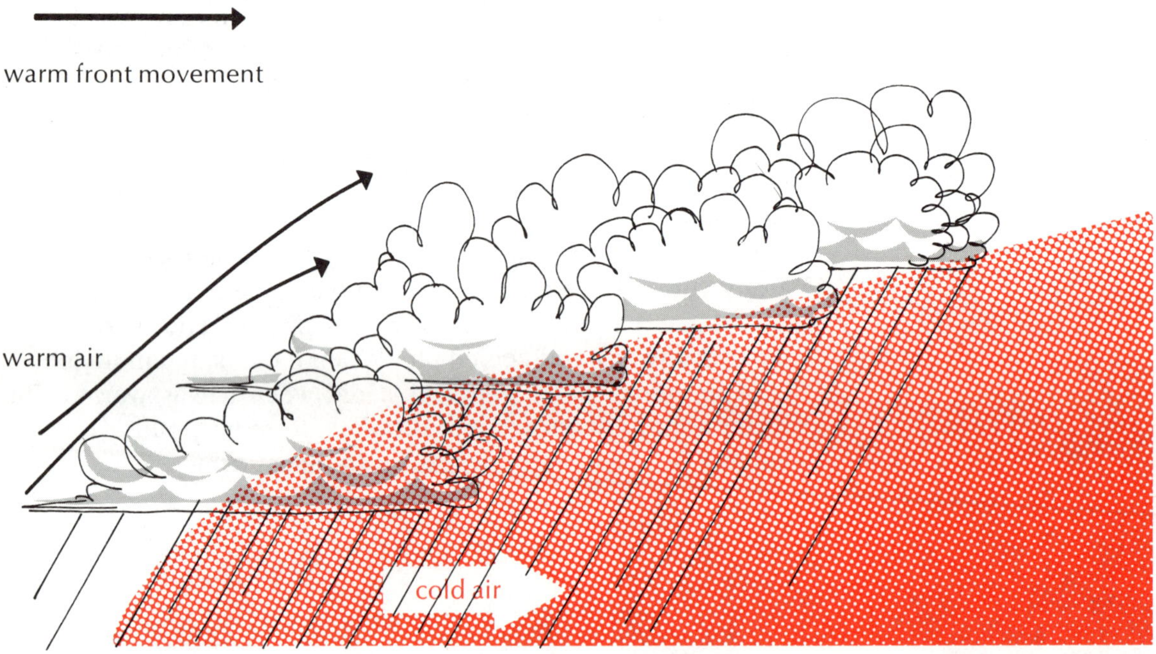

Figure 5–27
A warm front covers a larger area than a cold front, so it may bring a long light rain.

A **warm front** is formed when a low (which is warm) moves into an area of higher pressure (which is cool). The warm air slides over the cool air, as in Figure 5–27. As in a cold front, the warm air is cooled and rain may occur. However, the slope of a warm front is not as steep as the slope of a cold front. Therefore the rain from a warm front is spread over a larger area. When a warm front goes by, a long, light rain is more likely than a storm.

QUESTIONS

A

1. What is buoyancy?

2. A piece of steel and a piece of cork of the same volume are completely submerged in water. Are the buoyant forces on these objects the same or different? Explain.

3. (a) How is a solid different from a liquid? (b) How is a solid similar to a liquid?

4. Compare the two terms, *fluid* and *liquid*.

5. (a) List three substances that are soluble in water. (b) Name a substance that is insoluble in water.

6. Archimedes discovered how to find the volume of an irregular object. Explain his method.

7. Helium gas is used in blimps. State the property of helium that makes it useful for this purpose.

8. (a) What is a solution? (b) Give an example.

9. What is a saturated solution?

10. What is meant by the solubility of a solid?

11. What factor can change the solubility of a substance?

12. Why are the attractive forces between gas molecules much weaker than the forces between molecules of liquids and solids?

13. Where does the dissolved solid in seawater come from?

14. What is a solubility curve?

15. Why can a blimp, filled with helium, float in air?

16. Are the oceans becoming more salty or less salty? Explain.

17. State an important difference between the expansion of gases and the expansion of liquids.

18. What is meant by the term *absolute zero*?

19. What is the absolute zero of temperature on the Celsius scale?

20. A gas law concerns the volume and the ____________ of a gas. Supply the missing word for (a) Charles' Law, and (b) Boyle's Law.

21. What occupies the space above the mercury in a mercury barometer?

22. State Boyle's Law.

23. Express the pressure of the atmosphere in meters of mercury.

24. What is a low? How is it formed?

25. What is a high? How is it formed?

26. How is a cold front formed?

27. How does rain form along a cold front?

28. How is the weather along a cold front different from the weather along a warm front? Why?

B

29. A piece of cork is pushed under water and then released. (a) What happens to the cork? (b) Why?

30. A spring scale shows that a brick weighs 10 N in air but only 7 N in water. Is the loss of 3 N a real loss in weight? Explain.

31. A spring scale shows that the weight of an inflated loaded balloon is 10 N. How could you find the real weight of the balloon?

32. How would you change (a) a saturated solution to an unsaturated one? (b) an unsaturated solution to a saturated one?

33. When salt is dissolved in water, the temperature of the water drops slightly. Explain why.

34. A substance can never have a temperature of -1 K. Give a reason for this in terms of the motions of its molecules.

35. What would happen to the mercury column if a barometer were carried up a mountain? Give a reason for your answer.

36. What happens to the volume of an enclosed gas if the pressure on the gas is halved? Give a reason for your answer.

In Questions 37–41, write the letter of the word or expression that is not related to the others.

37. **a** water, **b** ethylene glycol, **c** potassium nitrate, **d** alcohol

38. **a** fluid, **b** solution, **c** soluble, **d** saturated

39. **a** Charles, **b** Newton, **c** Boyle, **d** Torricelli

40. **a** buoyant force, **b** Charles' Law, **c** Archimedes' Principle, **d** displaced water

41. **a** Kelvin, **b** Fahrenheit, **c** Celsius, **d** Torricelli

In Questions 42–45, write the letter (or letters) of the statement (or statements) that best completes the question. There may be more than one correct answer.

42. If a block of wood with a volume of 400 cm^3 and a mass of 320 g is placed in an overflow can filled with water,
a 400 g of water overflow
b 320 g of water overflow
c 400 ml of water overflow
d 80 g of water overflow

43. If a stone is submerged in an overflow can filled with water, and 100 g of water overflow,
a The stone has a volume of 100 cm^3
b The stone has a mass of 100 g
c The stone has a mass greater than 100 g
d The density of the stone is 1 g/cm^3

44. Ethylene glycol and water have different density values. If a wooden block floats in water and another block of identical mass and volume floats in ethylene glycol,
a The block in ethylene glycol floats higher
b The block in water displaces more liquid
c The block in water weighs more than the block in ethylene glycol
d Both blocks float at the same level

45. A tube about 80 cm long, sealed at one end, is filled with mercury and then inverted in a beaker of mercury. The column of mercury drops until it is about 76 cm in height. This happens because
a a narrow place in the glass tube keeps the mercury up
b there is more mercury in the beaker than in the tube
c air pressure exerted on the outside surface of the tube balances the weight of the mercury inside the tube
d air pressure exerting a force on the mercury surface in the beaker balances the weight of mercury in the tube.

PROBLEMS

1. A metal object, 2.0 cm by 5.0 cm by 10.0 cm, weighs 270 g in air. Compute its apparent weight when it is submerged in water.

2. A piece of wood, 2.0 cm by 5.0 cm by 10.0 cm, weighs 90 g in air. What is its apparent weight when it is placed in water?

3. A metal object, of density 1.70 g/cm^3, is 2.0 cm by 5.0 cm by 10.0 cm. Compute its apparent weight when it is submerged in water.

4. A stone weighs 450 g in air and 300 g in water. What is the density of the stone?

5. A piece of metal, 50 cm^3 in volume, is submerged in water. Compute the buoyant force of the water on the metal.

6. A block of wood, 10.0 cm^3 in volume, is submerged in gasoline of density 0.83 g/cm^3. Calculate the buoyant force of the gasoline on the wood.

7. A stone weighs 200 g in air and 120 g in water. Compute (a) the volume of the stone, (b) the density of the stone.

8. Calculate the apparent weight of a 150-g piece of copper (density of copper = 8.9 g/cm^3) when it is submerged in water.

9. A 75-cm^3 piece of wood displaces 60 g of water when it is placed in a filled overflow can. Calculate the density of the wood.

10. A 90-cm^3 block of metal weighs 190 g in water. How much does it weigh in air?

11. A saturated solution of a substance at 20°C contains 5.0 g of the substance and 15.0 g of water. Compute the solubility of the substance at 20°C.

12. If 25.0 g of a saturated water solution of a substance at 20°C is evaporated to dryness, 5.0 g of solid remain. Compute the solubility of the substance at 20°C.

13. Solubility values of a substance are: 40 g at 30°C, 60 g at 50°C, and 100 g at 70°C. (a) If 15 g of the substance are added to 25 g of water, at what temperature will the substance just dissolve? (b) If the solution is cooled to 30°C, what amount of the substance will leave the solution? (c) If the solution is warmed to 70°C, what additional amount of the substance can be dissolved?

14. The following data were obtained in a heat of solution experiment.

Mass of water	= 150 g
Mass of "hypo"	= 10.0 g
Initial temperature of water	= 23.6°C
Final temperature of solution	= 20.6°C

Compute the heat of solution of "hypo," that is, the heat involved when 1 g of "hypo" is dissolved.

15. If 12 g of "hypo" is dissolved in 100 g of water originally at 24.1°C, what is the temperature of the solution?

Chapter 6

WHAT ARE ATOMS?

From patterns like this, scientists can find out how atoms are arranged.

The meaning of the word *atom* has changed a great deal since it was first used by a group of Greek philosophers about 2000 years ago. The philosophers thought about the mysteries of the world about them. They were interested in the nature of things. Where did the earth come from? What is it made of? Is a piece of rock or metal really a continuous solid without parts, as it appears to be? Their answer to this last question was *No*. They argued that if a lump of copper is divided and subdivided into smaller and smaller pieces, a tiny particle would finally be reached that could be split no further. Thus they believed that matter is made up of extremely small "building blocks." They called these tiny particles atoms. The Greek word *atomos* means "cannot be cut."

What is the shape of an atom? How are atoms packed together? The early Greeks compared atoms with the variously shaped rocks used to build their temples. So they believed that atoms, even atoms of the same element, could have various shapes and sizes.

Since atoms were much too small to be seen, no experiment could be devised to test the truth of the ideas of the Greek philosophers. Nevertheless, these ideas lasted through the centuries and were generally accepted. Isaac Newton, who lived in the 17th century, certainly accepted them. In one of his scientific papers he wrote, "It seems probable to me that God in the beginning formed matter in solid, hard, movable particles, of such sizes and shapes, and with such other properties as most conduced to the end for which He formed them. These primitive particles were so hard as never to wear or break in pieces, no ordinary power being able to divide what God Himself made in the first creation." In short, Newton believed that atoms were indivisible particles created by God when he founded the universe.

PERSPECTIVES

perspective 6–1

Theories That Can Be Tested

There was no experimental evidence to support Newton's "theory" on atoms. In this respect it was very different from his law of gravitation which could be tested by experiment. However, in the early part of the 19th century (150 years after Newton's time) an English scientist named John Dalton expanded the old ideas about atoms. He expressed them in such a way that they could be tested by experiment. Dalton suggested that all atoms of the same element have the same mass and the same size. In other words, Dalton assumed that the atoms in every element had their own particular mass. In this respect Dalton's concept differed from the Greek idea. More important, Dalton's assumption led to the discovery of a new law in chemistry which could be tested by experiment. This was a great step forward. Following are the four main assumptions of his theory.

1. Atoms are the ultimate particles of matter. They cannot be split into smaller particles, nor can they be destroyed.
2. Atoms of the same element have the same mass, and also the same chemical properties.
3. Atoms of different elements have different masses and different chemical properties.
4. Atoms of different kinds combine chemically to form compounds.

As you learned in Chapter 1, a scientific theory must be able to predict the outcome of experiments related to the theory. Dalton's theory met this test, as you will prove experimentally in Chapter 7. At least some of his assumptions about atoms must have been true. Dalton's theory was accepted by scientists for almost a century.

Near the end of the 19th century, scientific investigators proved that atoms are themselves made up of tiny particles. This discovery meant that Dalton's theory had to be modified. Dalton was right in assuming that all atoms of the same element act alike and that an atom of an element is the smallest possible particle of that element. (See Figure 6–1.) But the new experimental work showed that he was wrong in assuming that atoms could not be split into smaller parts.

Scientists today are still interested in the nature of things. By doing experiments, they have learned many facts about the nature of matter. Also, they have developed theories that seem to explain many of these facts. But much is still unknown or unexplained. In this chapter you will learn some of these facts, theories, and unanswered questions. First, you will examine some evidence that suggests that matter is electric in nature.

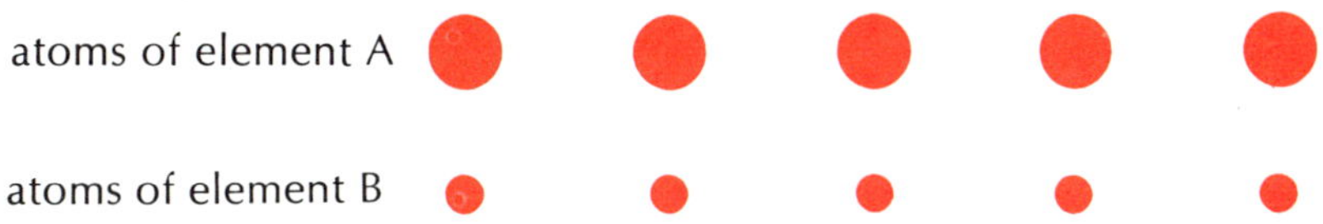

Figure 6–1
The atoms of an element are alike. Atoms of different elements are different.

perspective 6–2 Two Kinds of Electric Charge

You have probably noticed that when you comb your hair, particularly in winter when the air is dry, the hair spreads out and stands on end. (See Figure 6–2.) This is an electric effect. Both your hair and the plastic comb have been electrified by friction. The electricity that collects on your hair, or on the comb, is called an electric charge. Why does your hair become charged by friction, and why does hair stand on end? The answers to these questions will be apparent from the following demonstration.

One way to demonstrate frictional electricity is to rub a plastic rod with flannel and then hold the rod near a light object. The light object may be a small ball of wood pith or even a small piece of dried potato. It is suspended by a thread from a stand.

First rub the plastic rod with flannel and then hold the end of the rod near the pith ball. The pith ball is attracted to the rod, as in Figure 6–3A. But as soon as the ball touches the rod, it is repelled (pushed away) as in Figure 6–3B. Why? The rod is charged by friction with the flannel. When the ball touches the rod, it picks up some of the charge from the rod. An **electric charge** is an accumulation of electricity on an object. Both the rod and the pith ball then have the same kind of charge. We therefore conclude that objects with the same kind of electrical charge repel each other.

Figure 6–2
Why does hair become charged and spread out when it is brushed?

Figure 6–3A
When a plastic rod is rubbed with flannel, it attracts a pith ball.

Figure 6–3B
After the pith ball touches the rod, the ball is repelled.

Figure 6–4
Particles with like charges repel each other. Particles with unlike charges attract.

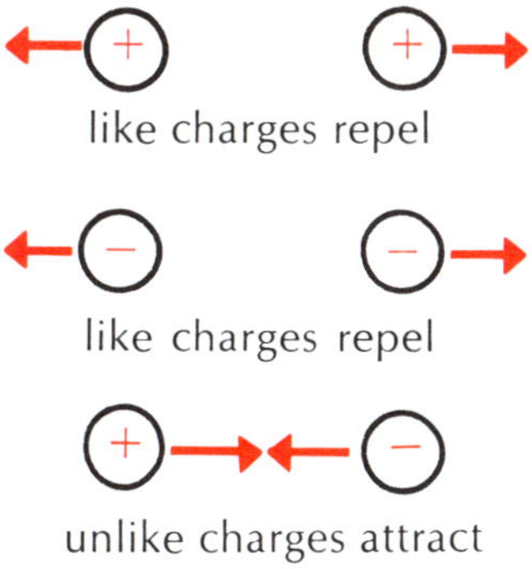

Now rub a glass rod with a piece of silk and hold the glass rod near the charged pith ball. This time the pith ball is attracted to the glass rod. Why? The glass rod is charged by friction. However, since the charged glass rod attracts the charged pith ball, the charge on the glass rod must be opposite in kind to that on the pith ball. So we conclude that objects attract each other if they have different kinds of charge. Particles with like charges repel each other, and particles with unlike charges attract each other. (See Figure 6–4.) This attraction or repulsion is called an **electric force.**

Actually it has been known for centuries that electricity could be generated by friction. Benjamin Franklin did many experiments with objects that were electrically charged by friction. He was the first to recognize that two different kinds of charge are formed in this way. The kind of charge on rubbed glass he called positive (+). The other kind of charge he called negative (−). These names introduced by Franklin are still used.

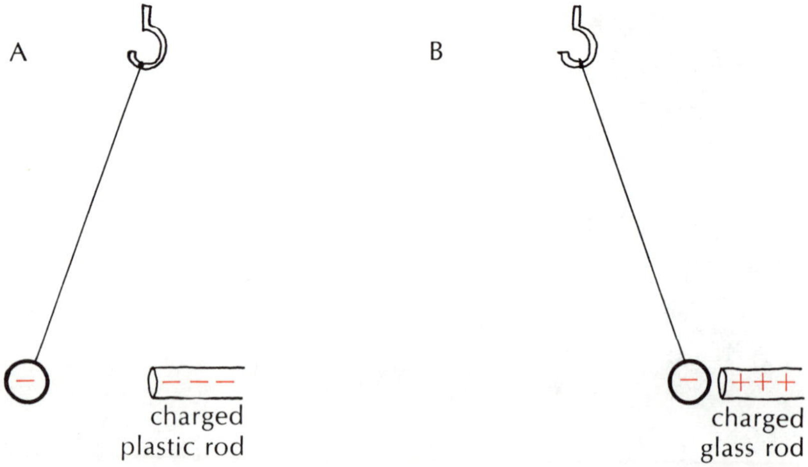

Figure 6–5
The charges on the rods and the pith ball.

Thus in the above demonstration the charge on the plastic rod is negative, the charge on the pith ball is negative, and the charge on the glass rod is positive. These charges are shown in Figure 6–5.

Franklin must have asked the two basic questions concerning objects charged by friction. Why does friction cause objects to become electrically charged? Why are some charged objects positive and others negative? Franklin could not answer these questions. The modern ideas on the structure of matter supplied the answers.

perspective 6–3

Particles in Atoms

There are about 100 different kinds of atoms. All atoms are built up from the same three basic particles. An **electron** (ee–LEK–tron) has a negative charge. A **proton** (PRO–ton) has a positive charge. A **neutron** (NEW–tron) has no charge at all. The charge on an electron is exactly equal, though opposite, to the charge on a proton. A proton is said to have a charge of +1. An electron has a charge of −1. Thus, in an electrical sense, an electron cancels out a proton. The different particles do not have the same mass, however. Protons and neutrons are much heavier than electrons. A proton is about 2000 times as heavy as an electron. A neutron has about the same mass as a proton. (See Figure 6–6.)

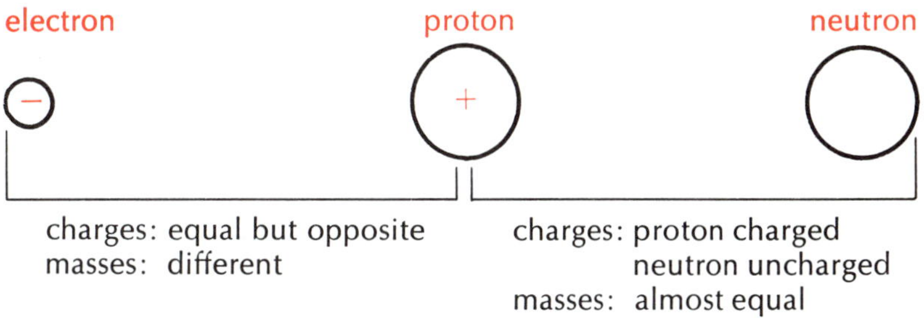

Figure 6–6
The three particles of which atoms are made.

Neutrons and protons are bound together very tightly and form the center or **nucleus** (NEW–klee–us) of every atom. Electrons, on the other hand, are in constant motion around the nucleus, forming a kind of spinning cloud. In all atoms the number of electrons equals the number of protons. So all atoms must be electrically neutral.

perspective 6–4

A Model of a Hydrogen Atom

A hydrogen atom has the simplest structure of all atoms. Its nucleus consists of only a single proton. Therefore the whole atom of hydrogen is made up of one proton and one electron. What kind of model will best represent this simple atom? At first it looks as though a picture like Figure 6–7 would be adequate. But this would be a misleading picture, because it places the electron at a definite spot in space. This could be true at a particular instant, but you must remember that an electron is in rapid motion all the time.

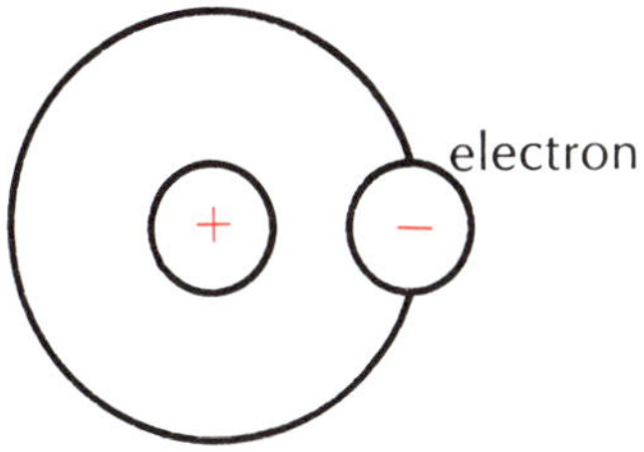

Figure 6–7
A possible picture of a hydrogen atom.

Then what is a reasonable picture or model of an atom of hydrogen? To answer this question, think of a mosquito buzzing around inside a basketball. What is its position? This question is hard to answer. At one instant it could be in position 1, the next instant in position 2, the next in position 3, the next in position 4, and so on. (See Figure 6–8.) The best picture for the position of the mosquito is probably that shown in Figure 6–9. But a note would have to be added, stating that the position could be represented by any one of the dots. How does this picture apply to a moving electron? The motion of the electron is affected by the attractive force of the proton in the nucleus. Therefore the electron is more likely to be found

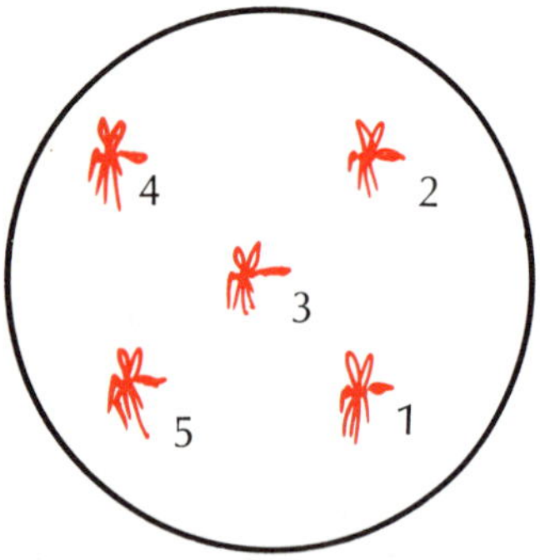

Figure 6–8
Like an electron, a mosquito in a basketball would move from one position to another very quickly.

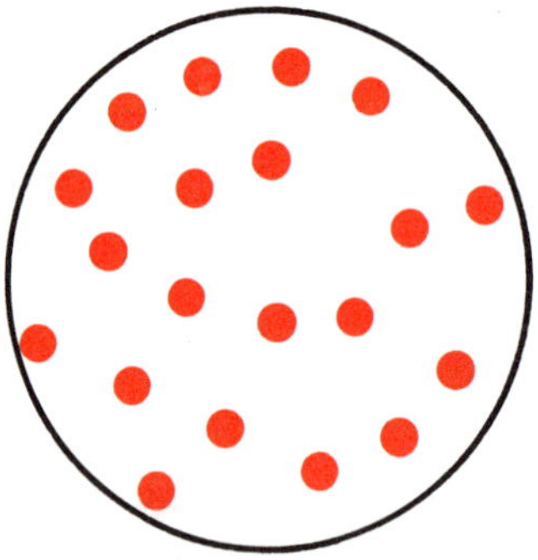

Figure 6–9
The mosquito could be at any one of these positions.

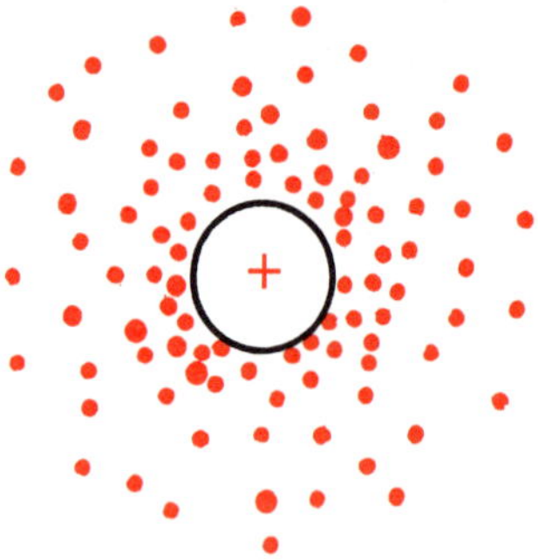

Figure 6–10
A better picture of a hydrogen atom. The electron could be in any one of the positions shown.

near the nucleus than far away from it. For this reason most of the dots are placed near the nucleus. Thus a reasonable model or picture for a hydrogen atom is shown in Figure 6–10, where ⊕ represents the nucleus.

Again it is important to understand that these dots represent the possible positions of a single electron. Thus, although the electron could be in any of these positions, the most likely position is near the nucleus. The space occupied by the electron is called an **electron cloud.**

perspective 6–5

Size of Atoms

Atoms are so very small that even the most powerful microscopes cannot bring them into view. However, you can get some idea of their size from the following facts. As you know, water is made up of atoms of hydrogen and oxygen. In one single drop of water there are about six thousand billion billion (6,000,000,000,000,000,000,000) atoms of hydrogen, a number so large that no one can understand it. If a drop of water could be magnified to the size of the earth, the atoms of hydrogen would become about as large as oranges.

perspective 6–6

Where Do Atoms Come From?

Figure 6–11
In the sun, a proton and a neutron may collide and stick together, forming a nucleus of heavy hydrogen.

Astronomers tell us that all atoms on earth — atoms in rocks, atoms in manufactured articles, and even atoms in your own body — may have come from the sun. How can this be explained? It is estimated that the temperature deep in the interior of the sun is about twenty million degrees Celsius (20,000,000°C). At such a high temperature there are no whole atoms, only parts of atoms (protons, neutrons, and electrons) moving about at exceedingly high speeds. Obviously there are frequent collisions between these particles. When a proton and a neutron collide they may stick together as shown in Figure 6–11. This proton-neutron pair forms the nucleus of a special kind of hydrogen atom called "heavy hydrogen."

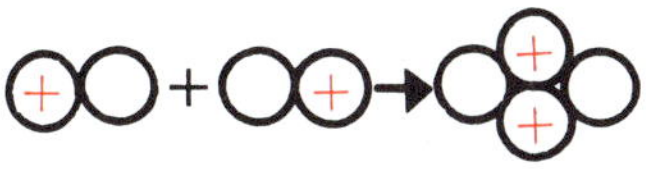

Figure 6–12
Two heavy hydrogen nuclei can combine to form a helium nucleus.

The surface of the sun is not as hot as the interior. The surface temperature of the sun is only about 6000°C. With lower temperatures, the particles slow down and it is easier for them to combine. One nucleus of heavy hydrogen can combine with another, as shown in Figure 6–12. When this happens a nucleus containing two protons is formed. It is the nucleus of a new element called helium. This combination is the first stage in the formation of heavier atoms.

If the nuclei (NEW–klee–eye) of three atoms of heavy hydrogen combine, the nucleus of the atom lithium is born, as shown in Figure 6–13. If the nuclei of three helium atoms combine, the nucleus of a carbon atom is formed, as in Figure 6–14.

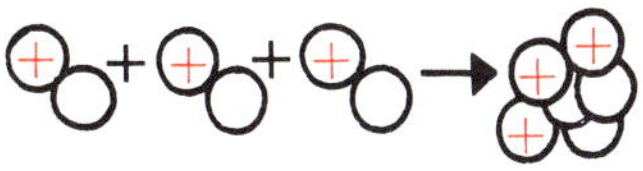

Figure 6–13
Three heavy hydrogen nuclei make a nucleus of lithium.

At the sun's surface, there are many different kinds of atoms. That is, the small nuclei have combined into larger and larger ones. At this temperature free electrons are attracted to the positively charged nuclei to form neutral atoms. However, most of the sun's atmosphere is hydrogen and helium.

Astronomers also tell us that the earth and the other planets may have come from the outer fringes of the sun. Thus the

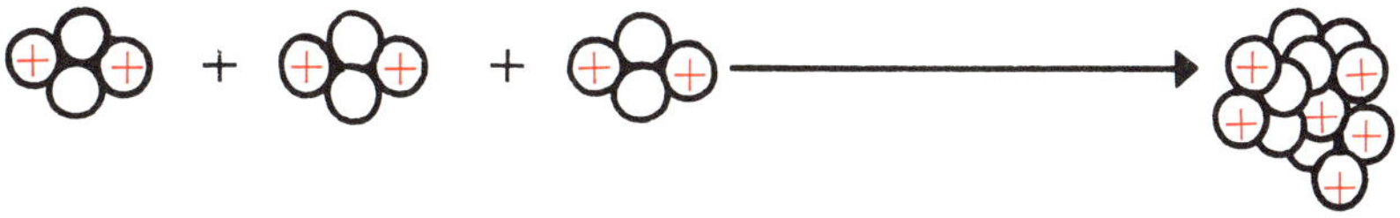

Figure 6–14
Three helium nuclei make a nucleus of carbon.

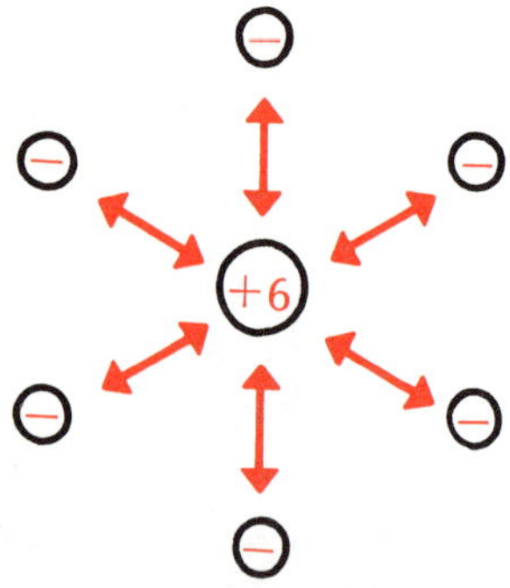

Figure 6–15
The electric force holds electrons to the nucleus.

original earth may have been an enormous cloud of extremely hot atoms. Gravitation pulled these atoms close together, and eventually the cloud condensed to a liquid ball. Then, through the ages, atoms combined chemically to form compounds. The ball cooled to form a solid crust.

If the earth came from the sun, how was the sun formed? The generally accepted theory is that it was formed from hydrogen atoms scattered throughout space. Then nuclear reactions occurred in the sun and all the known elements were made. This theory, however, leaves one big question unanswered. If the earth was made out of the sun, and if the sun was made out of hydrogen atoms in space, how were these hydrogen atoms made? No one knows the answer to this question. It is one of the mysteries of the universe.

perspective 6–7

Nuclear Forces

There is still another puzzling question about the structure of an atom. The presence of protons means that the nucleus of any atom must have a positive charge. Surrounding the nucleus, there are negatively charged electrons equal in number to the protons. Therefore there is an electric force of attraction between electrons and protons. This is the force that holds the atom together as a whole, as shown in Figure 6–15.

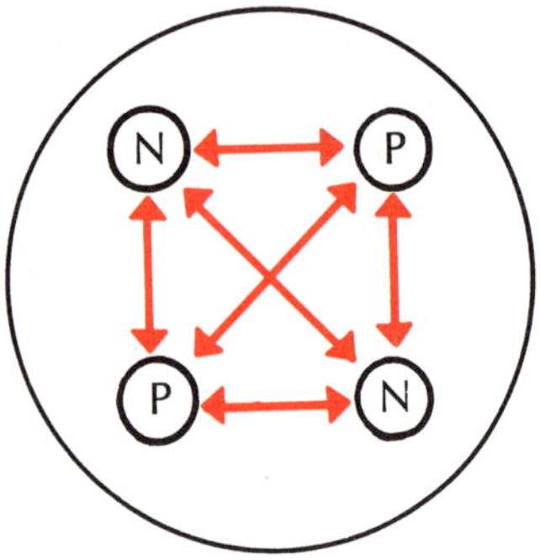

Figure 6–16
The nuclear force holds the protons and neutrons in the nucleus together.

Consider the electric force as it applies to the particles in the nucleus. In the nucleus of a carbon atom there are six protons tightly packed together. And, since the protons are all positively charged, they repel each other electrically. Then why does the nucleus not fly apart? This question can be asked of the nuclei of all atoms except hydrogen. The answer is that there must be some kind of binding force that is much stronger than the electric force. This binding force is called a **nuclear** (NEW-klee-ur) **force.** A nuclear force is a powerful force of attraction. It applies to both neutrons and protons in a nucleus. However, for this force to be effective, the particles must be extremely close together.

The nuclear force in the nucleus of an atom of helium is represented in Figure 6–16. Notice that proton attracts proton, neutron attracts neutron, and proton attracts neutron. The whole nuclear force is represented as made up of 12 separate forces. The electric force of repulsion between the protons is shown in Figure 6–17. The nuclear force is much stronger than the electrical force, so the helium nucleus is very stable. So far, scientists have not been able to split the nucleus of a helium atom.

The nature of these inner forces that hold a nucleus together is a mystery. It is another of the secrets of our universe.

perspective 6–8

The Three Forces in Nature

You now know about three different forces in nature: gravitational force, electric force, and nuclear force. All play their special and independent part in the nature of things.

As you know, the gravitational force was discovered by Newton in the 17th century. It is a force of attraction. Compared with the other two natural forces, it is a very weak force. It is too weak to overcome the force of friction and pull ordinary nearby objects together. Nevertheless it is an effective force through great distances, even through millions of miles of space. (See Figure 6–18.) In the vast regions of space it causes hydrogen particles to come together and condense to a cloud which then forms a star. Moreover, it is a force of great significance if we deal with enormous masses like moons, or planets, or stars.

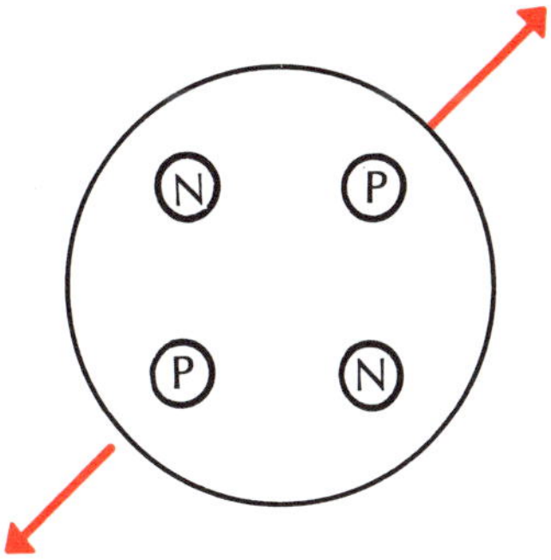

Figure 6–17
The protons in the nucleus repel each other with an electric force, but the nuclear force holding them together is stronger.

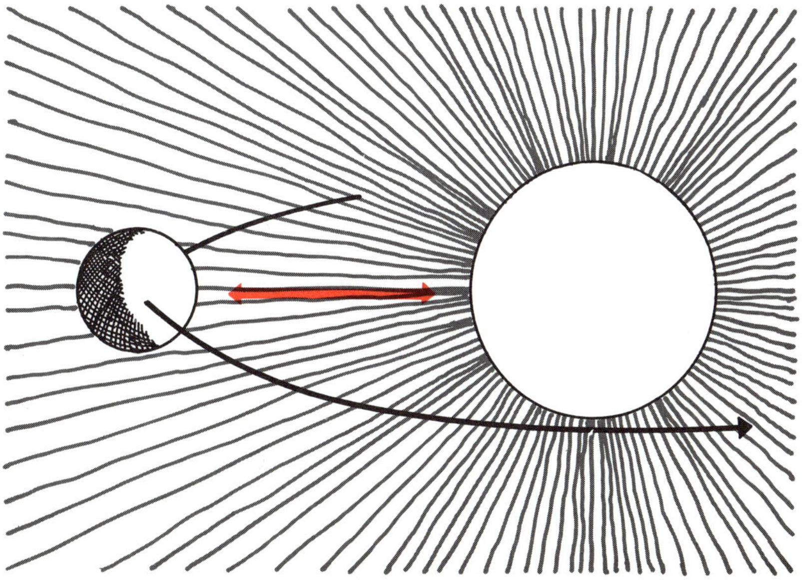

Figure 6–18
Gravitational force is weak, but it acts through great distances.

The way electric forces operate was not discovered until the middle of the 19th century. An electric force differs from a gravitational force in two ways. First, an electric force may be either a force of attraction or a force of repulsion. Second, it is a much stronger force than gravitation. For example, the electric force between two protons is billions of times stronger than the gravitational force between them. Even though it is gravitation that keeps us on earth, electric forces affect us much more than the gravitational force. Every atom is held together by the electric force between a positively charged nucleus and its negatively charged electrons. If there were no electric force, there would be no atoms, only loose electrons and nuclei. So this force accounts for the endless variety of substances in the world, even for plants and animals.

Nuclear forces were discovered in the early part of the twentieth century. A nuclear force is a very powerful force of attraction, at least 100 times as strong as an electric force. The most surprising thing about a nuclear force is the extremely short distance through which it acts. It does not begin to be effective until the nuclear particles are almost touching each other. Strangely enough, nuclear forces apply only to protons and neutrons, not to electrons. What would happen if this strange force became ineffective just for an instant? At that instant the electric force of repulsion would be the controlling force in the nuclei of all atoms. Atoms would all be blown apart. The whole universe would be reduced to a "sea" of neutrons, protons, electrons, and hydrogen atoms.

This summary of the three forces has raised some questions. (1) What causes a natural force? Nobody knows. (2) Why do protons repel protons? Why do electrons repel electrons? Why do protons attract electrons? There are no answers to these questions. Scientists merely state the facts that describe the way electric charges behave. These facts are called natural laws. (3) Why do nuclear forces apply only to protons and neutrons and not to electrons? Again, there is as yet no answer to this question. Scientists simply accept this strange selective behavior as part of the nature of things. (4) Finally, how did the three forces come about? Did they just happen, or were they created?

Some of these questions cannot be answered by experiments. Therefore they are outside the field of science. But some of these questions can be investigated. As the search for knowledge goes on, someone may eventually find the answers.

perspective 6–9

Atoms and Their Nuclei

As stated earlier, every atom has a central positively charged nucleus which is surrounded by just enough electrons to **neutralize** (NEW-truh-lyz), or cancel out, the positive charge. That is, the number of electrons always equals the number of protons. In some of the lighter atoms the number of neutrons in the nucleus equals the number of protons, but in most of the heavier atoms there are more neutrons than protons. For example, oxygen and nitrogen are light elements that make up almost 99 percent of the air. The nucleus of an oxygen atom consists of eight protons and eight neutrons, as shown in Figure 6–19. A nitrogen nucleus consists of seven protons and seven neutrons. The composition of the nuclei of a few other common elements is shown in Figure 6–20.

Figure 6–19
The nucleus of an oxygen atom.

Figure 6–20
The composition of the nuclei of some common elements.

Element	*Protons*	*Neutrons*	*Atomic Mass*
Hydrogen	1	0	1
Helium	2	2	4
Carbon	6	6	12
Sodium	11	12	23
Sulfur	16	16	32
Chlorine	17	18	35
Calcium	20	20	40
Iron	26	30	56
Silver	47	61	108
Mercury	80	120	200
Uranium	92	146	238

The sum of the numbers of protons and neutrons in a nucleus is called the **atomic mass** of the element. Uranium is the heaviest and largest of the natural elements. Its atomic mass is 238 and there are 92 electrons outside its nucleus. To get some idea of the size of a nucleus compared with the size of the whole atom, suppose an atom of uranium could be increased in size by one billion times. Then its diameter would be about one foot. But the nucleus at the center of this greatly magnified atom would be no bigger than a grain of sand. Because the nuclear material is concentrated in a very small volume, its density must be very high.

The volume occupied by the nucleus is less than one billionth of the whole atom. Yet about 99.9 percent of all the mass of the atom is concentrated into that tiny volume. Hence the density of matter in the nucleus of an atom is enormous, actually more than one billion grams per cubic centimeter. One billion grams is more than one thousand tons. Such a high density value is almost beyond our understanding. However, the density of uranium as a whole (that is, including the electron clouds) is only 18.9 g/cm^3, which is about 19 times as dense as water.

perspective 6–10

The Nature of a Chemical Reaction

As you already know, electrons are attracted to the nucleus by an electric force. However, electrons near the nucleus are attracted more strongly than those farther away. (See Figure 6–21.) Electrons on the outside of a large electron cloud are only weakly held. Some may be pulled out of one atom and into a nearby atom of a different kind. When this happens, a **chemical change** is said to have taken place. For example, let us see what happens if atoms of sodium and chlorine approach each other very closely. A sodium atom has a nucleus of 11 protons and 12 neutrons, surrounded by 11 electrons, as shown in Figure 6–22. A chlorine atom has a nucleus of 17 protons and 18 neutrons, surrounded by 17 electrons, as shown in Figure 6–23.

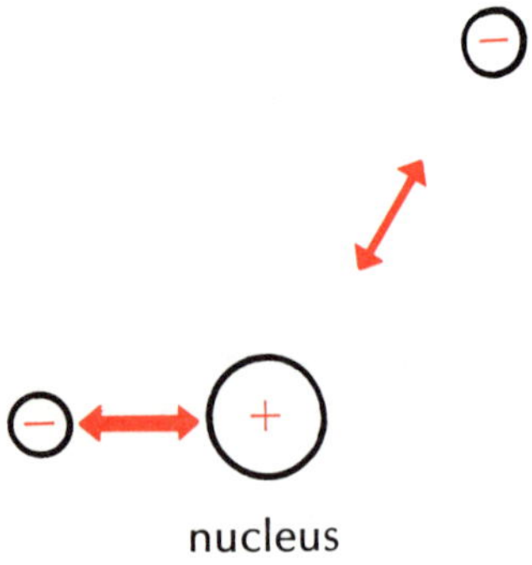

Figure 6–21
An electron near the nucleus is attracted more strongly than an electron farther away.

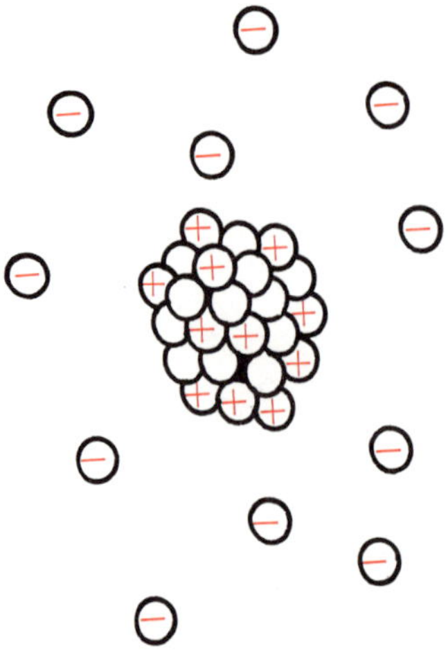

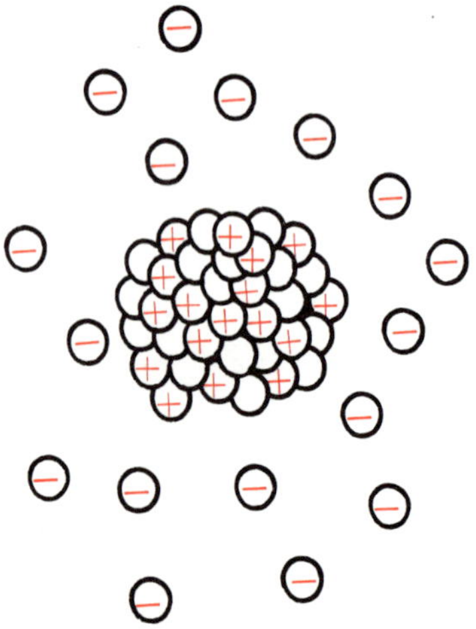

Figure 6–22 (left)
A sodium atom.

Figure 6–23 (right)
A chlorine atom.

A sodium atom has a nuclear charge of +11 surrounded by a large cloud of 11 electrons. A chlorine atom has a nuclear charge of +17 surrounded by a denser cloud of 17 electrons. If atoms of sodium and chlorine come very close together, the chlorine nucleus is able to pull an electron out of the sodium cloud and into its own. When an electron (or electrons) is transferred from one atom to another, ions are formed, in this case a sodium ion and a chlorine ion. An **ion** is an electrically charged atom. A chlorine ion is called a chloride ion. In summary,

sodium atom − electron = sodium ion
chlorine atom + electron = chloride ion

In an ion the number of electrons does not equal the number of protons in the nucleus. If the number of electrons is greater than the number of protons, the ion has a negative charge. If the number of electrons is less than the number of protons, the ion has a positive charge.

A chloride ion has 18 electrons and 17 protons. Therefore it has a negative charge of −1. On the other hand, a sodium ion has 10 electrons and 11 protons. Therefore it has a positive charge of +1. A diagram of a sodium atom and a sodium ion is shown in Figure 6–24. A diagram of a chlorine atom and a chloride ion is shown in Figure 6–25.

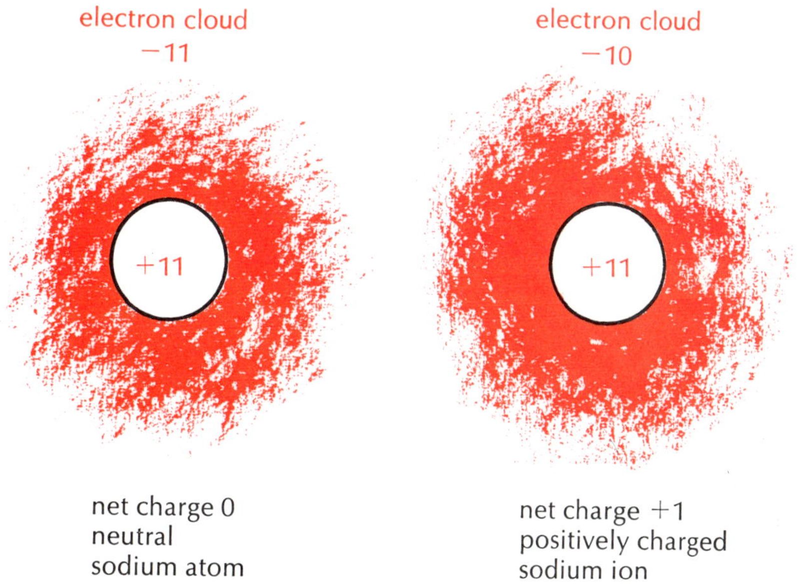

Figure 6–24
A sodium atom loses one electron to become a sodium ion.

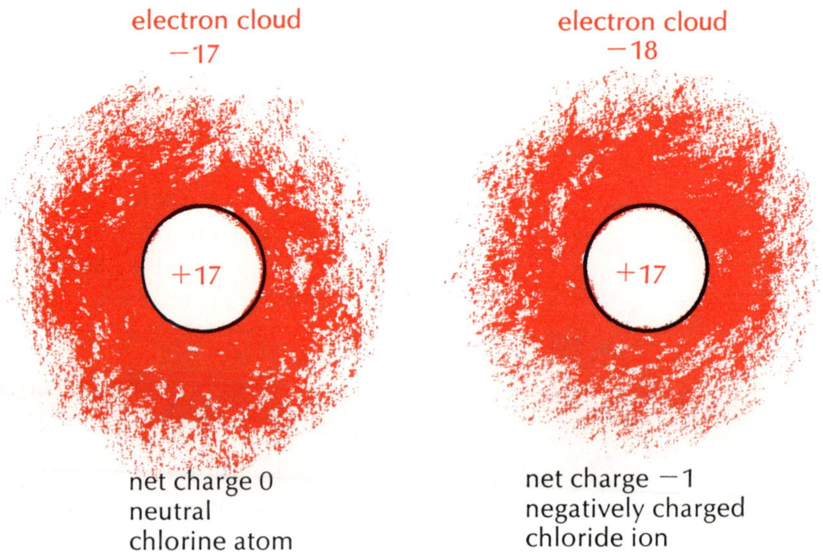

Figure 6–25
A chlorine atom gains one electron to become a chloride ion.

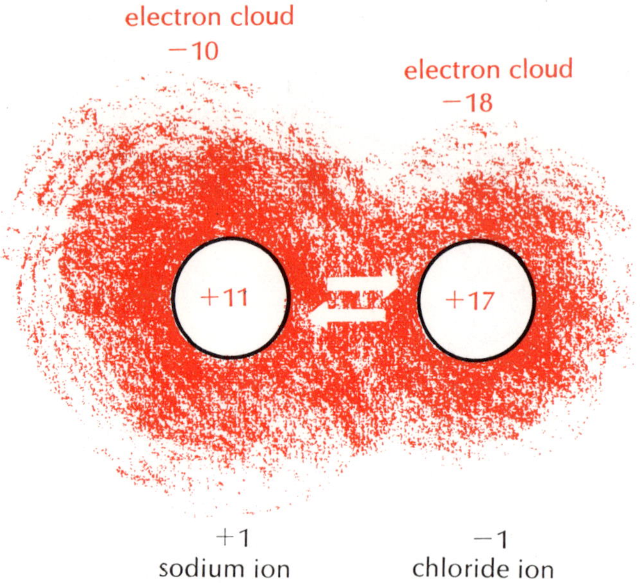

Figure 6–26
The sodium ion and the chloride ion attract each other. Why is the electron cloud of the chloride ion smaller?

Since sodium ions and chloride ions have opposite charges, they attract each other, as shown in Figure 6–26. This force of attraction binds together billions of oppositely charged ions into a solid called sodium chloride, or common salt.

perspective 6–11

Electric Charges Produced by Friction

Some atoms can be charged by friction. That is, in some atoms the outer electrons are so loosely held that they can be rubbed away from the nucleus. You remember that if glass is rubbed with silk, the glass becomes positively charged. Why? During the rubbing, some of the outer electrons of the atoms in the glass are dislodged. These loose electrons are attracted and captured by the nuclei of the atoms in the silk. As a result, the glass acquires a positive charge and the silk acquires a negative charge. Can you explain why the plastic rod becomes negatively charged when rubbed with flannel? Does the flannel become charged in the process?

This explanation of the charging of atoms by friction assumes that electrons are movable parts of atoms. This is a far cry from Dalton's theory that atoms are indivisible.

perspective 6–12

Atomic Numbers

As you know, atoms lose or gain electrons in a chemical reaction. The nuclei of atoms are never affected in chemical reactions. The nuclei are affected, however, when heavier nuclei are built up from lighter ones, as in the sun. When a lighter nucleus is changed into the next heavier nucleus, the number of protons in the nucleus increases by one. At the same time, the number of neutrons may increase by one or more than one. Thus the nucleus of a carbon atom is made up of 6 protons and 6 neutrons, whereas the nucleus of a sodium atom is made up of 11 protons and 12 neutrons.

Both the physical and chemical properties of an atom are determined by the number of electrons it has. Therefore this is an important number. The number of electrons equals the number of protons in the nucleus of an atom. The number of protons in the nucleus of an atom is called the **atomic number** of that particular element. In short, we can say that the properties of an element depend upon its atomic number.

Some atoms of an element may have a different number of neutrons from the rest. An example is heavy hydrogen. The number of neutrons does not affect the properties.

perspective 6–13

Symbols of Elements

A symbol is a kind of shorthand, often a substitute for a word. The most obvious symbol to use for an element is the first letter of the word. This is precisely what is used whenever possible. For example, H is the symbol for hydrogen, C for carbon, O for oxygen, S for sulfur, and so on. Notice that all these symbols are capital letters.

When a number of elements begin with the same letter, as in carbon, copper, calcium, cadmium, chlorine, chromium, and cobalt, the second or third letter of the word is added. Thus the symbol for calcium is Ca, for cadmium Cd, for chlorine Cl, for chromium Cr, and for cobalt Co. Notice that the second letter of a symbol is not capitalized; it is in small type. Other common elements with two-letter symbols are He for helium, Al for aluminum, Li for lithium, Mg for magnesium, and Zn for zinc.

Notice again that the symbol for cobalt is Co. Then what is the symbol for copper? Copper was discovered much earlier than cobalt and it is also more plentiful. It would seem reasonable then to use the symbol Co for copper. However, copper is one of a small group of elements which were known to the ancient Romans. The symbols of these elements come from their Latin names. These elements include copper (Cu), gold (Au), iron (Fe), lead (Pb), mercury (Hg), silver (Ag), and tin (Sn). The symbols for potassium (K) and sodium (Na) also come from their Latin names, even though these two elements were unknown in Roman times.

perspective 6–14

Arrangement of Elements According to Their Atomic Numbers

The first twenty elements are arranged in order of their atomic numbers in Figure 6–27.

If you examined and compared the chemical properties of these elements, you would find remarkable similarities among some of them. For example, lithium, sodium, and potassium are metals. Moreover, they are light metals (that is, metals of low density). They also enter into the same kinds of chemical reactions. Thus, lithium and potassium atoms behave like sodium atoms. They readily give up an electron to form a

Figure 6–27
The first twenty elements.

Atomic Number	*Element*	*Symbol*	*Pronunciation*
1	Hydrogen	H	HY-droh-jen
2	Helium	He	HEE-lee-um
3	Lithium	Li	LITH-ee-um
4	Beryllium	Be	buh-RILL-ee-um
5	Boron	B	BOR-on
6	Carbon	C	CAR-bun
7	Nitrogen	N	NY-tro-jen
8	Oxygen	O	OX-ih-jen
9	Fluorine	F	FLUR-een
10	Neon	Ne	NEE-on
11	Sodium	Na	SO-dee-um
12	Magnesium	Mg	mag-NEE-zee-um
13	Aluminum	Al	ah-LOO-mih-num
14	Silicon	Si	SIL-ih-kon
15	Phosphorus	P	FOS-for-us
16	Sulfur	S	SUL-fer
17	Chlorine	Cl	KLOR-een
18	Argon	Ar	AR-gon
19	Potassium	K	po-TAS-see-um
20	Calcium	Ca	KAL-see-um

positively charged ion. These three elements look like members of a chemical "family" that could be grouped together.

The elements helium, neon, and argon are also similar. These elements are gases. They take no part in chemical reactions. That is, these elements form no compounds. It looks as though these elements belong to a chemically inactive family that could also be grouped together.

The elements fluorine and chlorine are both gases. They also take part in the same kinds of chemical reactions. A chlorine atom readily gains an electron to form a negatively charged ion. Fluorine atoms do the same thing, even more readily than chlorine atoms. So again it looks as though fluorine and chlorine belong to a special chemical family of their own.

If you continued this kind of examination you would find that the elements beryllium, magnesium, and calcium have many similar properties and therefore belong to the same family.

There are also chemical and physical similarities between the following pairs of elements: boron and aluminum, carbon and silicon, and nitrogen and phosphorus.

perspective 6–15

A Classification of the First Twenty Elements

If the first twenty elements are arranged according to their atomic numbers and the families to which they belong, they form a table like Figure 6–28. In this table the atomic numbers increase from left to right, as you would read a book. The vertical columns are the families of elements.

Figure 6–28
The first twenty elements, arranged by atomic number and by family.

Column ("Family")

	I	II	III	IV	V	VI	VII	VIII
Row 1	1 H							2 He
Row 2	3 Li	4 Be	5 B	6 C	7 N	8 O	9 F	10 Ne
Row 3	11 Na	12 Mg	13 Al	14 Si	15 P	16 S	17 Cl	18 Ar
Row 4	19 K	20 Ca						

A hydrogen atom loses its electron to form a hydrogen ion with a charge of +1 (H^+). The atoms of the elements in column I form ions with charges of +1 too. For this reason hydrogen is placed above lithium in column I. In the same way, all the other elements have been arranged in families according to their properties. The complete classification table of all the elements is called the **periodic table.** The periodic table is shown in Appendix 2.

What is the meaning of a classification table based on atomic numbers? In a neutral atom, the number of protons in the nucleus is equal to the number of electrons outside the nucleus. Moreover, practically all properties of atoms depend upon the number and arrangement of their electrons. From left to right along the first horizontal row, Li, Be, and B are metals, while N, O, F, and Ne are non-metals. Carbon acts sometimes like a metal and sometimes like a non-metal. Hence, in going from left to right in the periodic table, there must be a changing pattern in the arrangement of the electrons of the atoms.

Elements 11 to 18, in the second horizontal row, have the same pattern of electrons as the elements in the first row. That is, Li and Na have similar properties because their electrons are arranged in a similar way. The same is true of F and Cl, and also of the other pairs.

perspective 6–16

The Arrangement of Electrons in Atoms

You know that there are two electrons in an atom of helium, ten electrons in an atom of neon, and eighteen electrons in an atom of argon. How are these electrons arranged? Is there a pattern? Scientists believe that electrons are arranged in different levels. The first level is filled when it contains two electrons. The second is filled with eight electrons. The third is also filled with eight electrons, and so on. The first three levels are shown in Figure 6–29.

Electrons can occupy positions in these levels but nowhere in between. If a ball is thrown into the air, its potential energy increases. The higher its position, the greater is its potential energy. Similarly, we can think of the electron levels as **energy levels.** The farther an electron is from the nucleus, the greater is its energy. Electrons in level 2 have more energy

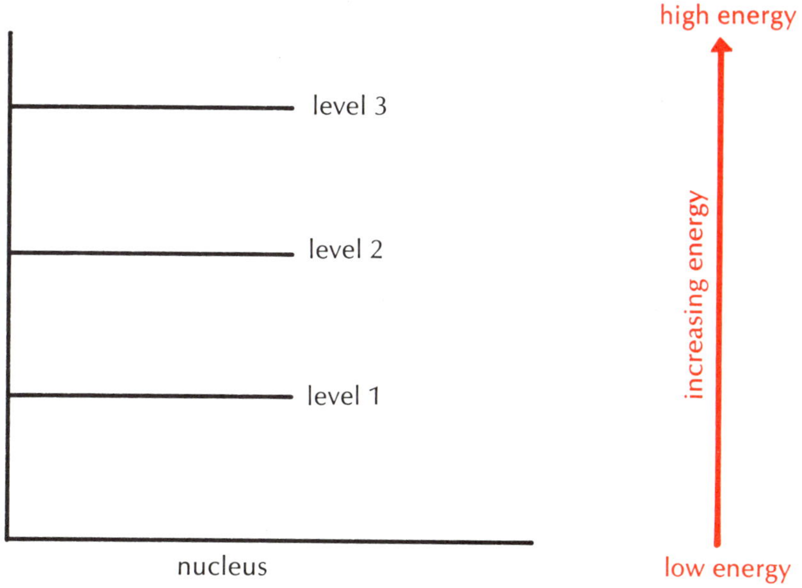

Figure 6–29
Electrons in higher energy levels have greater potential energy.

than electrons in level 1, and electrons in level 3 have more energy than electrons in level 2. Energy levels are also called **shells.**

Suppose atoms are formed by adding electrons one at a time. If an electron is added, a proton must be added to the nucleus. Otherwise the atom would not be neutral.

A hydrogen atom has a single electron in the first energy level. If a second electron and a second proton are added, a helium atom is formed, as shown in Figure 6–30. The first energy level is now filled. If a third electron is added, it will enter the second energy level. The next seven electrons will also enter this same energy level. When eight electrons occupy this level, it is filled and an atom of neon has been formed.

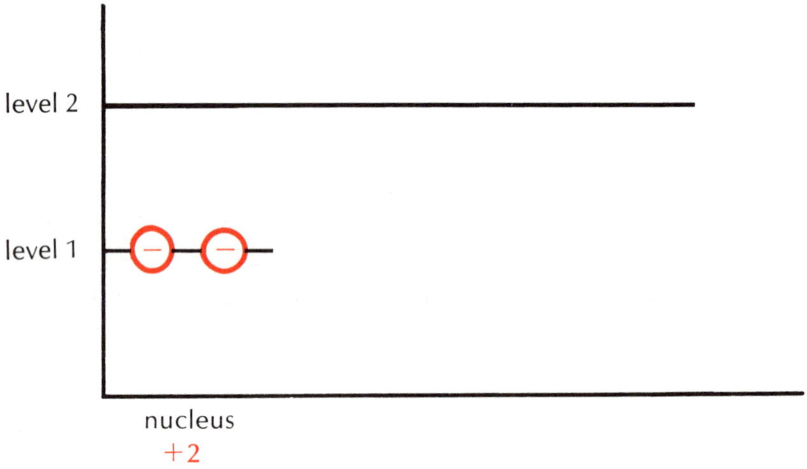

Figure 6–30
In a helium atom, the first energy level is filled.

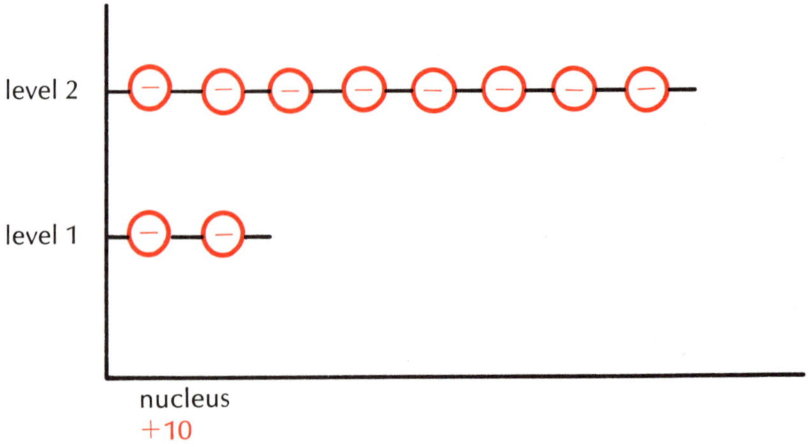

Figure 6–31
In a neon atom, the second energy level is filled.

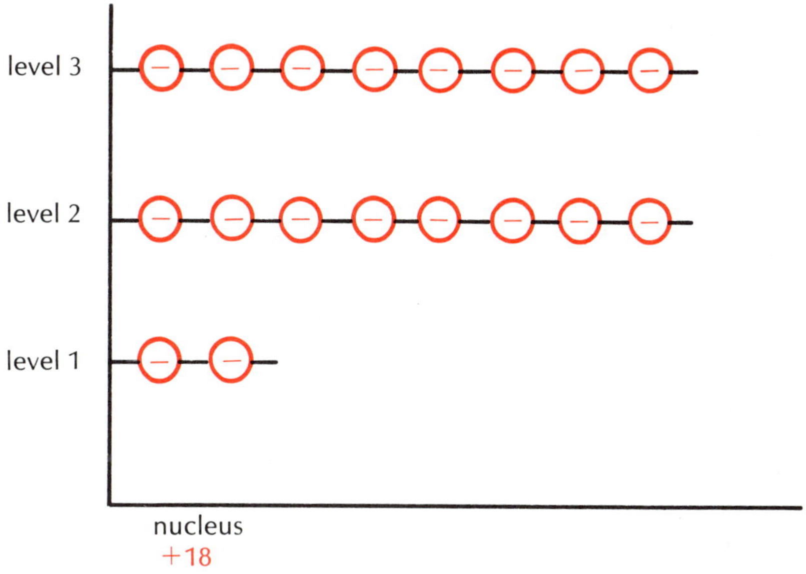

Figure 6–32
In an argon atom, the third energy level is filled.

The electron arrangement of an atom of neon is shown in Figure 6–31. An atom of neon has ten electrons and ten protons.

The eleventh electron enters the third energy level. If seven more electrons are added, this energy level becomes filled to capacity and an atom of argon is formed. The electron arrangement of argon is shown in Figure 6–32. An atom of argon has 18 electrons and 18 protons.

perspective 6–17

Electron Arrangement and Chemical Activity

A chemical change occurs if one or more electrons move from one kind of atom to another. Atoms of helium, neon, and argon have a full shell of electrons. Therefore there is no tendency for atoms of these elements to gain or lose electrons. That is why they are not active chemically and form no compounds.

Examine the electronic arrangement in atoms of sodium and chlorine as shown in Figure 6–33. In a sodium atom the electron in level 3 is farther away from the nucleus than the electrons in level 2. Thus the electron in level 3 is only held by a weak force and is easily displaced. When this happens, a chemical change takes place.

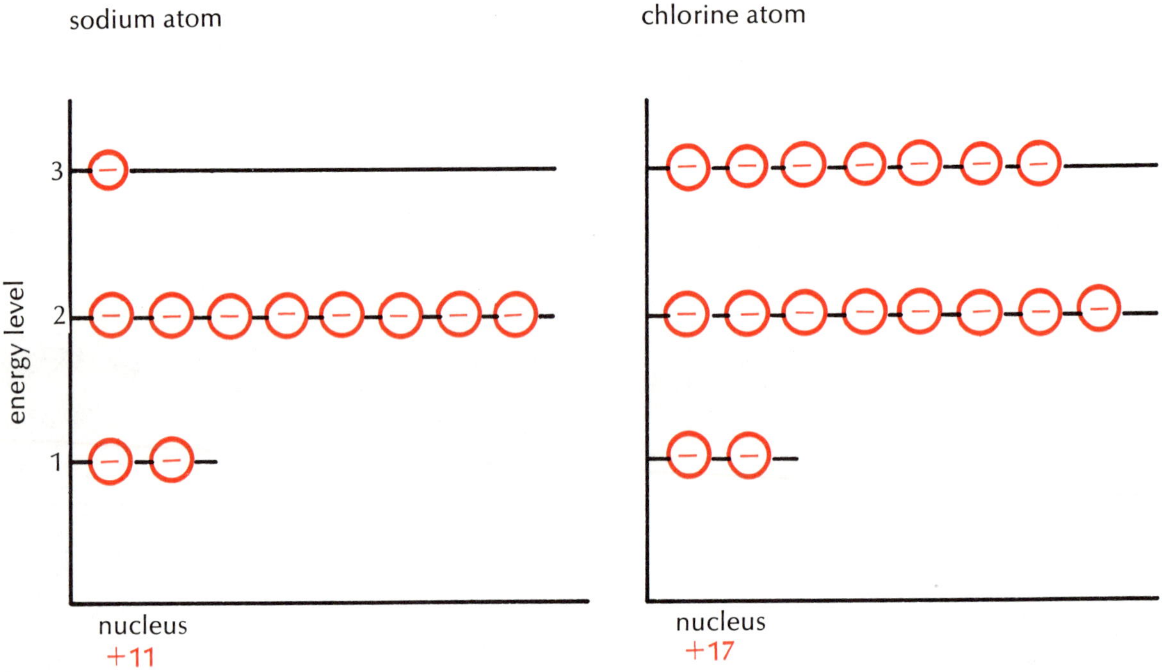

Figure 6–33
A sodium atom has only one electron in level 3. A chlorine atom has room for one more electron in level 3.

In a chlorine atom, the charge of the nucleus is +17. In a sodium atom, the charge of the nucleus is only +11. Thus the force available to hold electrons in the third level is much stronger in a chlorine atom than in a sodium atom. (See Figure 6–34.) In a chlorine atom the force is strong enough to attract an electron from outside into this level. The shell thus becomes filled with eight electrons. When this happens, a chemical change takes place.

The reaction between sodium and chlorine takes place in two steps.

Step 1: (a) A sodium atom loses an electron to form a positively charged sodium ion.

$$\text{sodium atom} - \text{electron} = \text{sodium ion}$$

Or, using symbols,

$$Na - e^- = Na^+$$

where e^- stands for an electron and Na^+ stands for a sodium ion. (See Figure 6–24.) Because a sodium atom loses one electron to form a positive ion, its combining capacity is said to be +1.

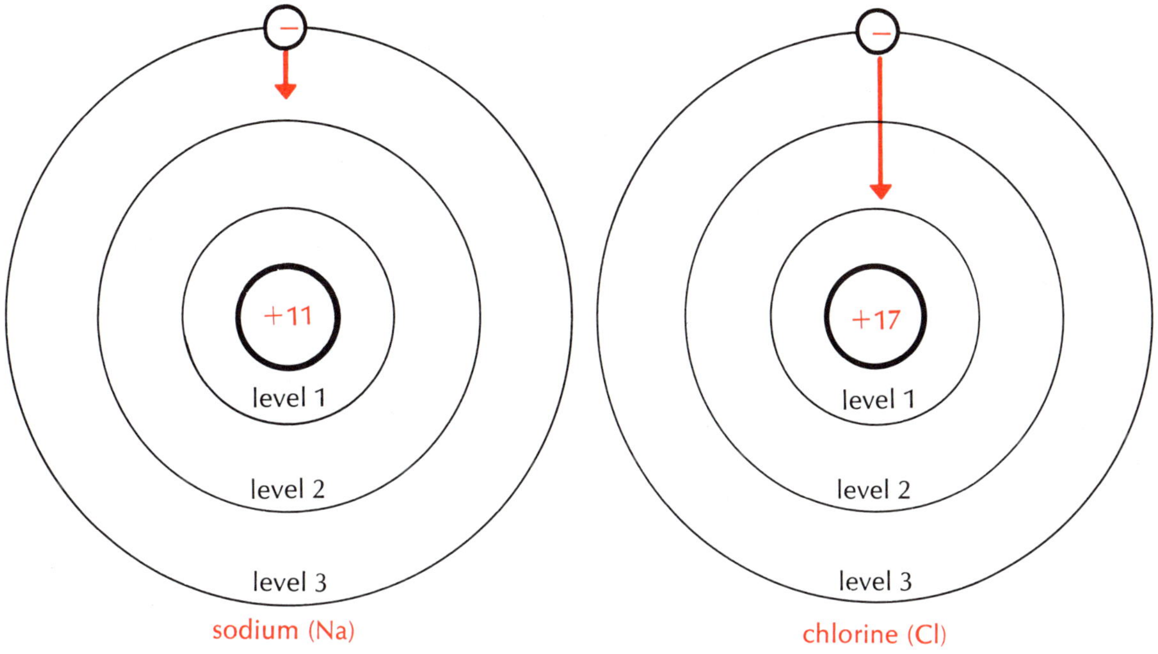

Figure 6–34
A chlorine atom has a greater charge in its nucleus, so it can hold an electron in the third level more strongly than a sodium atom.

(b) The electron lost by the sodium atom is gained by a chlorine atom to form a negatively charged chloride ion. (See Figure 6–25.)

chlorine atom + electron = chloride ion

Or, using symbols,

$$Cl + e^- = Cl^-$$

Because a chlorine atom gains an electron to form a negative ion, its combining capacity is said to be −1.

Step 2: The positively charged sodium ions then combine with the negatively charged chloride ions to form the compound sodium chloride.

$$Na^+ + Cl^- = NaCl$$

The **combining capacity** of an element is the number of electrons lost or gained when an atom of that element reacts with another atom. If it loses electrons, its combining capacity is positive. If it gains electrons, its combining capacity is negative.

perspective 6–18

The Periodic Classification and Chemical Activity

The classification of elements derived from atomic numbers is by far the most useful aid in the study of chemistry. If you know the properties of one particular element, you can predict the properties of other elements in the same family. And if you know which column an element is in, you can predict how it will behave. For example, the periodic classification gives information about the combining capacities of elements. All elements in the same column of the periodic table have the same combining capacity. Thus lithium and potassium both have a combining capacity of +1.

Notice that the elements helium, neon, and argon are in column VIII of the periodic table. All atoms in this column have filled shells, so these elements are not chemically active. The addition of another electron starts a new shell. The elements with a single electron in a new shell are lithium, sodium, and potassium in column I. All these elements are very active chemically.

The elements in column VII, which include fluorine, chlorine, and bromine, have seven electrons in their outer shells. Atoms of these elements need only a single electron to fill the shell. So these elements are also chemically active.

Chlorine combines with magnesium in column II to form magnesium chloride. An atom of magnesium loses two electrons in this reaction to form a magnesium ion with two positive charges. The symbol for a magnesium ion is therefore Mg^{2+}. This reaction can be written as

$$Mg - 2e^- = Mg^{2+}.$$

The two electrons released by a magnesium atom then combine with two chlorine atoms to form two chloride ions.

$$2Cl + 2e^- = 2Cl^-$$

The magnesium ions then attract the two chloride ions to form magnesium chloride ($MgCl_2$). In symbols this chemical reaction is

$$Mg^{2+} + 2Cl^- = MgCl_2$$

Notice that the two chloride ions in magnesium chloride are written as a subscript after the symbol in the lower corner. The combining capacity of magnesium is +2.

Finally, the combining capacity of elements in column III is +3. Hence, when aluminum combines with chlorine to form aluminum chloride, each atom of aluminum gives up three electrons and these three electrons are gained by three chlorine atoms. The two steps are

Step 1: (a) $Al - 3e^- = Al^{3+}$
(b) $3Cl + 3e^- = 3Cl^-$
Step 2: $Al^{3+} + 3Cl^- = AlCl_3$

Since the elements He, Ne, and Ar in Column VIII form no compounds, a combining capacity of zero is assigned to all elements in this column.

The combining capacities of the elements in the various columns are summarized in Figure 6–35.

Figure 6–35
The combining capacities of the elements in each column of the periodic table.

column number	I	II	III	IV	V	VI	VII	VIII
combining capacity	+1	+2	+3	+4 −4	−3	−2	−1	0

A knowledge of the combining capacities of the elements enables you to write formulas for chemical compounds. For example, potassium (K) combines with oxygen (O) in the air to form a white solid called potassium oxide. What is the formula for potassium oxide? The combining capacity of potassium is +1, so each atom of potassium will lose one electron. The combining capacity of oxygen is −2, so each atom of oxygen will gain two electrons. The total number of electrons lost by potassium atoms must equal the total number of electrons gained by oxygen atoms. Therefore two atoms of potassium combine with one atom of oxygen, so the formula for potassium oxide is K_2O.

perspective 6–19

Metals and Non-metals

Some elements are metals and some are non-metals. What do these terms mean? Examine some metals, such as copper, iron, and aluminum. You will see that metals have certain

Figure 6–36
Metals have many properties in common. Gold (above) is very valuable because it is shiny. Steel (below left), which is mostly iron, can be rolled into sheets for many uses. Copper (below right) is used for wire because it is a good conductor of electricity.

Figure 6–37
Non-metals are more varied than metals. Carbon (above left) comes in many forms, including graphite (pencil lead) and diamond. This is lampblack. Iodine (above right) is a shiny black solid that changes to a purple gas when heated. Sulfur (right) is a soft yellow rock.

characteristic properties. All metals except mercury are solids at ordinary temperatures. (See Figure 6–36.) They are shiny, they have a high density, and they are good conductors of heat and electricity. These are physical properties of metals. **Physical properties** of a substance are properties that do not involve any changes in the composition of its atoms or molecules.

Metals also have characteristic chemical properties. **Chemical properties** are those by which atoms or molecules can be changed so that one substance can be transformed into another. Atoms of all metals have some loosely held electrons which can easily be detached. This is a chemical property. Or, expressed another way, atoms of metals usually lose electrons in chemical reactions.

Some examples of non-metals are oxygen, chlorine, and sulfur. Non-metals may be gases, or solids, or even liquids at ordinary temperatures. (See Figure 6–37.) Moreover, they are poor conductors of heat and electricity. They are best defined in terms of the way they react chemically. Atoms of non-metals have no loosely held electrons, so in chemical reactions they gain rather than lose electrons.

In summary, metals usually form positively charged ions in chemical reactions. Non-metals usually form negatively charged ions.

QUESTIONS

A

1. In what important respect did Dalton's ideas on atoms differ from the ideas of the Greek philosophers?

2. In what important respect do the modern ideas on atoms differ from Dalton's?

3. (a) What effect do like charges have on each other? (b) What effect do unlike charges have on each other?

4. Compare the charge of a proton with the charge of an electron.

5. Compare the mass of a proton with the mass of an electron.

6. Compare (a) the nuclei and (b) the number of electrons in atoms of hydrogen and heavy hydrogen.

7. What is the marked difference between an electrical force and a nuclear force?

8. (a) Define *atomic number.* (b) Why is the atomic number of an element important to a scientist?

9. The nucleus of an atom of copper is made up of 29 protons and 35 neutrons. State (a) the atomic number and (b) the atomic mass of this element.

10. What is the combining capacity of an element?

11. (a) Write the formula for hydrogen oxide. (b) What is the common name for this compound?

12. The formula for aluminum oxide is Al_2O_3. Explain why.

13. (a) How does a metal usually behave chemically? (b) Give an example.

14. (a) How does a non-metal usually behave chemically? (b) Give an example.

A	B	C	D
E	F	G	H
I	J	K	L
M	N	O	P

15. The left column of Figure 6–38 contains words that complete or answer the statements in the column on the right. Select the number of the word that best answers each statement and place the number in the proper space in an answer box like the one at left. If you correctly complete the box, the total of your numbers in each horizontal row, vertical column, and diagonal will be the same.

B

16. If you comb your hair with a steel comb, rather than with a plastic comb, will your hair spread out? Give a reason for your answer.

17. Draw a diagram to represent the nucleus and the electron cloud of a hydrogen atom.

18. Explain how the nucleus of an oxygen atom (8 protons and 8 neutrons) could be formed from protons and neutrons in the surface region of the sun.

19. Compare the effects of an electrical force on (a) 2 protons, (b) 2 neutrons, (c) 1 proton and 1 electron.

20. What is a characteristic property of chlorine (atomic number 17)?

21. What is a characteristic property of argon (atomic number 18)?

22. Draw a diagram to show electrons in the energy levels of an atom of magnesium.

23. Draw a diagram to show electrons in the energy levels of an atom of sulfur.

24. (a) What is the formula for the compound of magnesium and sulfur? (b) Why have you written this particular formula?

Figure 6–38
Use this table for Question 15.

Answers	*Statements*
1. gravitational	A. an atom with an electrical charge
2. nuclear	B. the center of an atom
3. atomic number	C. an atom that gains electrons
4. electron	D. the number of protons in the nucleus
5. compound	E. the charge on a sodium ion
6. element	F. a negatively charged particle
7. combining capacity	G. a force of attraction or repulsion
8. atom	H. a group of atoms
9. electrical	I. a substance that is made of different kinds of atoms
10. ion	J. electrons pulled out of one atom and into a nearby atom of a different kind
11. proton	K. a positively charged elementary particle
12. electron cloud	L. a building block of matter
13. nucleus	M. space occupied by an electron
14. molecule	N. capacity of an atom to gain or lose electrons
15. weight	O. a substance that contains the same kind of atoms
16. positive	P. a particle with no charge
17. non-metal	
18. neutron	
19. chemical change	
20. negative	
21. metal	

25. The formula for magnesium nitride is Mg_3N_2. What are the combining capacities of magnesium and nitrogen?

26. Compounds of sulfur with a metal are called sulfides. Write the formulas of (a) calcium sulfide, (b) lithium sulfide, (c) aluminum sulfide.

27. Franklin called the charge on rubbed glass positive and the other charge negative. He could have called the charge on rubbed glass negative and the other charge positive. How would this change have affected the kinds of charge later discovered on electrons and protons?

28. In your notebook (not on this page) list the symbols of the first twenty elements under their respective combining capacities according to their position in the periodic table. Your table should have headings like Figure 6–39.

Figure 6–39
Make a table like this for Question 28.

Column	I	II	III	IV	V	VI	VII	VIII
Combining Capacity	+1	+2	+3	+4 −4	−3	−2	−1	0
Elements	?	?	?	?	?	?	?	?

29. Write the formula for the compound formed from each of the following combinations of elements.

(a) hydrogen and fluorine
(b) lithium and fluorine
(c) lithium and chlorine
(d) magnesium and oxygen
(e) magnesium and sulfur
(f) sodium and oxygen
(g) sodium and sulfur
(h) potassium and sulfur
(i) calcium and fluorine
(j) calcium and chlorine
(k) aluminum and fluorine
(l) aluminum and chlorine
(m) aluminum and sulfur
(n) aluminum and oxygen

30. Complete the table in Figure 6–40, summarizing the information on forces.

Figure 6–40
Make a table like this for Question 30.

Force	Type of Force (attractive or repulsive)	Effective Distance	Strength
gravitational	?	?	?
electric	?	?	?
nuclear	?	?	?

Chapter 7

CHEMICAL REACTIONS

A flame is a sign of a chemical reaction.

Have you ever looked really closely at a wooden match? A good way to start learning about chemical reactions is to observe a wooden match before it is lit, while it is burning, and after it is blown out. The match is made up of two parts, a thin stick of wood and a blob of chemicals. If you rub the tip of the match on the smooth surface of the matchbox, nothing happens. If you rub the tip along the strip of sandpaper on the box, the tip bursts into flame. The flame then ignites the wooden stick, which burns quietly. If you blow out the flame, you can compare the burned part of the stick with the unburned part.

These observations raise many questions. Why was heat released? Why was there a burst of flame at the tip of the match? Did the chemicals combine in some way with the air? Why did the wood burn quietly? Why did nothing happen when the match was rubbed on a smooth surface? Why did a flame start when the match was rubbed on sandpaper? Finally, why was the flame put out when you blew on it? Is there something in your breath that prevents wood from burning?

What is burning? What is air? What is a chemical reaction? The experiments in this chapter will help you answer some of these questions.

WHAT IS A CHEMICAL REACTION?

Reactants and Products

A flame is a sign of a chemical change. When a match burns, there are two different flames, burning match and burning wood. Thus there must be at least two different chemical changes in a burning match. A chemical change is usually called a **chemical reaction,** or simply a reaction.

In any chemical reaction there are two sets of substances. First there are the substances that enter into the reaction. These are called reacting substances or, more simply, **reactants** (ree-AK-tantz). Then there are the substances produced in the reaction. These are called **products.** Thus reactants are the starting substances, and products are the substances produced in the reaction.

There are, of course, reactants and products in the reaction in the flame at the tip of the match. As yet, you do not know what these reactants and products are. In the next experiment you will be given two known reactants and you will see at least one of the products.

experiment 7–1

A Visible Product in a Chemical Reaction

Materials

sodium sulfate solution; barium chloride solution; ordinary test tube

Procedure

In this experiment you are supplied with water solutions of sodium sulfate and barium chloride. Both these compounds are white solids that readily dissolve in water.

Pour about 5 ml of the sodium sulfate solution into a test tube. (The capacity of an ordinary test tube is about 30 ml, so the sodium sulfate solution should occupy about one-sixth of the volume of the tube.) Then pour a few drops of barium chloride solution into the tube. Observe that a white solid is formed, as shown in Figure 7–1. A solid that falls out of solution is called a **precipitate** (pree-SIP-uh-tayt). A precipitate, like a flame, is

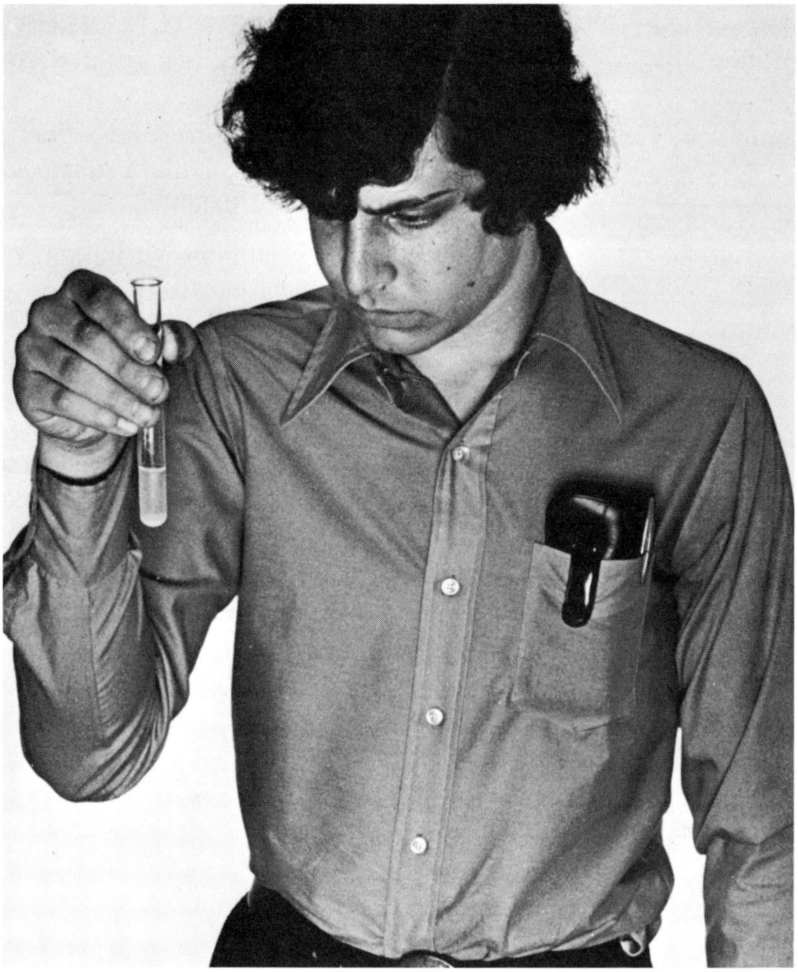

Figure 7–1
A precipitate is another sign of a chemical reaction.

evidence that a chemical change has taken place and that a new compound has been formed. The name of the visible compound formed in this reaction is barium sulfate.

Questions

1. Name the reactants in this reaction.

2. Name one of the products in this reaction.

3. (a) Is this product soluble or insoluble in water? (b) How can you tell?

4. What is the other product?

5. (a) Is this other product soluble or insoluble in water? (b) How can you tell?

experiment 7–2

To Compare the Masses of Reactants and Products in a Chemical Reaction

Materials

Erlenmeyer flask and solid rubber stopper to fit; small test tube with thread loop; sodium sulfate solution; barium chloride solution; equal-arm balance

Procedure

Pour a few milliliters of barium chloride solution into a small Erlenmeyer flask. The shape of an Erlenmeyer (UR-len-my-ur) flask is shown in Figure 7–2. Tie a loop of thread around the lip of a small test tube, small enough to be placed inside the flask. Almost fill the test tube with a solution of sodium sulfate. Holding the thread, lower the test tube into the flask. Then, to keep the thread in place and to prevent the test tube from falling, close the flask with a solid rubber stopper. The arrangement is shown in Figure 7–3. Find and record the mass of the flask, test tube,

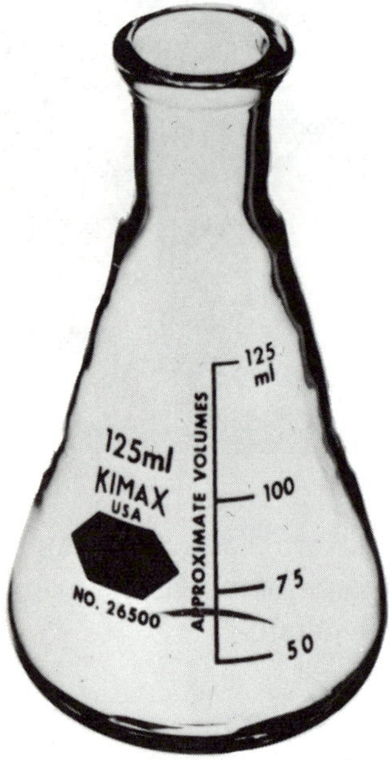

Figure 7–2 (above)
An Erlenmeyer flask.

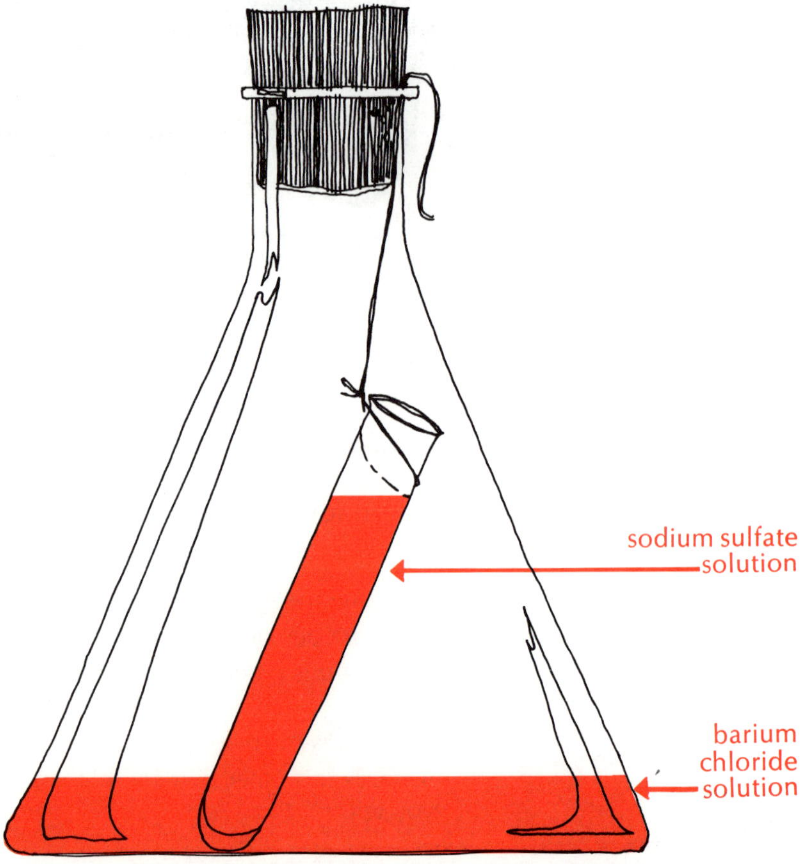

Figure 7–3 (right)
Setup for Experiment 7–2. An arrangement like this, with different chemicals, is used in many fire extinguishers.

and reactants. Then tilt the flask until the solutions mix. Find and record the mass of the flask, test tube, and products.

Questions

1. What evidence is there that a chemical reaction has taken place?

2. Is there any change in mass in this reaction? Support your answer by experimental data.

3. Does your experimental result support Dalton's atomic theory? If it does, explain why.

4. Your data on masses of reactants and products applies to all chemical reactions. This illustrates a scientific law. State this law in your own words.

REACTIONS WITH AIR

The Composition of Air

Air is a mixture of several gases, including oxygen and nitrogen. All animal life on earth depends upon oxygen. Oxygen is by far the most chemically active ingredient in air. By selecting a substance that will react with oxygen, you can remove oxygen from air. One such substance is iron. Iron will react with oxygen at ordinary temperatures, whereas many other substances will not. However, the iron must be damp, otherwise it will not react.

Oxygen in the air is free. That is, oxygen atoms are not attached to atoms of any other element in air. As a result, oxygen can be removed from air in such a simple reaction as the rusting of iron. In the next experiment oxygen will be removed from the air in a test tube.

experiment 7–3

To Find the Percentage of Oxygen in Air

Materials

800-ml beaker; 3 test tubes; glass marking pencil; iron filings; spatula; test tube rack; matches; centimeter ruler

Procedure

Write the numbers 1, 2, and 3 on the test tubes. Almost fill the beaker with water. Pour a little water into test tube 1. Close the

Figure 7–4
Use a spatula to spread iron filings over the inside of the test tube.

Figure 7–5
Invert the test tubes in the beaker of water.

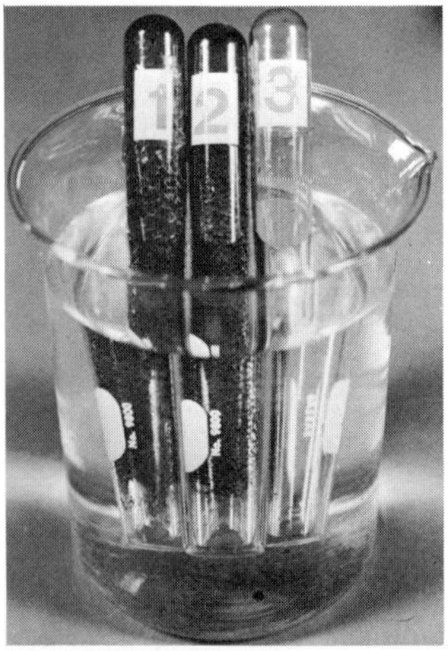

tube with your thumb. Shake the tube to wet the inside. Then pour the water out. Using a flat, knifelike tool called a spatula (SPACH-uh-luh) spread iron filings over the inside of the tube as in Figure 7–4. Use about enough filings to fill the rounded bottom of the tube. Now invert the tube and lower it into the beaker of water.

To get a second set of data, repeat the above procedure with test tube 2. Lower this tube into the beaker alongside tube 1. Finally, lower test tube 3 into the beaker. Test tube 3 should not contain iron filings. The arrangement is shown in Figure 7–5.

Put the apparatus aside until the next day or until the next laboratory period. Then examine it and describe any changes that have occurred.

Get a match and have it ready to light. With the mouth of tube 1 still below the water surface, close the tube with your thumb. Then take the tube out of the beaker and set it upright in the test tube rack. Without delay, light the match and thrust the flame just inside the tube. Now thrust the end of a burning match just inside test tube 3 and compare the observations for each of the tubes.

Measure the height of the water column in test tube 1. Also measure the length of the tube. Repeat this procedure for test tube 2, and record your readings as in Figure 7–6.

Figure 7–6
Data table for Experiment 7–3.

	Height of Water Column	Length of Test Tube
test tube 1	? cm	? cm
test tube 2	? cm	? cm

Questions

1. What observations did you make after the apparatus was set aside for a day?

2. The height (or length) of the test tube can be used as a measure of the volume of the air in the tube. Explain why.

3. How does the volume of water that enters the tube with the iron filings compare with the volume that enters the empty tube?

4. How does the amount of water that enters the iron filings tube compare with the amount of oxygen originally in the tube? Why?

5. What information is given by the test with the burning match in all three tubes?

6. Name at least two gases that remain in the test tube after a day.

Computations

1. Calculate the percentage of oxygen in air for each set of readings. Then find the average value. Remember that

$$\%\text{ of oxygen} = \frac{\text{height of water in tube}}{\text{length of tube}} \times 100\%$$

Water Vapor in the Air

Water vapor is always present in air. It can be detected in several ways. For instance, if the air contains a large amount of water vapor, the outside of a glass filled with ice water becomes coated

Figure 7–7
Water vapor from the air condenses on a cold glass.

with droplets. (See Figure 7–7.) Water vapor from the air condenses to liquid water on the cold surface of the glass. On the other hand, if the amount of water vapor is small, condensation does not take place.

The presence of water vapor can be proved by a chemical test. For example, some substances absorb water vapor from the air and then combine with it. Such substances are called **drying agents.** In the next experiment you will use a common drying agent called anhydrous (an-HY-drus) calcium chloride. The word **anhydrous** means "without water." That is, there is no water in combination with this particular form of calcium chloride. If the compound is exposed to water, it will at once combine with the water vapor.

experiment 7–4 Water Vapor in the Air

Materials

watch glass or evaporating dish; anhydrous calcium chloride; spatula; equal-arm balance

Procedure

Pour a small amount of anhydrous calcium chloride onto a watch glass and spread it into a thin layer as in Figure 7–8. Now find the mass of the watch glass plus calcium chloride to the nearest 0.1 g and record this value.

Leave the calcium chloride exposed to air until the end of the laboratory period. Again find the mass of the watch glass and its contents. Record the mass.

Leave the contents of the watch glass exposed to air until the next day (or until the next laboratory period). Again find the mass and record it.

Let the glass and its contents stand for at least a week and again record the total mass.

Questions

1. Compute the change in mass after an exposure of (a) a laboratory period, (b) a day, (c) a week.

2. How do you account for the change in mass?

3. Describe any change in the appearance of the calcium chloride.

4. How do you account for the change (or changes) described in Question 3?

Figure 7–8
Why will a thin layer of anhydrous calcium chloride work better than a thick layer?

Chemical Reactions with Oxygen

Oxygen will combine with most other elements, both metals and non-metals. A compound of oxygen and another element is called an **oxide** (ox-eyed). Iron oxide, or rust (Figure 7–9), is a metallic oxide, and sulfur dioxide is a non-metallic oxide. The

Figure 7–9
Rust is an oxide of iron.

formation of oxides is a chemical process, **oxidation** (ox-ih-DAY-shun). Some elements oxidize faster than others. For example, sodium oxidizes much faster than iron. The rate of oxidation also depends upon the temperature. The higher the temperature, the more rapid is the rate of oxidation.

However, even at a high temperature it is difficult to convert all of a metal into its oxide. This is because the oxide formed on the surface prevents air from making contact with the metal underneath the surface. In other words, the oxide on the surface "protects" the underlying metal from the oxidizing action of air.

In the next experiment you will study the reaction between the metal copper and the oxygen of the air. At ordinary temperatures the oxidation process is very slow. To speed up the reaction, the copper must be strongly heated.

experiment 7–5 The Reaction between Copper and Oxygen

Materials

evaporating dish; copper powder; stand; 4-inch ring; gauze; burner; equal-arm balance

Procedure

Find the mass of an evaporating dish to the nearest 0.1 g. Place about 5 g of copper powder in the dish and find the mass of dish plus copper.

Place a gauze on a ring clamped to a stand. Set the dish on the gauze as shown in Figure 7–10. Spread the copper in a thin layer. Heat the dish with a flame for five minutes. Let the dish cool, and then find the mass and record it.

Now reheat the dish and its contents for ten minutes. While the dish is being reheated, compute (1) the mass of the copper and (2) the change in mass during the first heating. Then compute the percentage change in mass during the first heating. When the dish has been heated for ten minutes, let it cool and record its mass. Then calculate the total percentage change in mass.

The percentage change in mass can be found as follows:

$$\%\ \text{change in mass} = \frac{\text{change in mass}}{\text{mass of copper}} \times 100\%$$

If there is time, heat again for another ten minutes, and again find the total percentage change in mass.

Questions

1. (a) Is the change in mass an increase or a decrease? (b) How do you account for the change?

2. (a) Is the total percentage change in mass the same after the second heating as the first? (b) If not, why not?

3. There are two different oxides of copper, called copper(I) oxide and copper(II) oxide. If copper is completely converted to copper(I) oxide the change in mass is 12%. If copper is completely converted to copper(II) oxide, the change in mass is 25%. How do you account for your own experimental results?

4. Copper(II) oxide is black and copper(I) oxide is red. Break up the product of the reaction in the dish. It readily crumbles in your fingers. Observe its color or colors. Does this observation support your answer to Question 3?

5. Write word equations for the formation of (a) copper(II) oxide, (b) copper(I) oxide.

6. Write symbolized equations for the formation of (a) copper(II) oxide, (b) copper(I) oxide.

Figure 7–10
Heat the dish of copper powder to make it oxidize faster.

TWO OPPOSITE KINDS OF CHEMICAL REACTIONS

Synthesis and Decomposition

In Experiment 7–5, you made copper oxide from copper and oxygen. This kind of process is called synthesis (SIN-thuh-sis). The word *synthesis* means "to put together." As applied to chemical reactions, **synthesis** means the building up of a compound from its elements. Synthesis is a common and important chemical process.

The opposite of synthesis is **decomposition** (dee-kom-poh-ZISH-un). Decomposition is the breaking down of a compound into the elements it is made of. You recall that some two-element compounds can be decomposed into their elements simply by heating them. If the mass of the compound and the mass of one of the products can be determined, the percentage composition of the compound can be computed. The decomposition of a compound in order to find its composition is called **analysis** (uh-NAL-uh-sis). A suitable compound for analysis is the red oxide of mercury. If this compound is heated, it decomposes into mercury and oxygen.

$$\text{mercury oxide} = \text{mercury} + \text{oxygen}$$

Mercury oxide is a red powder, mercury is a liquid, and oxygen is a gas. However, because the mercury oxide is heated, some of the mercury becomes a vapor. If it is heated in a test tube, the mercury vapor condenses to droplets on the relatively cool wall of the tube. If it is not heated too strongly, very little mercury escapes into the air. In other words, the only product that escapes from the test tube is oxygen. Therefore the mass of mercury can be found, as in the following experiment. The percentage of mercury in mercury oxide can be computed from

$$\%\ \text{mass of mercury} = \frac{\text{mass of mercury}}{\text{mass of mercury oxide}} \times 100\%$$

experiment 7–6

To Find the Percentage Composition of Mercury Oxide by Decomposition

Materials clean dry test tube; equal-arm balance; red oxide of mercury; stand; clamp; burner

Procedure

Find the mass of the test tube and record it. Add about 5 g of mercury oxide to the tube and find the mass of the tube plus oxide.

Clamp the tube almost horizontally to a stand as shown in Figure 7–11. Heat the closed end of the tube with a small flame. Notice that the red oxide turns black when it is hot. Notice also that gray droplets of mercury condense on the lower side of the

Figure 7–11
The test tube of mercury oxide should be almost horizontal. Be careful not to let mercury vapor escape.

tube. *Caution:* If you heat the tube too strongly, some mercury vapor will escape into the air. This loss will cause an experimental error. Moreover, mercury vapor is poisonous.

Continue the gentle heating until the powder is completely decomposed. This takes about 30 minutes. Then allow the tube to cool and find its mass.

Return the test tube and contents to your teacher so that the mercury (which is expensive) can be recovered. Keep in mind that mercury is poisonous and can be absorbed by the skin.

From your data determine (a) the mass of the mercury formed and (b) the mass of the oxygen released.

Computations

1. Compute the percentage by mass of mercury in mercury oxide.

2. Compute the percentage by mass of oxygen in mercury oxide.

3. (a) Add the percentage by mass of mercury and oxygen. (b) Is your sum of the percentages what you expect? (c) Why?

4. Your teacher will give you the correct value for the percentage by mass of mercury in the red oxide. Compute your percentage error.

Questions

1. Suppose that you lost some mercury vapor during the heating. How would this affect your value for the percentage by mass of oxygen? Explain.

Oxygen as a Product of Decomposition

Sometimes it is necessary to prepare samples of pure oxygen. One way to do this is to collect the oxygen released when mercury oxide is heated. However, mercury oxide is expensive and there are other more satisfactory sources of oxygen. One is a white crystalline solid called potassium chlorate. The ending *-ate* indicates that the compound is made up of three elements, one of which is oxygen. Thus, the name potassium chlorate indicates that the three elements in the compound are potassium, chlorine, and oxygen. The chemical formula for this compound is $KClO_3$.

If potassium chlorate is heated, it decomposes, releasing oxygen. Potassium chloride, a white solid, remains behind.

$$\text{potassium chlorate} = \text{potassium chloride} + \text{oxygen}$$

In the next experiment you will find the percentage by mass of oxygen in potassium chlorate.

experiment 7–7

To Find the Percentage of Oxygen in Potassium Chlorate

Materials

clean, dry test tube; potassium chlorate; equal-arm balance; stand; clamp; burner; wooden splint

Procedure

Caution: The test tube must be clean and dry. An impurity such as carbon on the wall of the test tube could be dangerous, since it might combine with hot oxygen and cause an explosion.

Find and record the mass of a clean, dry test tube. Add about 4 g of finely ground potassium chlorate and again find the mass. From these two mass readings find the mass of the potassium chlorate.

Clamp the test tube to a stand as shown in Figure 7–12. Notice that the tube is clamped near the open end and held in a sloping position. The bottom of the tube should be about 10 cm above the top of the burner.

Light the burner and heat the potassium chlorate with a small flame. At first the potassium chlorate crackles. Then it melts. Finally bubbles of gas escape, and the melted potassium chlorate appears to be boiling. The escaping gas is oxygen. The gas can be tested for oxygen by holding a glowing splint of wood just inside the mouth of the test tube. What happens to the glowing splint?

When the bubbling slows down, heat the bottom of the tube strongly until no more bubbles of gas escape. The remaining substance is potassium chloride. Allow the test tube to cool and then find its mass. From your data find the mass of potassium chloride.

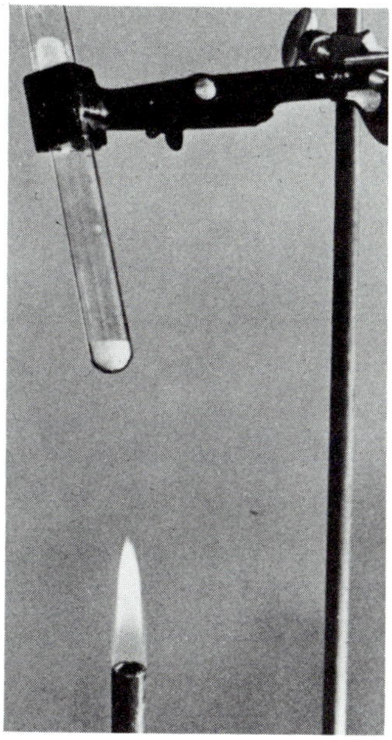

Figure 7–12
Start with a clean, dry test tube. To heat it, clamp it in a sloping position.

Computations

1. Compute the mass of oxygen that escaped during the heating.

2. Compute the percentage by mass of oxygen in potassium chlorate from your experimental data.

3. Ask your teacher for the correct percentage by mass of oxygen in potassium chlorate and then compute your percentage error.

Questions

1. Write, in symbols, the equation for the decomposition of potassium chlorate by heat.

experiment 7–8 To Find the Percentage by Mass of Sodium Chloride in Sodium Chlorate

Laboratory Test

Materials

same as for Experiment 7–7, except that sodium chlorate is substituted for potassium chlorate

Procedure

As you might expect, sodium chlorate behaves in the same way as potassium chlorate when heated.

Following the same procedure as in Experiment 7–7, find the percentage by mass of sodium chloride in sodium chlorate. Ask your teacher for the correct percentage of sodium chloride and then compute your percentage error.

CHEMICAL ACTIVITY

The Chemical Activities of Metals

The rate at which a metal reacts with another substance is called the **chemical activity** of the metal. One way to compare the chemical activities of metals is to observe how rapidly they react with an acid. All acids contain hydrogen. Some metals release the hydrogen from an acid, but others do not. Metals that release hydrogen are said to be more active chemically than hydrogen. On the other hand, metals that do not release hydrogen from acids are said to be less active than hydrogen.

Some of the active metals release hydrogen more rapidly than others. The rate at which bubbles of hydrogen are released is a measure of the activity of the metal. In the next experiment you will use hydrochloric acid and compare the chemical activities of six different metals. Remember that whenever hydrochloric acid reacts with a metal, a chloride of the metal is formed. For example, when hydrochloric acid reacts with sodium, hydrogen gas is released and sodium chloride is left behind.

experiment 7–9

The Chemical Activities of Metals

Materials

6 test tubes; rack; dilute hydrochloric acid; zinc; copper; aluminum; iron; magnesium; mercury

Procedure

Place six test tubes in the rack and pour dilute hydrochloric acid into each tube to a depth of about 1 cm. (See Figure 7–13.) Then add a small piece of one of the following metals to each of the test tubes: zinc, copper, aluminum, iron, magnesium, and mercury (one drop). Record the result of each test before proceeding to the next.

Do not dispose of the mercury in the sink or trash bucket. Place it in the container provided by your teacher.

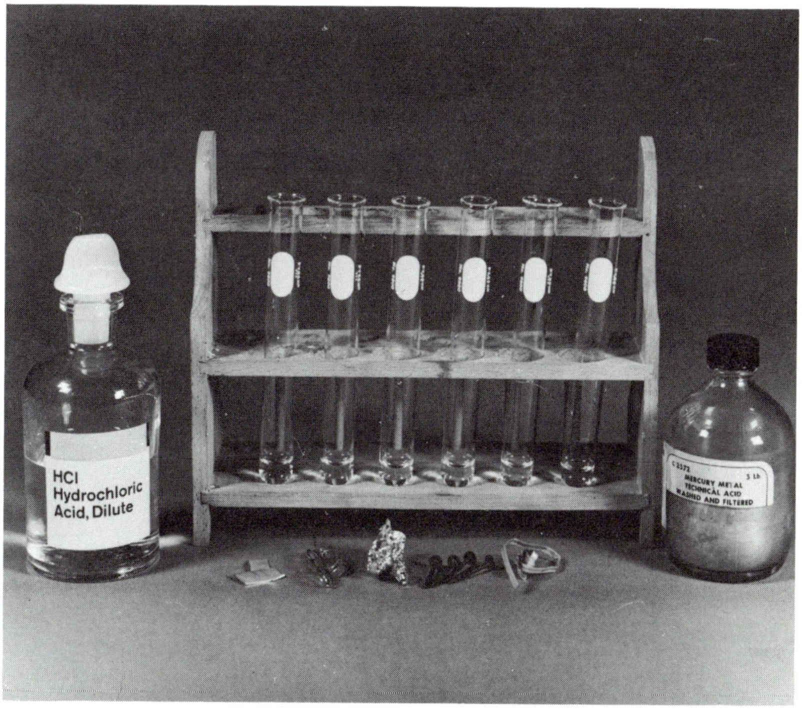

Figure 7–13
Setup for comparing the chemical activities of metals.

Questions

1. Which of the metals are more active than hydrogen?
2. Which of the metals are less active than hydrogen?
3. List the metals in Question 1 in order of their activity as indicated by the rate at which hydrogen is released.
4. The word equation for the reaction between aluminum and the acid is

aluminum + hydrochloric acid = aluminum chloride + hydrogen

Write word equations for the other reactions.

5. Select one of the reactions in Question 4 and explain why it takes place.

HEAT OF COMBUSTION

Determining the Heat of Combustion of a Substance

A candle is made of paraffin, which is composed of compounds of carbon and hydrogen. Such a compound is called a **hydrocarbon** (HY-droh-car-bun). When a candle burns, a chemical reaction takes place. The candle is used up and heat energy is released. The amount of heat released per gram of candle burned is called the **heat of combustion** (kum-BUS-chun). The mass of the candle used up can be found from its mass before and after burning. The amount of heat released can be found by using the heat of the flame to raise the temperature of a known mass of water. You will do this in the next experiment.

experiment 7–10 The Heat of Combustion of a Hydrocarbon

Materials

candle; glass plate; equal-arm balance; large can (coffee can, shortening can, or large peach can) fitted with wire supports; 100-ml graduate; stand; clamp; thermometer; matches

Procedure

Light a candle and hold it horizontally over a glass plate, as in Figure 7–14, so that a few drops of melted paraffin fall onto the center of the plate. Blow out the flame and press the base of the

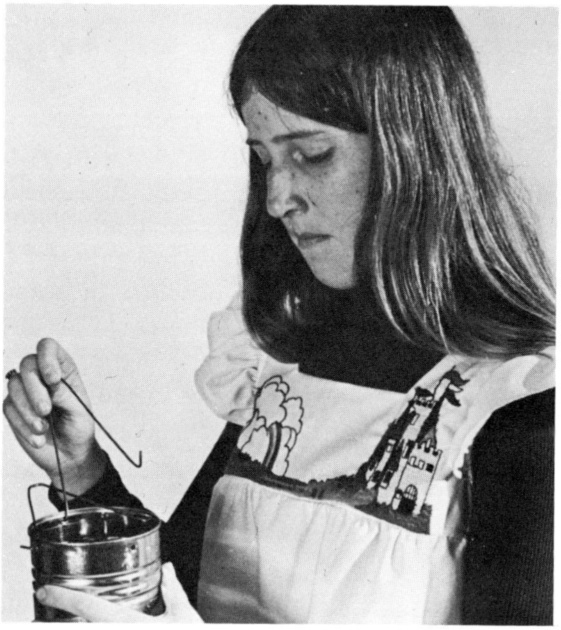

Figure 7–14 (above)
Drip paraffin onto a glass plate and stick the candle on it.

Figure 7–15 (above right)
Attach bent wire supports to the coffee can.

Figure 7–16 (right)
Adjust the clamp so the flame does not touch the can.

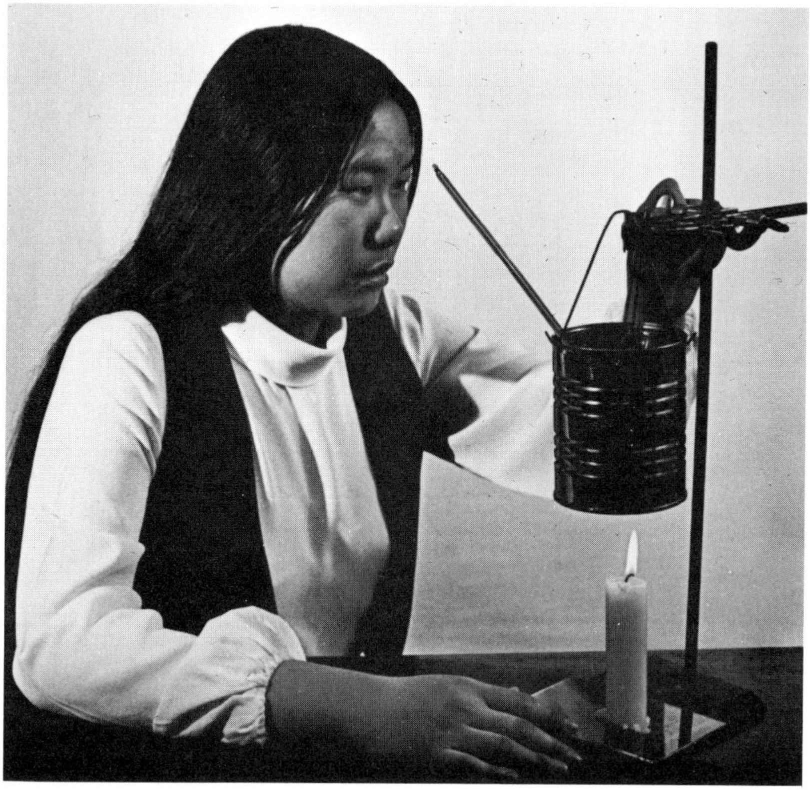

candle into the melted paraffin until it solidifies and the candle is stuck to the plate. Find the mass of candle and plate to the nearest 0.1 g.

Attach two wire supports to a coffee can as in Figure 7–15. Find the mass of the can and wire to the nearest gram. Pour about 500 ml of cold water (at about 5°C to 10°C) into the can and again find the mass.

Suspend the can from the stand and clamp so that the bottom of the can is about 10 cm above the wick of the candle. Record the room temperature by reading the thermometer in air. Place the thermometer in the water and record the temperature to the nearest 0.5°C. Light the candle and then adjust the height of the clamp so that the flame of the candle does not touch the bottom of the can. The arrangement is shown in Figure 7–16.

Stir the water with the thermometer. When the temperature of the water is as far above room temperature as it was below room temperature at the beginning, gently blow out the flame. Record the temperature of the warm water to the nearest 0.5°C. Then find the mass of the candle and plate.

Computations

1. Compute the mass of candle burned.

2. Find the rise in temperature of the water.

3. Compute the heat energy gained by the water from the equation

$$\text{heat gained by water} = m \times c \times (t_2 - t_1)\ \text{cal}$$

where m = mass of water, c = heat capacity of water, t_2 = temperature of warm water, t_1 = temperature of cold water.

4. Compute the heat energy gained by the can from the equation

$$\text{heat gained by can} = m_1 \times c_1 \times (t_2 - t_1)\ \text{cal}$$

where m_1 = mass of can, c_1 = heat capacity of iron $\left(0.12 \dfrac{\text{cal}}{\text{g} \times \text{C}^\circ}\right)$, and t_2 and t_1 are the same as in Question 3.

5. Compute the total heat gained by the water and can.

6. Compute the heat of combustion of 1 g of paraffin.

$$\text{heat of combustion} = \frac{\text{heat gained by water and can}}{\text{mass of candle burned}}$$

Questions

1. The correct value for the heat of combustion of paraffin is 10.3 kcal/g. Compare this with your computed value and then give reasons for the difference between the two values.

2. How would you modify the apparatus and the procedure in order to get a more reliable value for the heat of combustion of paraffin?

PERSPECTIVES

perspective 7–1

What is Fire?

Ancient man was afraid of fire. To him it often meant the destruction of his home. Sometimes it meant death, as when a forest fire got out of hand. The cause of fire was a mystery to him. He thought a destructive fire was a punishment heaped upon him by an angry god. But a campfire under control was useful and friendly. (See Figure 7–17.) It kept him warm, it enabled him to cook, and it kept wild animals away. No wonder early man developed skill in making campfires. To start a fire he struck two different rocks together. One was a

Figure 7–17
Ancient man learned to make a fire by striking two rocks together. There was no easy way to start a fire until the 1800s.

hard rock called flint and the other a softer rock called pyrites (py-RY-teez). When the rocks were struck together, a shower of white-hot particles was released from the pyrites. A heap of dry grass or powdered bark was prepared beforehand. The fire was started when the hot sparks fell into the grass or bark.

Fire remained a mystery until relatively recent times. It was finally solved about 200 years ago by a French scientist named Lavoisier (lah-vwaz-ee-AY). According to Lavoisier, fire is the result of a chemical reaction between a combustible (burnable) substance and the active part of air, which he called oxygen. In this reaction, enough heat is released to cause the substance to burn.

perspective 7–2

Matches

Once the nature of fire was understood, men learned how to control it, and to use it to do useful work such as driving a steam engine. But how could a fire be started easily and conveniently? This problem was solved in the 1800s with the invention of the match. The match operates on the same principle as the two rocks and the heap of dry grass or bark. However, the combustible material in the match is the wooden or paper stick. One end of the stick is dipped into a paste of a substance called phosphorus sulfide. This substance is even more readily combustible than wood. When the tip of the match is rubbed on a rough surface, the heat of friction is enough to ignite the phosphorus sulfide. The burning phosphorus sulfide then ignites the wooden stick.

perspective 7–3

The Composition of Air

The ancient Greek philosophers thought about the composition of air as early as 500 B.C. They concluded that air was a single substance. For almost 2000 years air was thought of as an element. It was not until 1770 that a Swedish scientist named Scheele (SHEEL-uh) proved by experiment that air is a mixture of at least two gases. In doing this he rusted iron in a closed bottle and followed the same procedure as in your Experiment 7–2. He called the two gases "active air" and "inactive air." Like you, he found the percentage composition of each of these gases.

About twenty years later the French scientist Lavoisier first used the term **oxygen** for active air and **nitrogen** for inactive air. Near the end of the 19th century it was found that the "nitrogen" of Lavoisier contained about one percent of a very inactive gas which was called argon. (The name *argon* is well deserved because it comes from a Greek word that means "lazy.") The analysis of air, begun about 500 B.C., was not completed until the year 1898.

The atmosphere contains a small percentage of carbon dioxide. Where does the carbon dioxide come from? Carbon dioxide is formed whenever such fuels as coal, gasoline, and natural gas are burned. In this way thousands of tons of carbon dioxide enter the atmosphere every day. Moreover, food is burned in the cells of animals. This process is called **respiration** (res-puh-RAY-shun). In respiration, carbon dioxide is formed. The exhaled breath of all animals is at least four percent carbon dioxide.

If the carbon dioxide from combustion and respiration were allowed to accumulate, the atmosphere would not be able to support animal life. Human beings, for example, would die of suffocation if the air were 10 percent carbon dioxide.

Fortunately there is a natural process called **photosynthesis** (foh-toh-SIN-thuh-sis) which uses up carbon dioxide and so allows animals to live. The word *photo* is from a Greek word that means "light." Photosynthesis occurs only in the presence of sunlight. Photosynthesis is a complicated process that occurs in green plants. Water is taken in through the roots of plants. Carbon dioxide is taken out of the air through the pores of the leaves. The water and carbon dioxide combine to form starch or sugar, plus oxygen. The reaction is

carbon dioxide + water = starch (or sugar) + oxygen

Notice that, in this reaction, not only is carbon dioxide removed from air, but oxygen is released into the air. The processes of combustion and respiration remove oxygen from air, but the process of photosynthesis puts oxygen back into the air. (See Figure 7–18.) Thus as a result of photosynthesis the air contains much oxygen and very little carbon dioxide, actually only 4 parts of carbon dioxide in 10,000 parts of air, or 0.04 percent. No wonder photosynthesis is called the most vital of all chemical reactions.

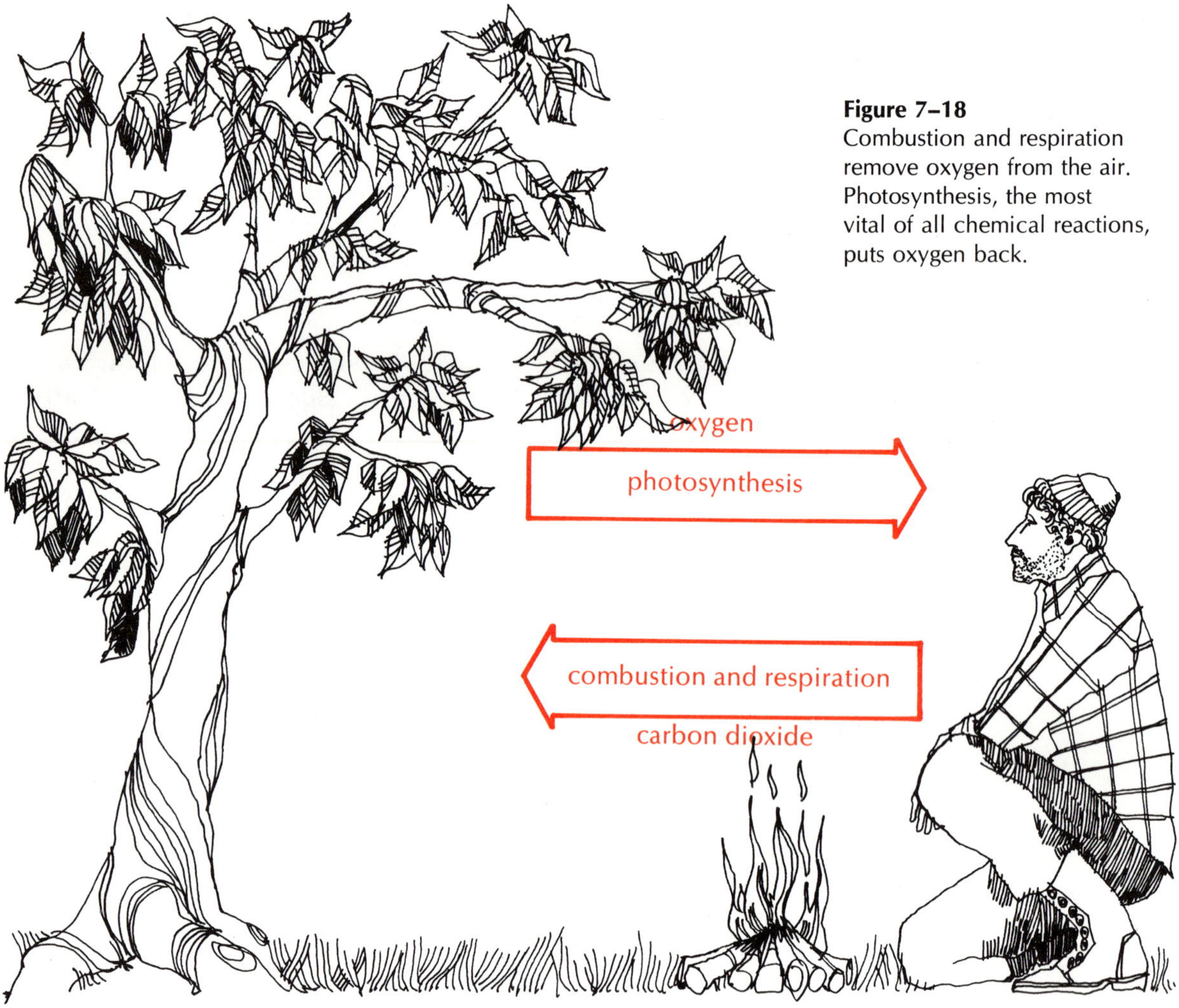

Figure 7–18
Combustion and respiration remove oxygen from the air. Photosynthesis, the most vital of all chemical reactions, puts oxygen back.

perspective 7–4

Water Vapor in the Air

Evaporation occurs at the surfaces of all bodies of water such as ponds, rivers, and seas. The warmer the water is, the faster it evaporates. The warmer the air is, the more water it can hold. The ratio of the amount of water in the air to the amount of water the air can hold at a certain temperature is called the **relative humidity.**

$$\text{relative humidity} = \frac{\text{amount of water in the air}}{\text{amount of water the air can hold}} \times 100\%$$

When the relative humidity is high, the air is holding nearly all the water it can. Then water will not evaporate very fast. If you work hard and perspire, you will feel "sticky."

On a cold winter day, the air cannot hold much water. If there is snow on the ground, the relative humidity may be high because of evaporation. But there is really not much water in the air. If the air enters a house, it is warmed up, and so is able to hold much more water. Since there is little water in it, its relative humidity is low.

Indoors in the winter, the relative humidity may be so low that you can cause electric sparks by combing your hair, or stroking the cat, or scuffing your feet on a rug. The comb and hair are charged by friction. The tiny sparks are caused by electric charges jumping from one to the other. If the air is damp, the electric charges escape through the moisture. If the relative humidity is high, charges do not accumulate on the comb and hair, so sparking does not occur.

Before air is analyzed, it is passed over a drying agent such as anhydrous calcium chloride to remove water vapor. Air is dried before it is analyzed because the amount of water vapor in the air varies from day to day, whereas the percentages of the other gases remain fairly constant. The analysis of air gives the percentages of the ingredients of "dry" air.

perspective 7–5

Combining Capacities and Formulas

In Chapter 6 the combining capacities of a number of elements were discussed. The combining capacity of an element is sometimes called its valence (VAY-lentz). Knowing the combining capacities of elements, you can write the formulas of their compounds. Figure 7–19 is a list of the combining capacities of some metals and non-metals.

Notice that a positive sign is placed before the combining capacities of metals and a negative sign before the combining capacities of non-metals. This is because metals usually lose

Figure 7–19
The combining capacities of some metals and non-metals.

Metal	Combining Capacity	Non-metal	Combining Capacity
Hydrogen (H)	+1	Chlorine (Cl)	−1
Potassium (K)	+1	Bromine (Br)	−1
Sodium (Na)	+1	Oxygen (O)	−2
Silver (Ag)	+1	Sulfur (S)	−2
Barium (Ba)	+2	Nitrogen (N)	−3
Calcium (Ca)	+2		
Mercury (Hg)	+2		
Lead (Pb)	+2		
Magnesium (Mg)	+2		
Zinc (Zn)	+2		
Aluminum (Al)	+3		

electrons in chemical reactions, whereas non-metals usually gain electrons. Notice also that hydrogen is listed with the metals and that its combining capacity is +1. This is because hydrogen may lose its single electron in a chemical reaction. That is, hydrogen may behave like a metal.

To write the formula of a compound bear in mind two rules. (1.) The first symbol in the formula is the symbol of the element with the positive combining capacity. (2.) Combining capacities must balance each other so that the net electric charge in a molecule of the compound is zero.

To show how these rules apply, try writing the formula for barium chloride. Because barium has a positive combining capacity and chlorine a negative combining capacity, the symbol for barium is written first. The combining capacity of Ba is +2 and the combining capacity of Cl is −1. Therefore two atoms of chlorine are needed to electrically balance one atom of barium. The formula for barium chloride is thus $BaCl_2$. Notice that the number 2 (the number of atoms of chlorine) is written after the symbol as a subscript (that is, in the lower right corner).

Barium chloride is made up of two elements, barium and chlorine. All compounds composed of two elements have names ending in the letters *-ide*. For example, a compound of iron and oxygen is called iron oxide and a compound of copper and sulfur is called copper sulfide.

perspective 7–6

Elements with More than One Combining Capacity

Some metals have more than one combining capacity. Two of these metals are iron and copper. Iron can have a combining capacity of +2 or +3, depending upon the conditions of the reaction. Copper can have a combining capacity of +1 or +2. Thus there are two chlorides of iron, $FeCl_2$ and $FeCl_3$. What are the names of these compounds? In naming a compound, the combining capacity of the variable element is mentioned, and when it is written a Roman numeral is used to show its combining capacity. Thus $FeCl_2$ is called "iron two chloride," written as iron(II) chloride. Similarly, $FeCl_3$ is called "iron three chloride," written as iron(III) chloride.

There are two oxides of copper. If the combining capacity of copper is +1, the formula of the oxide is Cu_2O. The name of this compound is "copper one oxide," written as copper(I) oxide. If the combining capacity of copper is +2, the formula for the oxide is CuO. The name of this compound is "copper two oxide," written as copper(II) oxide.

There is also an older method of naming compounds, which is still in use. In this older system, the combining capacity of the metal is indicated by adding an ending to the name. The ending *-ous* indicates the lower combining capacity of the metal. The ending *-ic* indicates the higher combining capacity. Thus the older name for $FeCl_2$ is ferrous chloride and for $FeCl_3$ it is ferric chloride. Similarly, the older name for Cu_2O is cuprous oxide and for CuO it is cupric oxide.

perspective 7–7

Radicals—Their Combining Capacities and Formulas

The formula for sulfuric acid is H_2SO_4. This means that one molecule of sulfuric acid contains two atoms of hydrogen, one atom of sulfur, and four atoms of oxygen. If a metal like zinc is dropped into sulfuric acid, a vigorous reaction takes place in which hydrogen gas escapes and the metal zinc disappears.

Thus it looks as though the zinc has replaced the hydrogen of the acid and that the equation for the reaction is

$$Zn + H_2SO_4 = ZnSO_4 + H_2$$

Or, in words,

$$\text{zinc + sulfuric acid = zinc sulfate + hydrogen}$$

A diagram of this reaction is shown in Figure 7–20. The diagram shows that the SO_4 group stays together in the reaction.

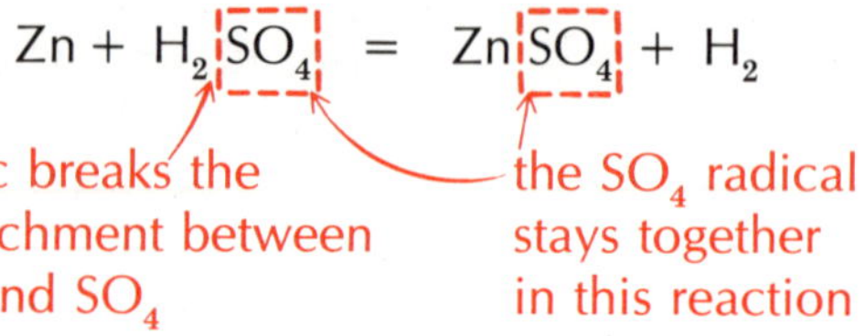

Figure 7–20
In the reaction between zinc and sulfuric acid, the sulfate radical stays together.

A group of atoms that remains together in a chemical reaction is called a *group*, or a **radical.** Thus the SO_4 group is the sulfate radical. A radical is always part of a compound. It can never exist by itself. Radicals, like atoms, have combining capacities. You can see from the formulas for sulfuric acid and zinc sulfate, H_2SO_4 and $ZnSO_4$, that the combining capacity of the sulfate radical must be −2. The formula for potassium nitrate is KNO_3. The formula for the nitrate radical is NO_3. Since the combining capacity of potassium is +1, the combining capacity of the nitrate radical must be −1. Another common radical is hydroxide (OH), with a combining capacity of −1. Still another is carbonate, with a combining capacity of −2. These radicals and their combining capacities are listed in Figure 7–21.

Figure 7–21
The combining capacities of some common radicals.

Radical	*Formula*	*Combining Capacity*
hydroxide	OH	−1
nitrate	NO_3	−1
carbonate	CO_3	−2
sulfate	SO_4	−2

In formulas of compounds that contain radicals, the negative combining capacity of the radical must be balanced by the positive combining capacity of the metal. Thus, to write the formula for silver carbonate, remember that the combining capacity of silver is +1 and the combining capacity of the carbonate radical is −2. Therefore two atoms of silver attach themselves to one carbonate radical. The formula of silver carbonate is Ag_2CO_3.

Notice that although the symbol for hydrogen is H, the formula for hydrogen gas is H_2. Molecules of hydrogen gas each contain two atoms of hydrogen. Such a molecule is called a **diatomic** (*dy*-ah-TOM-ik) **molecule.** A diatomic molecule may also consist of atoms of two different elements. A molecule of hydrochloric acid (HCl) is diatomic.

Zinc sulfate is soluble and so remains in solution. Notice the ending *-ate* in the word sulfate. This ending always means that the compound contains three elements, one of which is oxygen. Thus the compound $ZnSO_4$ contains the three elements zinc, sulfur, and oxygen. The word *nitrate* also has the ending *-ate*. Therefore all nitrates contain the elements oxygen and nitrogen. The compound potassium nitrate contains the three elements potassium, nitrogen, and oxygen. The formula of this compound is KNO_3.

Notice that two of the three elements are always included in the name of an *-ate* compound. The third element, oxygen, is not included in the name, since it is always present in an *-ate* compound. In the compound calcium carbonate the elements calcium and carbon are in the name, but oxygen is not. The formula for calcium carbonate is $CaCO_3$.

Calcium hydroxide is a common compound. What is its formula? The combining capacity of calcium is +2. The combining capacity of the hydroxide radical is −1. Therefore two hydroxide radicals are needed to balance one atom of calcium. Two hydroxide radicals are written as $(OH)_2$, not as OH_2, since the number 2 applies to both the atoms O and H. Hence the formula is $Ca(OH)_2$.

This method of writing radicals can be expressed as the following rules: If a formula consists of more than one radical of the same kind, the radical is placed in parentheses. A number is then added at the lower right side of the parentheses to show the number of radicals in the formula. Thus the formula for lead nitrate is $Pb(NO_3)_2$, and the formula for aluminum sulfate is $Al_2(SO_4)_3$.

perspective 7–8

Equations

A chemical reaction can be represented most briefly by an equation in which chemical symbols and formulas are used. For example, the gas hydrogen combines with the gas chlorine to form hydrogen chloride, which is also a gas. What is the equation for this reaction?

The formula for a molecule of chlorine is Cl_2. That is, there are two atoms per molecule as in hydrogen (H_2). Also, the formula for hydrogen chloride is HCl. So, using symbols, the reaction can be represented as

$$H_2 + Cl_2 = HCl$$

However, in this expression there are four atoms on the left and only two on the right. Since atoms are never destroyed in a reaction, the total number of atoms in the product or products must be the same as the number in the reactants. This can be achieved by placing the number 2 before the formula HCl. The expression then becomes

$$1\ H_2 + 1\ Cl_2 = 2\ HCl$$

which is read as "one molecule of hydrogen plus one molecule of chlorine yields two molecules of hydrogen chloride." However, the number 1 is not necessary, so the expression becomes

$$H_2 + Cl_2 = 2HCl$$

Notice that the number of atoms in the reactants (four) now equals the number in the product. Moreover, the number of atoms of each element is the same on both sides. Thus the equation is said to be **balanced.**

We know that hydrogen reacts with oxygen to form water. What is the equation for this reaction? In oxygen, as in hydrogen and chlorine, there are two atoms per molecule. So the formula for oxygen is O_2. The formula for water is H_2O. The first step in writing the equation is

$$H_2 + O_2 = H_2O$$

But this is not a balanced expression, since there are two atoms of oxygen on the left and only one on the right. However, the oxygen atoms can be balanced by using two molecules of water. To show two molecules of water, place the number 2 in front of the H_2O formula.

$$H_2 + O_2 = 2H_2O$$

This expression is still unbalanced, since there are four atoms of hydrogen on the right but only two on the left. The hydrogen atoms can be balanced by using two molecules of hydrogen. Place the number 2 in front of the H_2 formula. The expression is now a balanced equation:

$$2H_2 + O_2 = 2H_2O$$

Remember three factors when you write an equation:

1. Use correct formulas for all reactants and products.
2. Write reactants on the left and products on the right. Connect them by an equal sign (=).
3. Make the number of atoms of each element in the products equal the number of atoms of that element in the reactants.

Now write the equation for the reaction of sodium sulfate with barium chloride to yield sodium chloride plus barium sulfate. The formulas are: sodium sulfate, Na_2SO_4; barium chloride, $BaCl_2$; sodium chloride, NaCl; barium sulfate, $BaSO_4$. An unbalanced expression for the reaction is

$$Na_2SO_4 + BaCl_2 = NaCl + BaSO_4$$

The balanced equation for the reaction is

$$Na_2SO_4 + BaCl_2 = 2NaCl + BaSO_4$$

perspective 7–9

The Oxidation of Metals

Whenever oxygen reacts with a metal (or a non-metal), an oxide is always formed. The more chemically active the metal, the more readily will the metal form an oxide. For example, sodium is an active metal. It combines with oxygen even at ordinary temperatures to form the oxide, which is a white solid. The equation for the reaction is

$$4Na + O_2 = 2Na_2O$$

The metal magnesium is less active than sodium. At ordinary temperatures it reacts with oxygen very slowly. However, if it is heated in a flame, oxidation is so rapid that a brilliant white

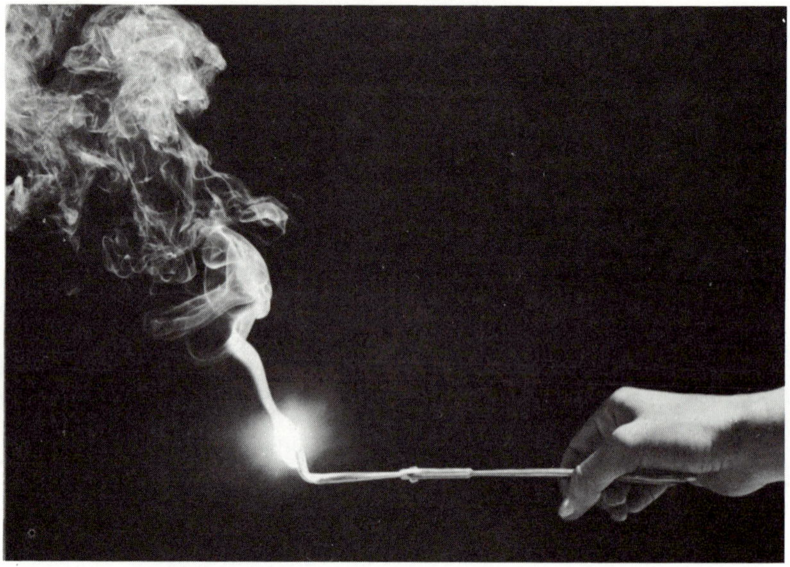

Figure 7–22
Magnesium oxidizes very rapidly. It is used in photographic flash bulbs.

light is emitted (See Figure 7–22.) The equation for the reaction is

$$2Mg + O_2 = 2MgO$$

Aluminum is even less active than magnesium. Its oxidation is also very slow at ordinary temperatures and rapid at high temperatures. The equation for the oxidation reaction is

$$4Al + 3O_2 = 2Al_2O_3$$

In writing the formula for aluminum oxide, remember that the combining capacity of aluminum is +3 and that the combining capacity of oxygen is −2. Since an atom of aluminum gives up three electrons and an atom of oxygen gains two, the smallest number of electrons that can share in the exchange is six. Two atoms of aluminum will give up six electrons, and three atoms of oxygen will gain six electrons. The formula of the oxide is thus Al_2O_3.

perspective 7–10

Synthesis and Decomposition

Water can be made from its elements. It can also be decomposed into its elements. That is, the opposing processes of synthesis and decomposition can be used with a compound and the elements that make it up.

A. The Synthesis of Water

At ordinary temperatures, hydrogen does not react with oxygen. However, if a mixture of these gases is warmed in a closed container (even in a soap bubble), the gases react with explosive violence. The explosive force is due to the heat released during the reaction. The heat of reaction causes so much expansion of the water vapor that it bursts its container. The water vapor could, of course, be condensed to liquid water, so the reaction can be expressed in words as

$$\text{hydrogen} + \text{oxygen} = \text{water}$$

B. The Decomposition of Water

Heat energy is released when water is formed from hydrogen and oxygen. Therefore, water molecules must have less potential energy than hydrogen and oxygen molecules. Before water can be decomposed, the energy that was lost as heat must be put back. The most suitable form of energy for this purpose is electric energy. Pure water does not conduct an electric current. Water can become a good conductor, however, if a little sulfuric acid is added to it. Thus, if an electric current flows through a solution of sulfuric acid, water molecules are decomposed. Hydrogen gas is released at one of the terminals and oxygen at the other. A word equation for this reaction is

$$\text{water} = \text{hydrogen} + \text{oxygen}$$

These opposing reactions can be shown as a single equation by using reversible arrows.

Reactants	Synthesis	Product
$2H_2 + O_2$	$\rightleftarrows$	$2H_2O$
Products	Decomposition	Reactant

perspective 7–11

Potassium Chlorate as a Source of Oxygen

A match has two parts, a flammable tip and a wooden stick. There are usually two compounds in match tips, potassium chlorate and phosphorus sulfide. The end of a match stick is first dipped into a mixture of potassium chlorate and glue. It is then dipped into a paste of phosphorus sulfide. (See Figure 7–23.)

When the match is struck, the heat of friction ignites the phosphorus sulfide. The flame of the burning phosphorus sulfide is not hot enough to ignite the wood. However, the flame is hot enough to decompose the potassium chlorate into potassium chloride and oxygen, as you did in Experiment 7–7. This oxygen makes the phosphorus sulfide burn with a hotter flame, hot enough to ignite the wood.

perspective 7–12

Reactions of Metals with Hydrochloric Acid

As stated earlier, hydrogen chloride (HCl) is a gas. This gas is very soluble in water. When it dissolves in water the molecules of HCl break up into hydrogen ions (H^+) and chloride ions (Cl^-). That is, a hydrogen atom loses its electron to form a positively charged hydrogen ion.

$$\underset{\text{atom}}{H} - e^- = \underset{\text{ion}}{H^+}$$

At the same time a chlorine atom gains the electron to form a negatively charged ion.

$$\underset{\text{atom}}{Cl} + e^- = \underset{\text{ion}}{Cl^-}$$

In a chemical reaction a hydrogen ion tends to gain an electron and re-form its neutral atom. As you learned in Chapter 6, the atoms of metals tend to lose electrons. If a metal gives up electrons to hydrogen ions, it is said to be more active than hydrogen. If it does not give up electrons to hydrogen ions, it is said to be less active than hydrogen.

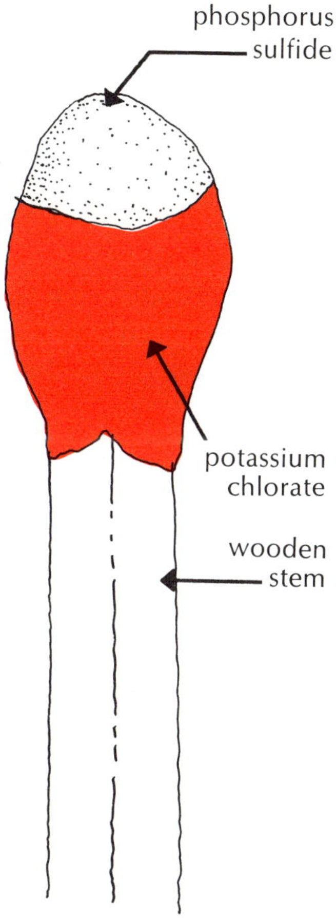

Figure 7–23
Why are there two different flames when a match is struck?

Consider the reaction of the metal magnesium with hydrochloric acid. This is really a reaction of magnesium atoms with hydrogen ions in the acid solution. Let us see why. Magnesium is in column II of the periodic table. Therefore its combining capacity is +2. A magnesium atom has two electrons in its outer shell. These electrons are easily lost. A hydrogen ion will attract an electron to form a neutral hydrogen atom.

$$H^+ + e^- = H$$

Hence, in this reaction two hydrogen ions will remove two electrons from every atom of magnesium. In symbols this reaction is

$$Mg + 2H^+ = Mg^{2+} + 2H$$

However, the two hydrogen atoms combine to form a hydrogen molecule, H_2. A more correct equation for the reaction is

$$Mg + 2H^+ = Mg^{2+} + H_2$$

Two hydrogen ions are provided by two molecules of hydrogen chloride.

$$2HCl = 2H^+ + 2Cl^-$$

So the ionic equation for the reaction between magnesium and hydrochloric acid is

$$Mg + 2H^+ + 2Cl^- = Mg^{2+} + 2Cl^- + H_2$$

It is usually written as

$$Mg + 2HCl = MgCl_2 + H_2$$

perspective 7–13

Slow and Rapid Oxidation

The combination of oxygen with another element is an example of a chemical process called oxidation. The rusting of iron is an oxidation process in which iron is oxidized to iron(III) oxide.

At high temperatures oxygen is a powerful oxidizer. It combines with many elements producing both heat and light, as in a flame. Sulfur does not react with oxygen at ordinary temperatures, but at high temperatures it burns rapidly to form the gas sulfur dioxide (SO_2) as in Figure 7–24. A reaction in which a flame is produced is called **rapid oxidation** or **combustion.** If a flame is not produced in an oxidation reaction, the process is called **slow oxidation.** The rusting of iron is an example of slow oxidation.

Paraffin, or wax, belongs to a group of chemical compounds known as hydrocarbons. A hydrocarbon is a compound made up of atoms of hydrogen and carbon. When a candle wick is ignited, some of the wax melts. The melted wax seeps up the wick and burns. In burning, the hydrogen of the wax combines with oxygen of the air to form water vapor. The carbon combines with oxygen to form carbon dioxide (CO_2). Some of the carbon is released as free carbon (soot).

As the candle burns, the tiny particles of carbon become white hot. This makes the flame white and luminous. As you know, a candle flame is used more as a source of light than a source of heat. However, a candle flame is smoky. If it touches the underside of the tin can in Experiment 7–10, some of the particles of carbon will be cooled enough to form a black deposit of soot. This would cause an experimental error.

A word equation for the reaction in the candle flame is

hydrocarbon + oxygen = carbon dioxide + water vapor
+ carbon + heat + light

Figure 7–24
Sulfur burns to produce sulfur dioxide, a choking gas. Sulfur in oil is a source of pollution.

Figure 7–25A
To test for oxygen, light a wooden splint and blow it out. The wood glows red hot.

Figure 7–25B
If you plunge the glowing splint into a bottle of oxygen, the splint bursts into flame.

As would be expected, oxidation takes place much more rapidly in pure oxygen than in air. A simple test can distinguish between air and pure oxygen. To do this test, first light a wooden splint. Blow out the flame and observe the glow of red-hot wood at the end of the splint. (See Figure 7–25A.) The red glow indicates that the wood is combining with oxygen of the air to form CO_2 and H_2O, and the heat released in the reaction is enough to cause the wood to glow. If the glowing splint is plunged into a bottle of pure oxygen, it instantly bursts into flame as in Figure 7–25B. The reaction is speeded up in the higher concentration of oxygen, and more heat is produced. The extra amount of heat causes the wood to burn rather than glow.

perspective 7–14

The Importance of Metals

Have you ever thought how much you depend upon metals? Think of the many uses of metals in your home. The furnace and radiators are made of steel. Water pipes and electrical wiring are made of copper, and cooking pans of aluminum. The refrigerator and dishwasher are made of metals, chiefly zinc, copper, and chromium. Most of the bathroom fixtures are made of zinc, copper, aluminum, and magnesium.

Outside the home, metals are just as plentiful and useful. A skyscraper looks like a concrete structure, but it has a framework of steel. Even a wooden house is held together by nails and bolts made of steel. Airplanes are made of aluminum, and trains are built of steel. The body of an automobile is made of steel, but zinc, copper, lead, nickel, and chromium are also used in various parts of the automobile.

Most metals used in making machines or in any construction work are not pure metals. Instead, they are mixtures of metals, called **alloys.** Probably the only pure metals you have seen outside a laboratory are copper in wire and mercury in thermometers. Alloys can be made to have more useful properties than pure metals. For example, very few people have seen pure iron. Pure iron is soft, it has little strength, and it rusts easily. It would be useless in construction work. However, if small amounts of other metals such as manganese, nickel, tungsten, and chromium are added to iron, it can be made hard, tough, elastic, and almost rustproof. An alloy of iron is called steel. There are many kinds of steel, each with special properties designed for its own special use.

Active metals (metals that are more active chemically than hydrogen) react with acids, oxygen, and other substances. Thus it is not surprising that active metals are not found as pure metals in nature. Instead, they are found as ores. On the other hand, inactive metals like silver and gold (and sometimes copper) are often found as pure metals in nature. That is, they are found in the uncombined or free state.

Some metals have been known for thousands of years. The early Egyptians used metals as early as 1000 B.C. Gold was relatively plentiful, and cheaper than silver. Many articles of gold have been found in the tombs of Egyptian kings. The early Egyptians also learned to extract iron, copper, and tin from their ores to make cooking utensils, mirrors, and daggers.

The Egyptians also made some alloys. They learned by experience that copper could be made harder by adding a small quantity of tin. The alloy of copper and tin is called bronze. The alloy of copper and zinc is called brass. There are references to both of these alloys in the Bible. For example, in a letter written about 2000 years ago, Paul tells the Corinthians that without love "we become as sounding brass or a tinkling cymbal."

Our modern way of living depends upon an abundance of metals. Every factory needs metals for pipes, boilers, and machinery. Without metals we would have to live under primitive conditions. Are there enough metals for our future needs?

Metal ores are unevenly distributed in the earth. All countries are short of some of them. The United States has more metals than many other countries. Yet we import such metals as aluminum, copper, tin, chromium, nickel and manganese.

Remember that there is no second crop of metals. Once they are taken from the earth, they cannot be renewed. Some metals are already in short supply. Others will run out in a few years. Our supplies of metals will last longer if we recycle metals. For example, aluminum cans can be melted, and the aluminum can be used to make more cans.

A possible new source of metals is the sea. (See Figure 7–26.) The sea collects everything carried from the land by rivers. Seawater contains practically every known element, even gold. As you learned in Perspective 5–4, if 100 kg of seawater are evaporated, about 3.5 kg of solid remains. At least two thirds of this, about 2.3 kg, is sodium chloride. This in turn yields about 1 kg of the metal sodium. However, the large deposits of salt in the earth will provide all the sodium and chlorine needed for centuries to come.

The other substances in seawater are less concentrated. A large volume of seawater must be treated to recover a small amount of metal. Thus, it is usually more expensive to extract a metal from seawater than from its ore. That is why the sea is almost untapped as a source of metals. Magnesium, however, is an exception. Magnesium ions are relatively abundant in the sea and it is almost as cheap to get the metal from seawater as from its ores. Nevertheless, even for magnesium, large quantities of seawater yield only a small amount of metal. It has been estimated that to obtain 1 kg of magnesium, at least 800 kg of seawater must be treated.

Figure 7–26
This factory extracts magnesium from seawater. A compound of magnesium is obtained, and electricity is used to release the metal.

QUESTIONS

A

1. Lavoisier gave a satisfactory explanation of combustion. How would he have explained the fact that wood burns in air?

2. (a) Name four gases present in air. (b) Which is the most abundant gas in air?

3. How does photosynthesis affect the amount of (a) oxygen, (b) nitrogen, (c) carbon dioxide in air?

4. (a) What is relative humidity? (b) Why is the relative humidity usually low indoors in winter?

5. Why do metals have positive combining capacities?

6. What does the ending *-ide* indicate?

7. (a) What is a radical? (b) Give an example of a radical. (c) What does the ending *-ate* indicate?

8. The equation for the reaction of hydrogen with chlorine is

$$2H + 2Cl = 2HCl$$

Is this equation right or wrong? Why?

9. What is (a) synthesis, (b) photosynthesis?

10. What is the difference between the decomposition of a compound and its analysis?

11. Write the (a) ordinary equation, (b) ionic equation for the reaction between sodium and hydrochloric acid.

12. (a) What is slow oxidation? Give an example. (b) What is rapid oxidation? Give an example.

13. (a) What is an alloy? (b) Give an example of an alloy.

14. Compare the terms *reactant* and *product*.

15. How is air different from oxygen?

16. How can the presence of water vapor in air be demonstrated?

17. Compare the terms *synthesis* and *decomposition*.

18. What element is present in all acids?

19. What two processes are responsible for the balance of carbon dioxide and oxygen in the atmosphere?

20. Why is hydrogen usually listed as a metal?

21. What is the difference between a sulfide and a sulfate?

22. Compare the terms *symbol* and *formula*.

23. Write the names of the elements represented by the following symbols: Na, P, N, F, S, Cl, K, Ag, Hg, Au, and He.

B

24. What are the main differences between the combustion of sugar and the photosynthesis of sugar?

25. If you comb your hair on a very dry day you may hear a crackling noise and see tiny electric sparks near the comb. Why?

26. On a humid day there is no crackling or sparking when your hair is combed. Why not?

27. In each of items (a), (b), and (c), only one of the formulas is correct. Choose the correct formula and name the compound. (a) HCl_2, Na_2Cl, $ZnCl_2$, Al_3Cl (b) NaS, AgS, ZnS, AlS (c) KN, CaN, MgN, AlN.

Questions 28–30 refer to the reaction between zinc and sulfuric acid, for which the equation is

$$Zn + H_2SO_4 = ZnSO_4 + H_2$$

In each question, choose the correct answer.

28. The particle that increases its combining capacity in the reaction is

a zinc atom

b sulfate radical

c hydrogen ion

29. The particle that does not change its combining capacity is
a zinc atom
b sulfate radical
c hydrogen ion

30. A sulfate radical is an ion because
a it has a charge
b it has lost two electrons

31. If mercury is gently heated in air, it forms mercury oxide. If mercury oxide is strongly heated, it decomposes to mercury and oxygen. Set up these reactions as a single equation that shows both synthesis and decomposition.

32. Sulfuric acid in water solution is present as ions. Write an equation to show the formation of these ions, and name the ions that are formed.

33. What kinds of substances can people use instead of metals? Give several examples.

In each of Questions 34–38, three symbols are listed. Write the names of the elements and state what each three elements have in common.

34. H, N, O

35. Mg, Ca, Ba

36. K, Na, Hg

37. F, Cl, Br

38. Li, Be, B

39. How many atoms are present in each of the following? (a) $Ca(OH)_2$ (b) Na_2SO_4 (c) KNO_3 (d) H_3PO_4 (e) Li_2O (f) HF (g) $Al_2(SO_4)_3$ (h) Fe_2O_3 (i) H_2CO_3 (j) $AlCl_3$

40. Copy the following table and fill in the correct formulas. Each formula represents a compound made up of two elements. The first one is filled in for you.

	Cl (−1)	Br (−1)	O (−2)	S (−2)
Na (+1)	NaCl	?	?	?
Ag (+1)	?	?	?	?
Ca (+2)	?	?	?	?
Mg (+2)	?	?	?	?
Al (+3)	?	?	?	?

41. Copy the following table and fill in the correct name of each compound. The first one is filled in for you.

	Cl	Br	O	S
Na	sodium chloride	?	?	?
Ag	?	?	?	?
Ca	?	?	?	?
Mg	?	?	?	?
Al	?	?	?	?

42. Copy the following table and fill in the correct formulas. The first one is filled in for you. Note that radicals are used in this table.

	OH (−1)	NO_3 (−1)	CO_3 (−2)	SO_4 (−2)
K (+1)	KOH	?	?	?
Ba (+2)	?	?	?	?
Pb (+2)	?	?	?	?
Fe (+2)	?	?	?	?
Fe (+3)	?	?	?	?

43. Copy the following table and fill in the correct name of each compound. The first one is filled in for you.

	OH	NO_3	CO_3	SO_4
K	potassium hydroxide	?	?	?
Ba	?	?	?	?
Pb	?	?	?	?
Fe(II)	?	?	?	?
Fe(III)	?	?	?	?

44. Write a balanced equation for each of the following reactions.
(a) $AgNO_3 + NaCl = AgCl + NaNO_3$
(b) $CO + O_2 = CO_2$
(c) $Na_2CO_3 + CaCl_2 = NaCl + CaCO_3$
(d) $NH_3 + H_2O = NH_4OH$
(e) $KClO_3 = KCl + O_2$

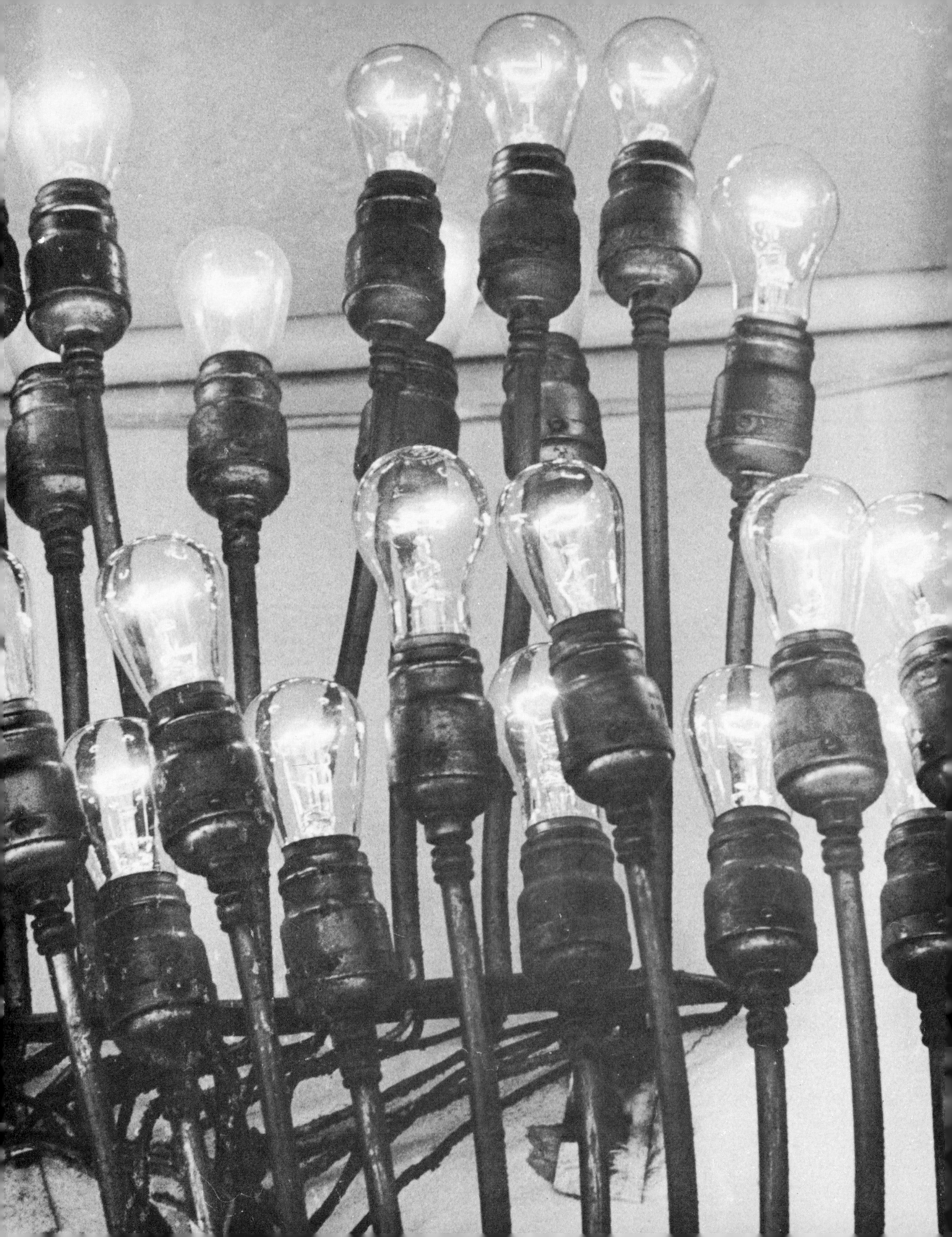

Chapter 8

ELECTRICITY

Light bulbs, and many other devices you use, use electricity.

Because you have grown up using electric appliances, it is probably difficult for you to realize that in Benjamin Franklin's time there were none at all. Indeed, very little was known about electricity. Franklin knew that some objects could be electrically charged if they were rubbed. In many of his experiments with charged bodies, Franklin produced electric sparks. He discovered that if a piece of metal with a sharp point was attached to a charged body, the charges readily escaped into the air through the point.

Lightning was still a mystery, and people were afraid of it. The Romans had believed that a flash of lightning was a bolt of fire hurled by an angry god at sinners on earth. Some people believed this even in Franklin's time. Franklin, however, thought there must be a scientific explanation of lightning. He suspected that a lightning flash was nothing more than a gigantic electric spark. How could he prove his theory? How could lightning be caught? Then he had the idea of using a kite. Could a kite with a spike draw electricity from a thunder cloud?

Franklin made a kite and fastened an iron spike to it. Then he tied a long string to the kite and tied a silk ribbon to the string. He fastened a metal key at the knot joining the silk and string.

During a thunderstorm in 1752, Franklin tried his experiment. He flew the kite, holding the silk ribbon in his hand. To be sure that the silk would not get wet from the rain, he stood under a shed. When the rain had wetted the string, he noticed that loose threads in the string stood on end. To Franklin it looked as though the string had become electrically charged. To test this idea he placed his knuckle near the key. Instantly sparks jumped from the key and Franklin received a painful shock. (See Figure 8–1.)

Franklin's experiment was highly dangerous. Fortunately for him, his kite went into a cloud that was not heavily charged. A French scientist who repeated Franklin's experiment a year or so later was killed by the charge that he received. Franklin knew that a charge from the cloud would run down the wet string. He also knew that the charge would not flow through the dry silk. Thus the dry silk ribbon protected him.

The sparks from the key were similar to the sparks from an electrically charged object. Franklin was convinced that thunder clouds are electrically charged and that a flash of lightning is a discharge of electricity. This was a great discovery. Because of the work of Franklin and other scientists, much is now known about electricity. You will investigate electricity in this chapter.

ELECTRIC CURRENTS

What Is an Electric Current?

All the experiments in this chapter involve an electric current. An **electric current** is a flow of electrons. But how can electrons flow? You can find out by observing what happens if an active metal such as zinc and an inactive metal such as copper are connected by a wire and dipped into a solution of sulfuric acid.

In Experiment 7–9 you observed that zinc reacts with hydrochloric acid and releases hydrogen gas. Zinc reacts the same way with sulfuric acid. You learned the reason for this reaction in Chapter 7. Zinc is an active metal because every atom of zinc has two loosely held electrons. Sulfuric acid, like hydrochloric acid, contains hydrogen ions. The hydrogen ions capture the loosely held electrons from the zinc to form molecules of hydrogen gas.

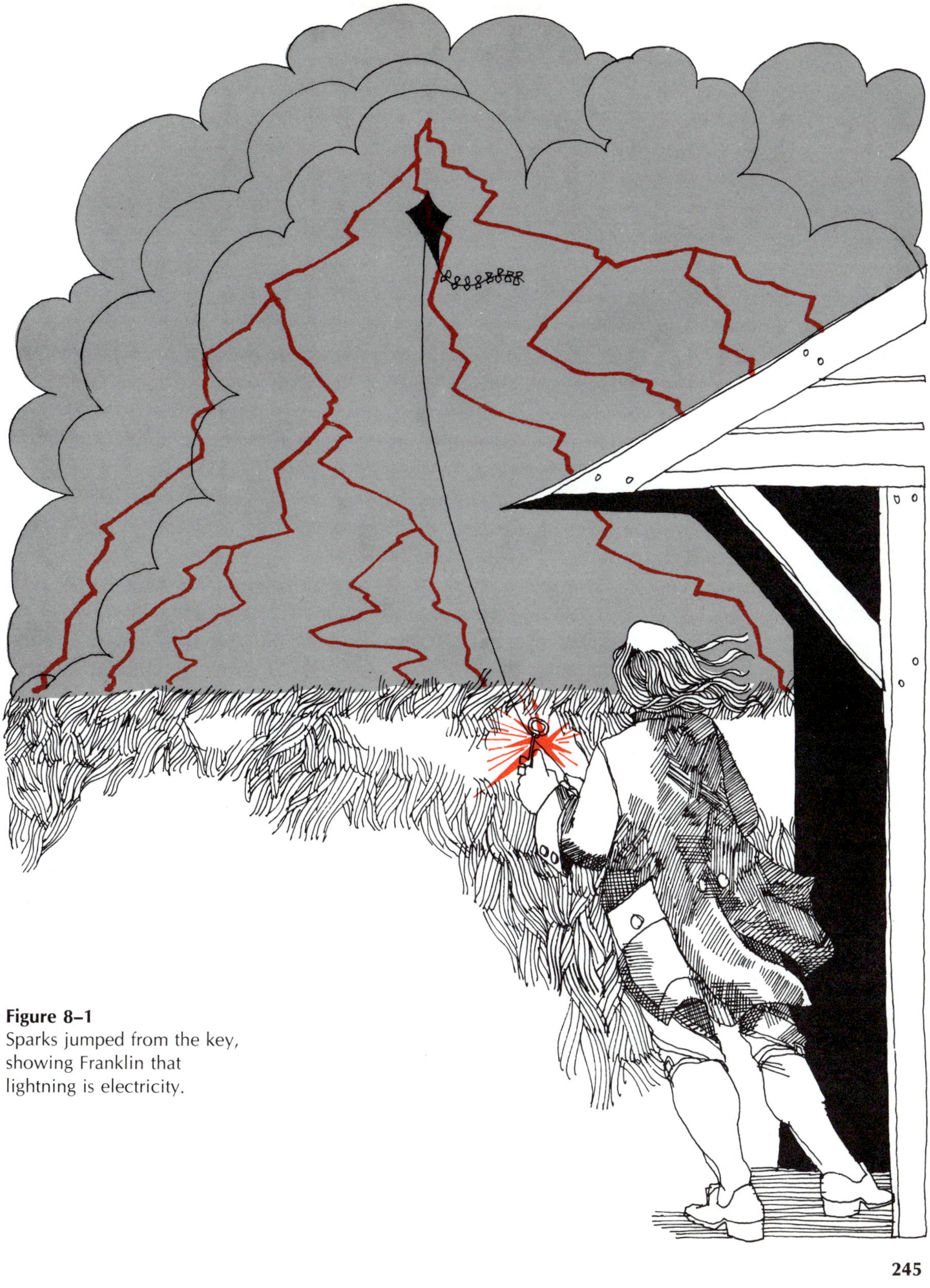

Figure 8–1
Sparks jumped from the key, showing Franklin that lightning is electricity.

2 H ions + 2 electrons = 2 H atoms = 1 H_2 molecule

In symbols the reaction is

$$2H^+ + 2e^- = 2H = H_2$$

On the other hand, copper, like mercury, is an inactive metal. That is, all the electrons in a copper atom are strongly attracted by the nucleus. As a result, hydrogen ions do not capture electrons from copper atoms.

An Electric Current from a Chemical Reaction

The difference in the chemical activity between zinc and copper in the presence of dilute sulfuric acid can be used to generate electric current. That is, chemical energy is changed into electric energy. A device which changes chemical energy to electric energy is called a cell or, more usually, a **voltaic** (vol-TAY-ik) **cell.** It is named after an Italian scientist named Volta, who made the first chemical cell about the end of the 18th century.

A copper-zinc chemical cell is only one of many different kinds of voltaic cells. This cell will be shown as a demonstration. Watch this demonstration closely, since there are a number of effects that call for explanation.

A **circuit** (SUR-kit) is the path followed by an electric current. In the demonstration the current will leave the zinc-copper cell, flow through the circuit, then return to the cell. You cannot see an electric current, but you can detect it with an instrument called an **ammeter** (AM-mee-tur). If a current flows through an ammeter, the needle of the ammeter moves away from the zero position. This movement is called deflection. An ammeter has two **terminals,** places to which wires are attached. One is marked + and is called the positive terminal. The other is marked − and is called the negative terminal.

A useful device in an electric circuit is a **switch.** It can control the flow of current by making or breaking a circuit. A switch has a movable arm. When the arm is in one position, there is a gap in the circuit so that there is no current. In this case the switch is said to be open. But if the arm is in the opposite position, the gap is closed. Then a current flows through the switch and, therefore, through the circuit.

experiment 8–1

A Zinc-Copper Electric Cell

Demonstration

Materials

zinc strip, 2 cm × 10 cm; copper strip, 2 cm × 10 cm; 2 alligator clips; copper wire; switch; ammeter; 150-ml beaker; dilute sulfuric acid

Procedure

Attach an alligator clip to each of the metal strips. Connect the clip on the zinc strip to the switch by a piece of wire. Attach the other end of the switch to the negative terminal of the ammeter. Connect the positive terminal of the ammeter to the clip on the copper strip. Fill the beaker to a depth of about 1 cm with dilute sulfuric acid. Be sure the switch is open. Lower the metal strips into the acid. The arrangement is shown in Figure 8–2.

Figure 8–2
A zinc-copper electric cell.

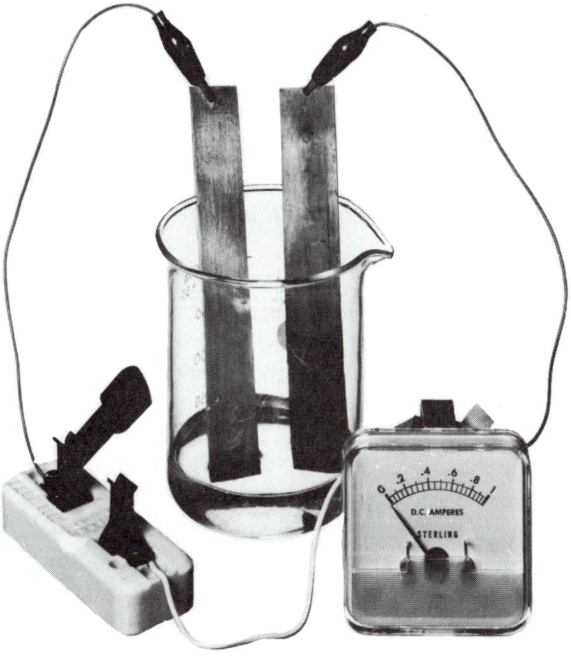

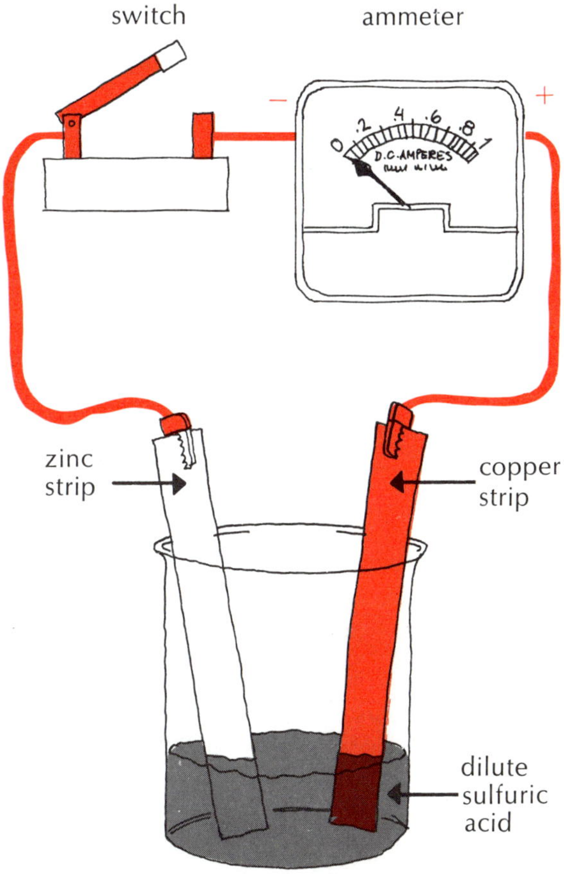

Notice that bubbles of gas (hydrogen) form at the surface of the zinc and escape. No gas bubbles appear on the surface of the copper strip.

Now close the switch. The needle of the ammeter is deflected. Bubbles of gas (hydrogen) now collect on and escape from the copper strip. But few bubbles escape from the zinc strip.

Finally, open the switch again. Gas bubbles no longer appear on the copper, but bubbles now escape freely from the zinc. Moreover, the ammeter needle returns to the zero mark.

These observations raise a number of questions, some of which will be asked and also answered.

Questions and Answers

1. Why does hydrogen gas escape from the zinc when the switch is open?
Answer: This reaction is precisely the same as if zinc is dropped into a test tube containing dilute sulfuric acid. Hydrogen ions attract and combine with the outermost electrons of zinc atoms to form hydrogen gas molecules.

$$2H^+ + 2e^- \text{ (from Zn atom)} = H_2$$

2. Why is hydrogen not formed on the surface of the copper when the switch is open?
Answer: Copper is an inactive metal and so does not give up its electrons to hydrogen ions.

3. Why is hydrogen gas formed on the surface of the copper when the switch is closed?
Answer: Electrons combine with the hydrogen ions in solution.

$$2H^+ + 2e^- = H_2$$

Zinc metal is the only source of these electrons. But in this case electrons flow from the zinc along the wire and then combine with the hydrogen ions that surround the copper. Hence the reaction on the surface of the copper is

$$2H^+ + 2e^- = H_2$$

This situation and the direction of the electron flow are shown in Figure 8–3. As a result of the reaction on the copper surface, the formation of hydrogen gas on the zinc strip is slowed down and may be stopped completely.

4. What is the electric circuit in this experiment?
Answer: A circuit is the path followed by electrons. In this experiment electrons leave the cell at the zinc strip, then flow through the connecting wires, switch, and ammeter, and back to the cell by way of the copper strip.

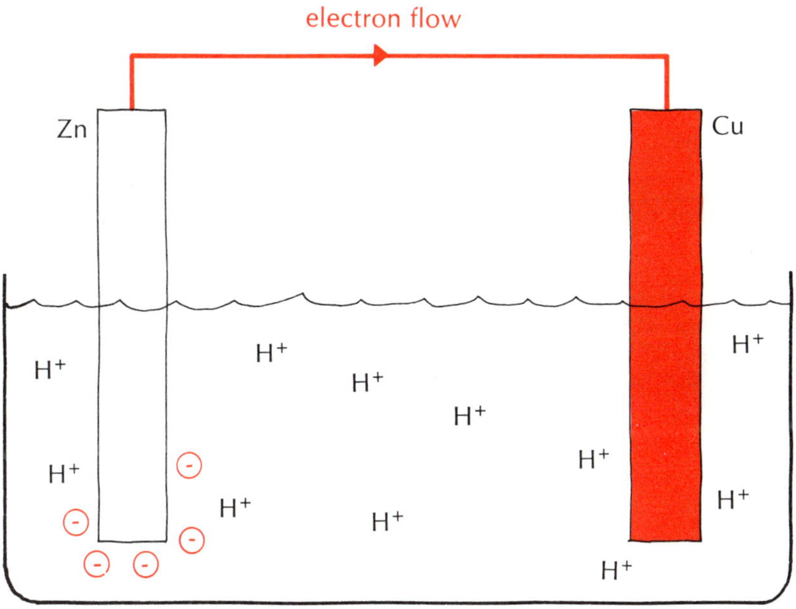

Figure 8–3
The electron flow in a zinc-copper electric cell.

ELECTRIC CIRCUITS

Parts of an Electric Circuit

A circuit consists of a source of current such as a dry cell, and of various other parts, or **components** (KOM-poh-nentz). The components may include connecting wires, a resistor (ree-ZIS-ter), a switch, and either an ammeter, a voltmeter (VOLT-mee-ter), or both. At this time we will discuss only those parts needed for the circuit in Experiment 8–2.

A dry cell is a source of electric energy. It works like the zinc-copper cell, except that it has a zinc terminal on the outside and a carbon terminal in the center. The terminal in the center of the dry cell is the positive (+) terminal. The terminal on the outside is the negative (−) terminal. A chemical reaction occurs in a dry cell when it is made part of a circuit. Electrons are forced through the circuit from the zinc to the carbon terminal, and chemicals in the cell are used up. To conserve the cell, maintain the current just long enough to obtain a reading of the ammeter (or voltmeter). When you are ready to make a reading, close the switch, read the meter, and then open the switch.

An ammeter measures the current in a circuit. The greater the number of electrons that flow through the ammeter per second, the higher the ammeter reading. An ammeter can be easily damaged and it should therefore be used with great care. One of the terminals of the ammeter is marked + and the other is marked −. The + terminal of the ammeter should always be connected to the + terminal of the dry cell.

A switch is shown in the open position in Figure 8–2. To complete the circuit, the two metal sections of the switch must be in good contact with each other.

A **resistor** slows down the flow of electrons. It is used to control the current in a circuit.

Connecting wires are lengths of copper wire, usually covered with rubber or plastic. Copper wire does not offer much resistance to the flow of electrons and therefore does not slow down the current very much. Thus, a copper wire is a good **conductor** of electricity. The rubber or plastic covering prevents the current from leaking away. Rubber and plastic are called **insulators.** A substance that is a good insulator is a bad conductor.

The various parts of a circuit can be represented by symbols as shown in Figure 8–4.

Figure 8–4
Symbols used for the parts of an electric circuit.

Symbol	*Meaning*
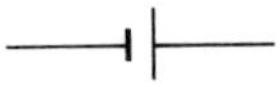	dry cell (long thin vertical line is +)
	switch
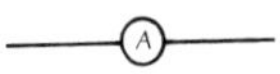	ammeter
	voltmeter
	resistor (the number of zigzags is not related to the strength of the resistor)
	rheostat (variable resistor)
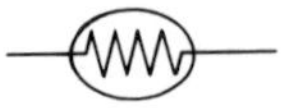	flashlight bulb

In each experiment in this chapter, you will assemble an electric circuit from components. You may be provided with a kit which includes all the components needed for these experiments. It includes a circuit board on which the various components can be mounted. The circuit board is shown in Figure 8–5.

The plug-in positions for components and connecting wires are numbered on the board. The solid lines from 1 to 2, 3 to 4, 5 to 6, 11 to 12, and 13 to 14 are sections of wire mounted underneath the board. Notice that there are two plug-ins for most of the positions. For position 4, for example, there are two plug-ins, both of which are attached to the wire leading from position 4 to position 3.

Resistors R_1 and R_2 are permanently mounted on the board. Connections to R_1 can be made at positions 7 and 8. Connections to R_2 can be made at positions 9 and 10.

Figure 8–5
The circuit board from the kit. It includes two resistors and a rheostat. Other components can be attached to any of the posts.

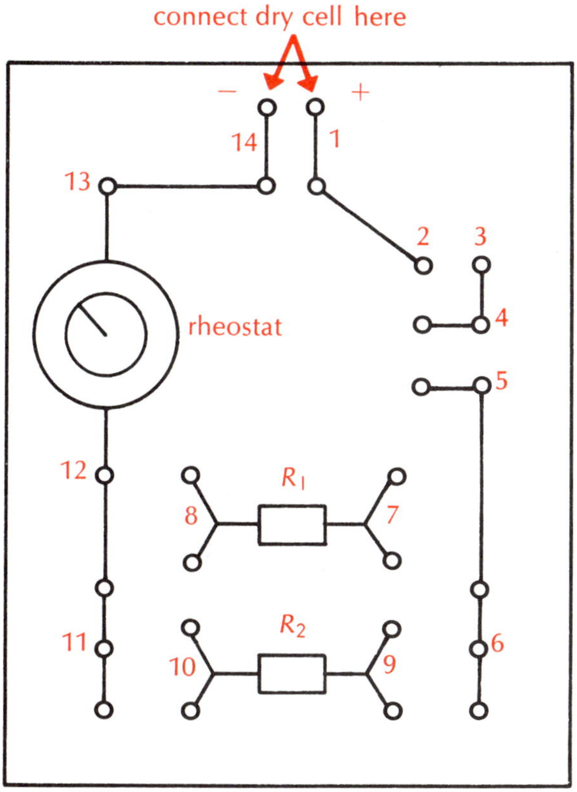

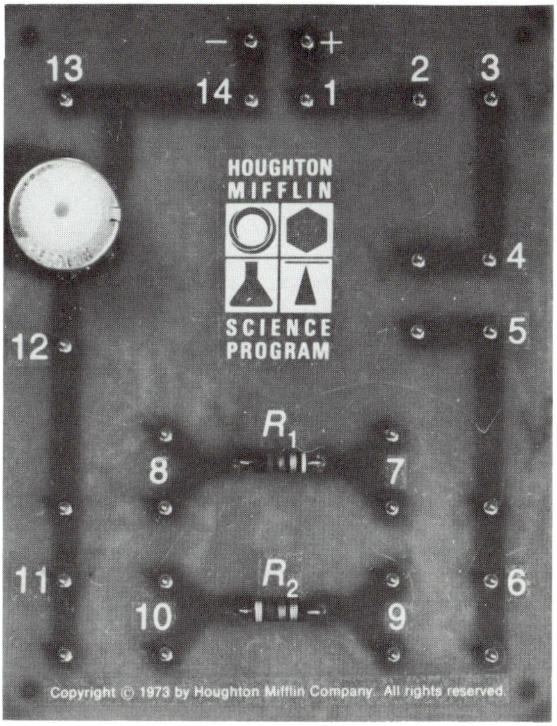

The numbering of the plug-in positions makes the assembly of a particular circuit quite simple. For example, suppose you are asked to assemble a circuit that includes a dry cell, an ammeter, and a fixed resistor (R_1). A diagram of this circuit and a picture of the circuit board arrangement are shown in Figure 8–6. The components would be attached to the circuit board as follows:

Dry cell: + terminal to 1, – terminal to 14
Ammeter: + terminal to 2, – terminal to 3
Resistor (R_1): connecting wire from 6 to 7
connecting wire from 8 to 11
Connecting wire from 12 to 13
Connecting wire from 4 to 5

The connecting wire from 12 to 13 bridges the rheostat. This produces, in effect, one long connecting wire from 11 to 14. The connecting wire from 4 to 5 serves as a switch.

Figure 8–6
Below: A simple circuit. Right: How to set up this circuit on the circuit board.

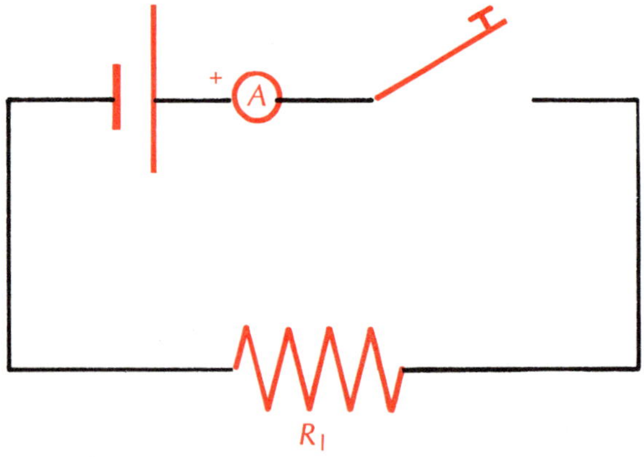

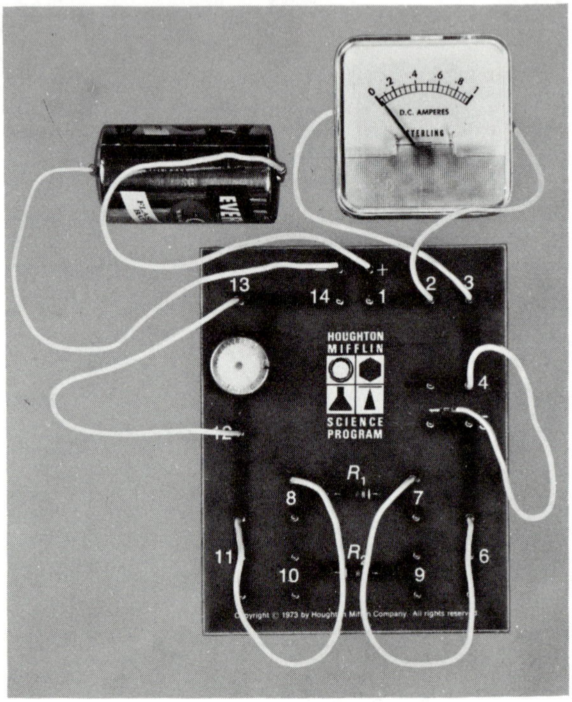

experiment 8–2 — Current in a Circuit

Materials

circuit board (or 2 resistors, R_1 and R_2); dry cell; ammeter; connecting wires

A. Different Positions of the Ammeter

Procedure

(1) Set up the circuit as shown in Figure 8–7, with the ammeter in position A_1. Attach the components to the circuit board as follows:
Dry cell: + terminal to 1, − terminal to 14
Ammeter: + terminal to 6, − terminal to 7
Resistor: (R_1): connecting wire from 8 to 11
Connecting wire from 12 to 13
Connecting wire from 2 to 3

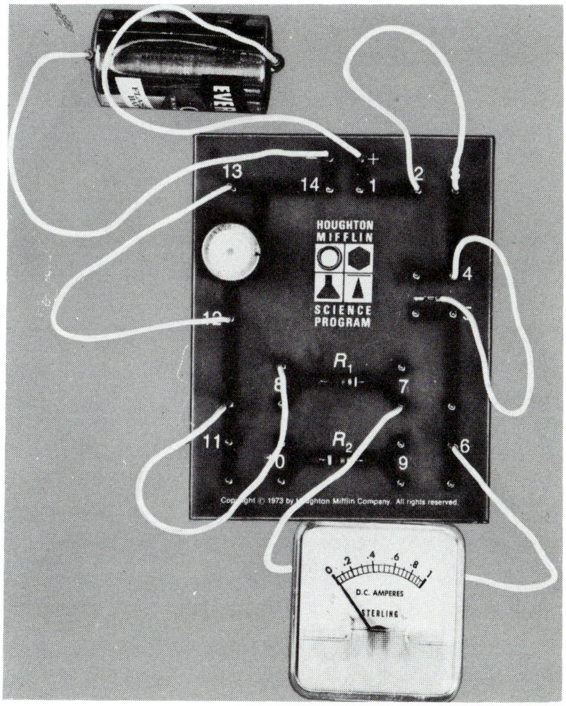

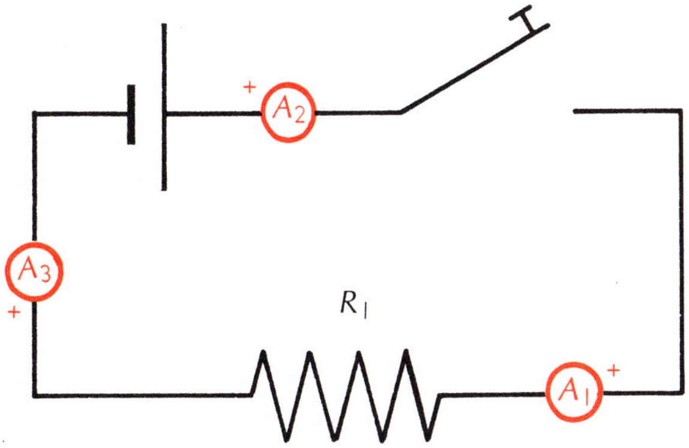

Figure 8–7
Below: Circuit for Experiment 8–2. Right: The circuit board arrangement, with the ammeter in position A_1.

Ask your teacher to inspect the circuit. Attach a connecting wire from 4 to 5 and read the ammeter. Disconnect the wire from 4 to 5. Record the ammeter reading you made when the wire was connected.

(2) Transfer the ammeter to position A_2 in Figure 8–7. The connections are the same as before except for:
Ammeter: + terminal to 2, − terminal to 3
Resistor (R_1): connecting wire from 6 to 7

Draw a circuit diagram for this arrangement. Record the ammeter reading.

(3) Transfer the ammeter to position A_3. The connections are the same as before except for:
Ammeter: + terminal to 8, − terminal to 11
Resistor (R_1): connecting wire from 6 to 7
Connecting wire from 2 to 3

Draw a circuit diagram for this arrangement. Record the ammeter reading.

B. Using a Different Resistor

Replace resistor R_1 by another resistor, R_2, and find the value of the current through R_2.

The only change in the circuit for part B is to connect 6 with 9 (instead of with 7) and 11 with 10 (instead of with 8) by means of connecting wires. R_1 is now out of the circuit and R_2 is in the circuit.

Questions

1. Compare the three ammeter readings in part A. What do you conclude?

2. Does it matter where you put the ammeter in the circuit? Why or why not?

3. Compare the current through R_1 with the current through R_2. What do you conclude?

4. Which resistor has the greater resistance, R_1 or R_2? Give a reason for your answer.

Resistors in Series and Parallel

Two resistors can be arranged in either of two ways, one behind the other or alongside each other. The first arrangement is shown in Figure 8–8A. In this arrangement the resistors are said

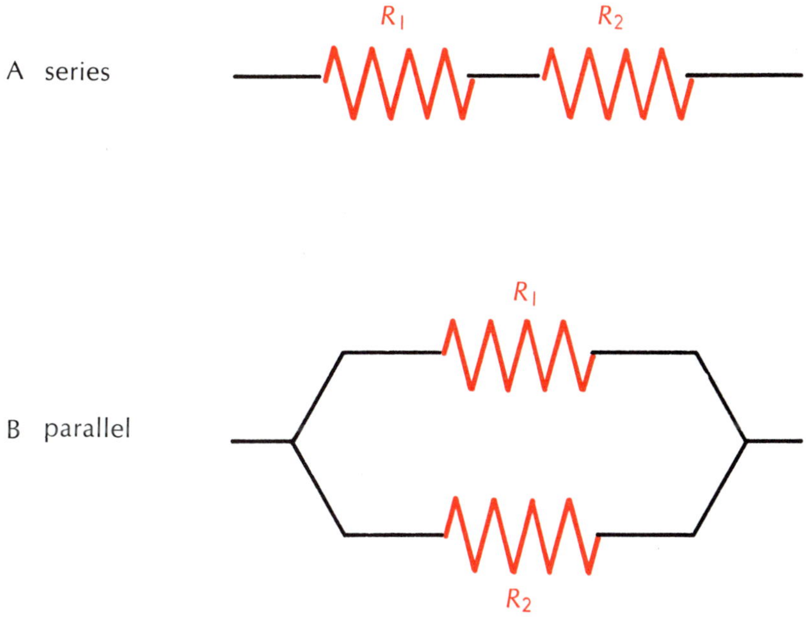

Figure 8–8
A. Resistors in series.
B. Resistors in parallel.

to be in **series.** In the series arrangement there is only one path for the flow of electrons.

The other arrangement is shown in Figure 8–8B. Here the resistors are said to be in **parallel.** In this parallel arrangement the total current is divided into two separate paths. Some electrons flow through resistor R_1 and the rest flow through R_2.

A circuit that has two or more resistors in series is called a **series circuit.** A circuit with two or more resistors in parallel is called a **parallel circuit.**

experiment 8–3 The Currents in Series and Parallel Circuits

Materials

circuit board (or 2 resistors, R_1 and R_2); dry cell; ammeter; connecting wires

A. Series Circuit

Procedure

To connect resistors R_1 and R_2 in series, arrange the circuit as shown in Figure 8–9. Place the ammeter in position A_1.

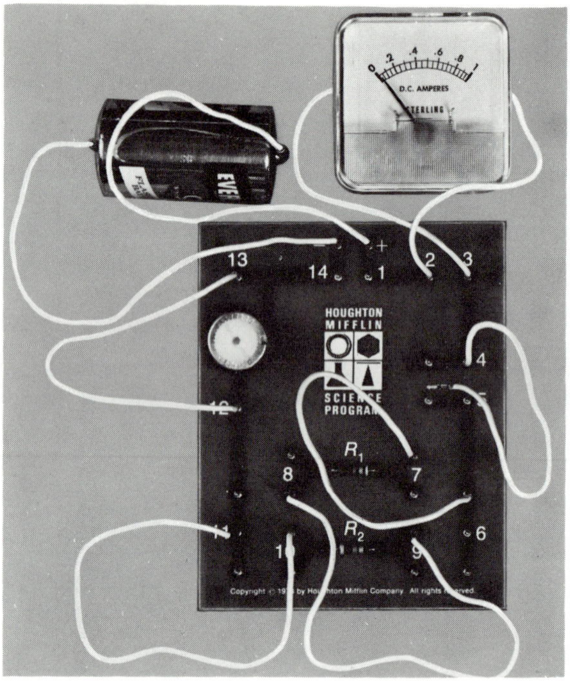

Figure 8–9
Circuit for Experiment 8–3A. The resistors are in series. The ammeter is in position A_1.

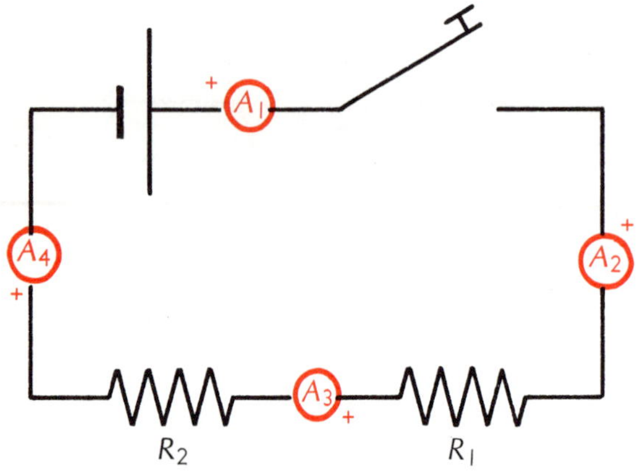

Attach the components to the circuit board as follows:
Dry cell: + terminal to 1, – terminal to 14
Ammeter: + terminal to 2, – terminal to 3
R_1: connecting wire from 6 to 7
connecting wire from 8 to 9
R_2: connecting wire from 10 to 11
Connecting wire from 12 to 13
Connecting wire from 4 to 5

Read the ammeter, and then disconnect the wire from 4 to 5. Record the ammeter reading.

Disconnect the ammeter and insert it in position A_2 in Figure 8–9. Now, with the ammeter in position A_2, the connections are the same as before except as follows:
Ammeter: + terminal to 6, – terminal to 7
Connecting wire from 2 to 3

Record the current.

Repeat the above with the ammeter in position A_3. With the ammeter in position A_3, the connections are the same as before except for:
Ammeter: + terminal to 8, – terminal to 9
Connecting wire from 6 to 7

Repeat with the ammeter in position A_4. With the ammeter in position A_4, the connections are the same as before except for the following:
Ammeter: + terminal to 10, − terminal to 11
Connecting wire from 8 to 9

1. Compare the current values for the four ammeter positions. What do you conclude?

2. Compare the current through R_1 and R_2 in series with the current values through R_1 and R_2 in Experiment 8–2. What do you conclude?

3. Why does the current act the way it does in a series circuit?

4. Write a statement that expresses the total current through a combination of several resistors in series.

B. Parallel Circuit

Connect resistors R_1 and R_2 in parallel. Arrange the circuit as shown in Figure 8–10 with the ammeter in position A_1.

Attach the components to the circuit board as follows:
Dry cell: + terminal to 1, terminal to 14
Ammeter: + terminal to 2, − terminal to 3
R_1: connecting wire from 6 to 7
connecting wire from 8 to 11
R_2: connecting wire from 6 to 9
connecting wire from 10 to 11
Connecting wire from 4 to 5
Connecting wire from 12 to 13

Record the ammeter reading.

Disconnect the ammeter. Insert it in position A_2 of the diagram. With the ammeter in position A_2, the connections are the same as before except for:
Ammeter: + terminal to 6, − terminal to 7
Connecting wire from 2 to 3

Find the ammeter reading and record it.

Insert the ammeter in position A_3 of the diagram. With the ammeter in position A_3, the connections are the same as before except for:
Ammeter: + terminal to 6, − terminal to 9
Connecting wire from 6 to 7

Find the ammeter reading and record it.

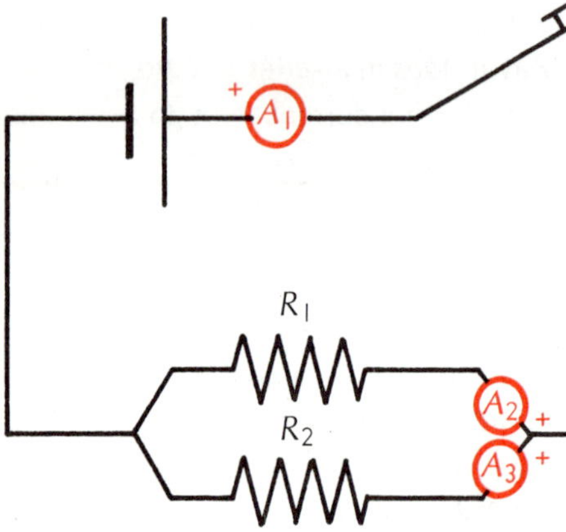

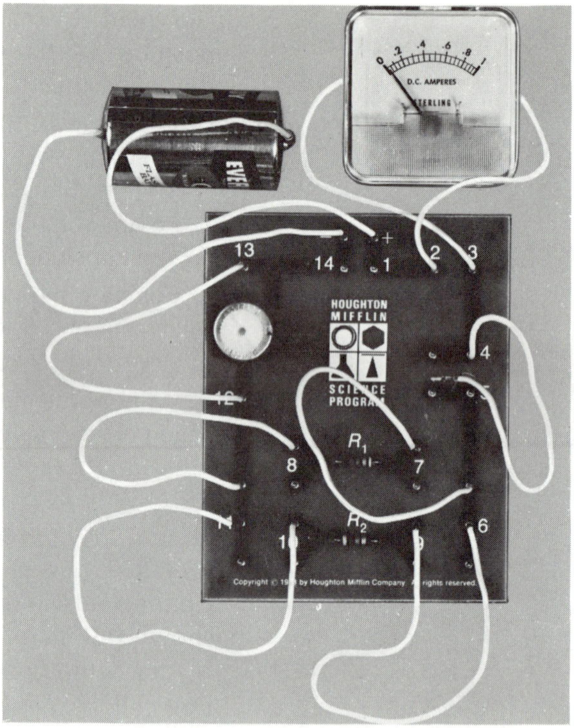

Figure 8–10
Circuit for Experiment 8–3B. The resistors are in parallel. The ammeter is in position A_1.

Questions

1. (a) What is the total current through R_1 and R_2 as indicated by the ammeter reading in position A_1? (b) What is the sum of the separate currents through R_1 and R_2 as indicated by ammeter readings in positions A_2 and A_3? (c) What do you conclude?

2. (a) Compare the current through R_1 in this experiment with the current in Experiment 8–2. (b) Compare the current through R_2 in this experiment with the current in Experiment 8–2. (c) What conclusions can you draw from these comparisons?

3. Write a statement that expresses the total current through a combination of several resistors in parallel.

Voltmeter

A voltmeter looks like an ammeter, but there are important differences between the two instruments. A **voltmeter** measures the voltage between two points in a circuit. But what is voltage?

To help answer this question, consider what happens when you turn the handle of the faucet at the kitchen sink. As you know, water flows from the faucet. But water flows only if a pressure is exerted on it. What causes the water pressure? In many towns, water is stored in a tank like that shown in Figure 8–11. Water is pumped into the tank from a reservoir. The water level in the tank is well above the height of all buildings in the town. The vertical distance between the faucet and the water level in the tank is called the "head" of water. The head of water provides the pressure that makes the water flow.

Similarly, an "electric pressure" is needed to force electrons through a wire. The electric pressure is due to the differences in chemical activity between the elements that form the terminals of a cell. This difference in activity between the terminals is called the **voltage** of the cell.

The voltage of a cell is shared among the various parts of the circuit. The difference in electric pressure at the two ends of a resistor is usually called **voltage drop.** A voltmeter measures either the voltage of a cell or the voltage drop in any part of the circuit.

Figure 8–11
Water is forced through pipes by water pressure. Electrons are forced through wires by "electric pressure," called voltage.

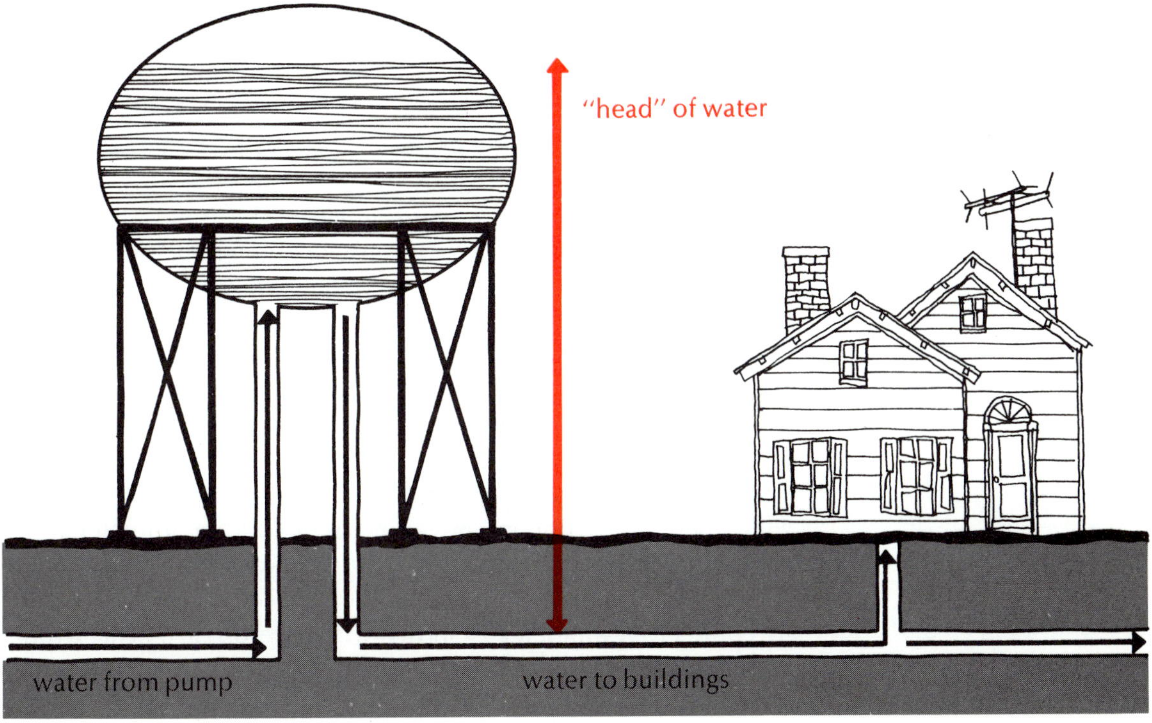

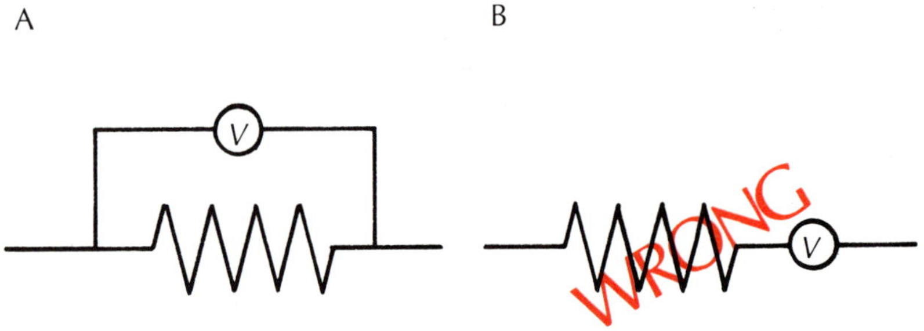

Figure 8–12
A. The right way to connect a voltmeter. **B.** The wrong way. Remember that you measure the voltage *across* a component.

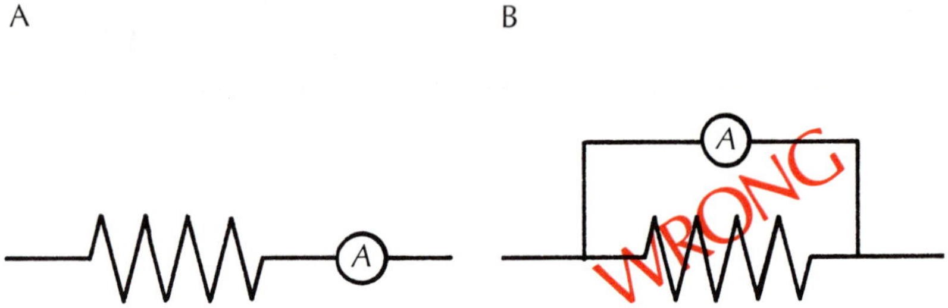

Figure 8–13
A. The right way to connect an ammeter. **B.** The wrong way. Remember that you measure the current *through* a component.

One of the terminals of a voltmeter is marked +. The other is marked −. The wire attached to the + terminal must be connected to the + terminal of the dry cell.

An important difference between a voltmeter and an ammeter is the way these instruments are connected in a circuit. A voltmeter is always connected across the electrical component such as a resistor. That is, *a voltmeter is always placed in parallel with the component.* The correct way to connect a voltmeter and resistor in a circuit is shown in Figure 8–12A. The wrong way is shown in Figure 8–12B.

On the other hand, *an ammeter must always be placed in series* with a component, never in parallel. If an ammeter were connected in parallel, a very large current would pass through it. This current might burn out the ammeter coil. The correct way to connect an ammeter is shown in Figure 8–13A. The wrong way is shown in Figure 8–13B.

experiment 8–4 Voltage Drops in a Circuit

A. The Voltage of a Dry Cell

Materials

circuit board; voltmeter; dry cell; connecting wires

Procedure

Connect the terminals of a voltmeter to the terminals of a dry cell by means of two connecting wires. The arrangement is shown in Figure 8–14. Notice that the positive terminal of the cell is connected to the positive terminal of the voltmeter.

Attach the components to the circuit board as follows:
Dry cell: + terminal to 1, − terminal to 14
Voltmeter: + terminal to 1, − terminal to 14

Record the voltmeter reading in your notebook.

Figure 8–14
Circuit for Experiment 8–4A

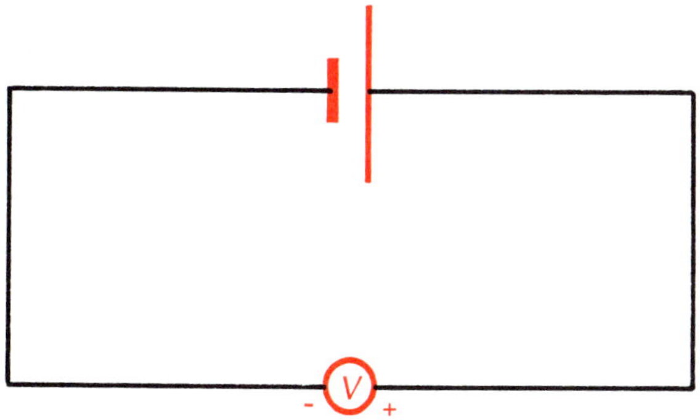

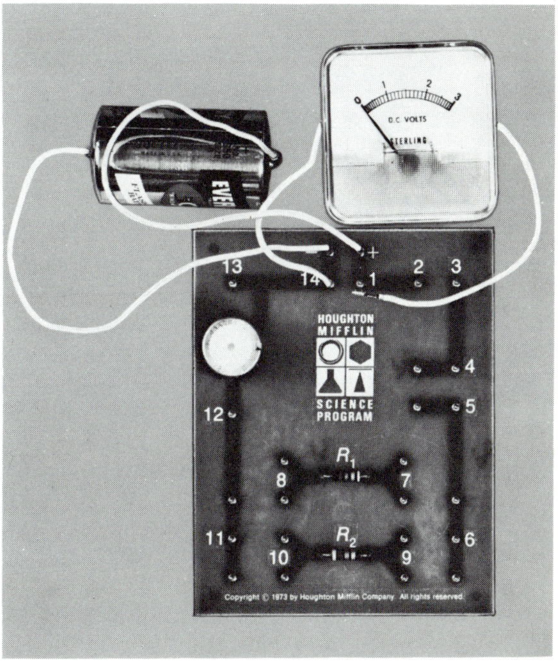

B. The Voltage Drop across the Parts of a Circuit

Materials

circuit board (or a resistor, R_1); voltmeter; dry cell; connecting wires

Procedure

Set up a circuit which includes a dry cell and a fixed resistor (*EF*) as shown in Figure 8–15. The voltmeter is shown across points *A* and *B*, the terminals of the dry cell. This will be the first voltage you measure.

Attach the components to the circuit board as follows:

Dry cell: + terminal to 1, – terminal to 14

Resistor (R_1): connecting wire from 6 to 7
connecting wire from 8 to 11

Connecting wire from 12 to 13

Connecting wire from 2 to 3

Connecting wire from 4 to 5

Voltmeter: across *AB*: + terminal to 1, – terminal to 14
across *EF*: + terminal to 7, – terminal to 8
across *BC*: + terminal to 1, – terminal to 4
across *CD*: + terminal to 4, – terminal to 5
across *FA*: + terminal to 8, – terminal to 14

Figure 8–15
Circuit for Experiment 8–4B. The voltmeter is across *AB*.

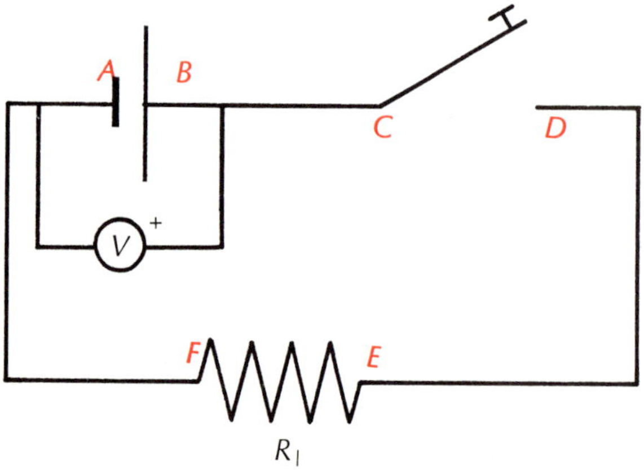

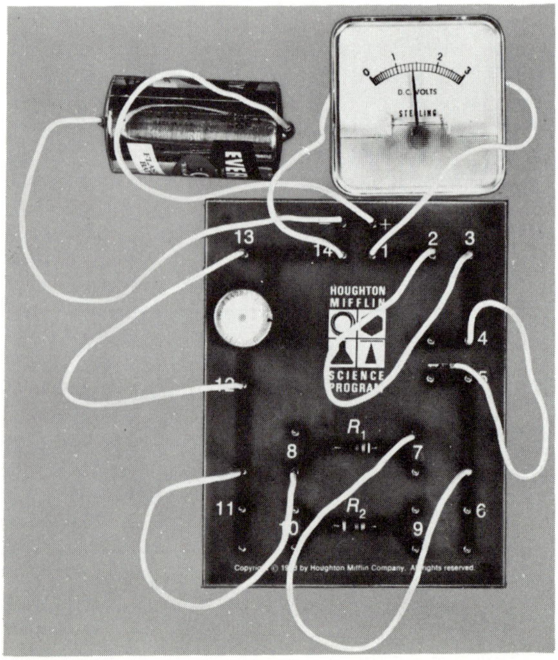

Measure and record the voltage drop across *AB*. To do this, connect the + terminal of the voltmeter to point *B* and the − terminal of the meter to point *A*. Now measure the voltage drop across *AB*.

Measure and record the voltage drop across resistor *EF*. Remember that the voltmeter must be connected in parallel with *EF*.

Now find and record the voltage drops across *BC, CD,* and *FA*.

Questions

1. Compare the voltage drop across *AB* with the voltage of the dry cell in part A of this experiment. Does the comparison make sense? Explain.

2. Compare the voltage drop across *AB* with the voltage drop across *EF*. What do you conclude?

3. How do you account for the size of the voltage drops across the sections *BC, CD,* and *FA*?

experiment 8–5

Voltage Drops in a Series Circuit

Materials

circuit board (or 2 resistors, R_1 and R_2); voltmeter; dry cell; connecting wires

A. A Series Circuit

Procedure

Set up a circuit containing resistors R_1 and R_2 in series, as shown in Figure 8–16.

The components should be attached to the circuit board as follows:

Dry cell: + terminal to 1, − terminal to 14
R_2: connecting wire from 6 to 9
connecting wire from 10 to 7
R_1: connecting wire from 8 to 11
Connecting wire from 2 to 3
Connecting wire from 12 to 13
Connecting wire from 4 to 5
Voltmeter: across *AB*: + terminal to 1, − terminal to 14
across *CD*: + terminal to 9, − terminal to 10
across *DE*: + terminal to 7, − terminal to 8
across *CE*: + terminal to 9, − terminal to 8

Measure and record the voltage drops across *AB, CD, DE,* and *CE*.

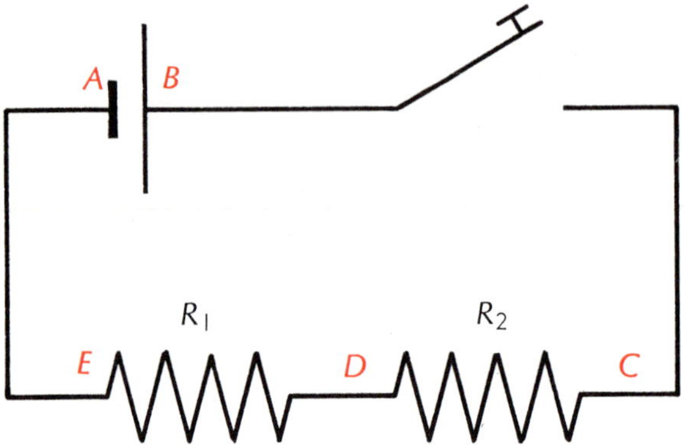

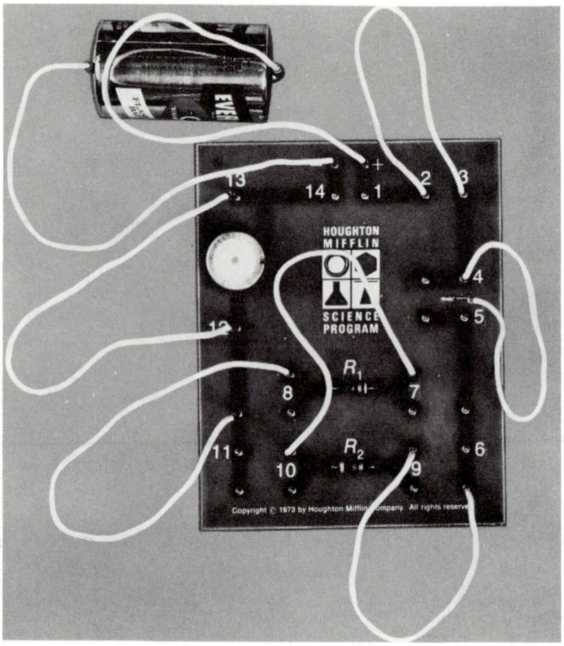

Figure 8–16
Circuit for Experiment 8–5A.

Questions

1. Compare the voltage drop across *CE* with the sum of the voltage drops across *CD* and *DE*. What do you conclude?

2. Compare the voltage drops across *AB* and *CE*. How do you account for the results?

B. Interchanging the Resistors

Procedure

Suppose the positions of the two resistors, R_1 and R_2, in Figure 8–16 were interchanged. Predict the voltage drops across R_1 and across R_2.

Now set up the circuit and check your prediction. To find the voltage across R_1, attach the components to the circuit board as follows:

Dry cell: + terminal to 1, – terminal to 14
R_1: connecting wire from 6 to 7
connecting wire from 8 to 9
R_2: connecting wire from 10 to 11
Connecting wire from 2 to 3
Connecting wire from 4 to 5
Connecting wire from 12 to 13
Voltmeter: + terminal to 7, – terminal to 8

To find the voltage across R_2, attach the + terminal of the voltmeter to 9, and the – terminal to 10.

Questions

1. Does it matter where the resistors in a series circuit are placed? That is, is their order of arrangement of any importance? Give a reason for your answer.

experiment 8–6

Voltage Drops in a Parallel Circuit

Materials

as in Experiment 8–5.

Procedure

Place the two resistors, R_1 and R_2, in parallel as shown in Figure 8–17.

The components should be attached to the circuit board as follows:

Dry cell: + terminal to 1, – terminal to 14

R_1: connecting wire from 6 to 7
connecting wire from 8 to 11

R_2: connecting wire from 6 to 9
connecting wire from 10 to 11

Connecting wire from 2 to 3

Connecting wire from 12 to 13

Connecting wire from 4 to 5

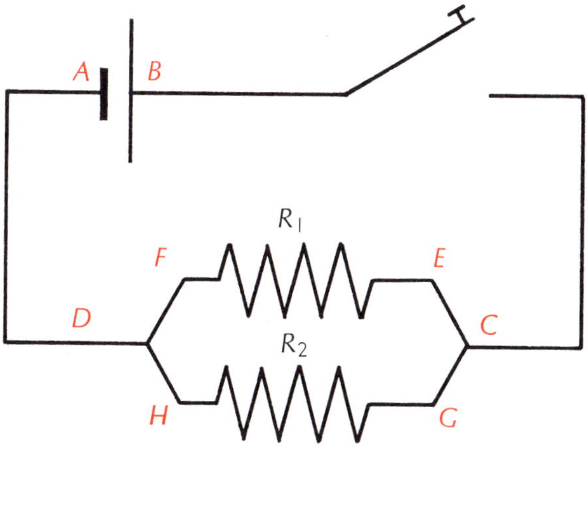

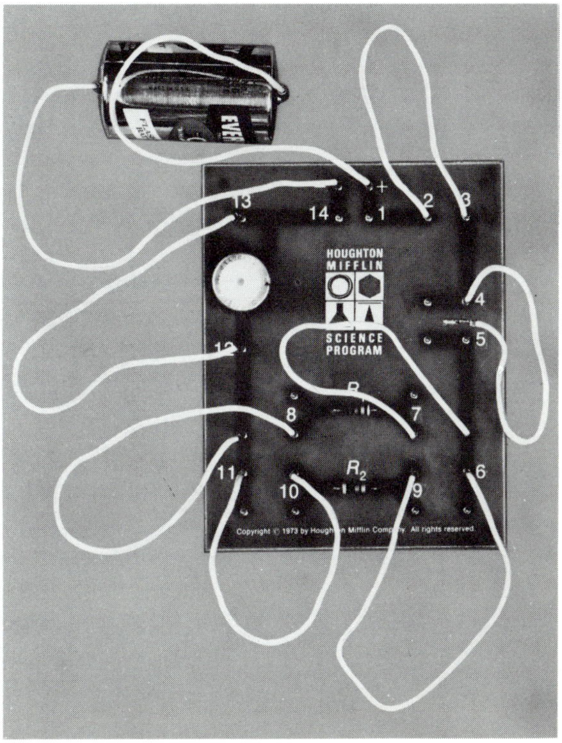

Figure 8–17
Circuit for Experiment 8–6.

Voltmeter: across *AB*: + terminal to 1, – terminal to 14
across *CD*: + terminal to 6, – terminal to 11
across *EF*: + terminal to 7, – terminal to 8
across *GH*: + terminal to 9, – terminal to 10

Measure and record the voltage drops across *AB, CD, EF,* and *GH.*

Questions

1. What can you say about the voltage drops across different resistors in parallel?

Rheostat

A **rheostat** (REE-oh-stat) is a resistor whose resistance value can be varied. Hence, a rheostat is used to vary the resistance to the flow of electrons through a circuit. Many rheostats consist of a coil of wire with a high resistance. A movable part makes contact with any desired point on the resistance wire. The position of the movable contact determines the length of the resistance wire in the circuit. The length of resistance wire determines the resistance to the current in the circuit.

experiment 8–7 Using a Rheostat

Materials

circuit board (or rheostat and resistor, R_1); voltmeter; ammeter; dry cell; connecting wires

Procedure

Assemble the electric circuit as shown in Figure 8–18. It includes a dry cell, an ammeter, a voltmeter, a rheostat, and a fixed resistor (R_1).

The components should be attached to the circuit board as follows:
Dry cell: + terminal to 1, – terminal to 14
Ammeter: + terminal to 2, – terminal to 3
R_1: connecting wire from 6 to 7
connecting wire from 8 to 11
Voltmeter: + terminal to 7, – terminal to 8

The ammeter is connected in series in the circuit. The voltmeter is connected in parallel across the resistor whose potential difference it measures. Turn the knob of the rheostat so that it offers maximum resistance to the current.

Figure 8–18
Circuit for Experiment 8–7.

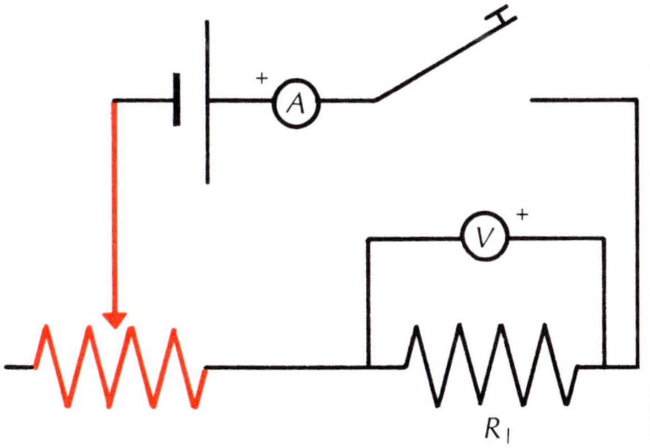

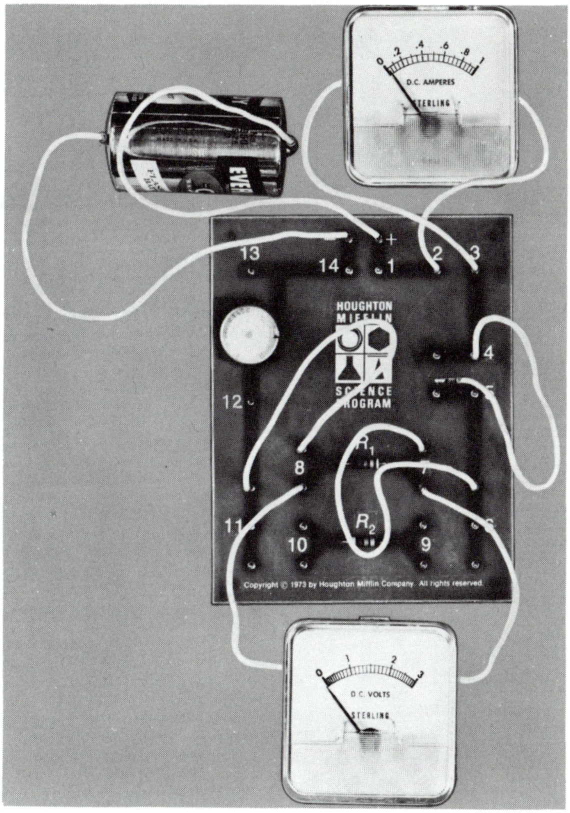

Now ask your teacher to check your circuit. When it is approved, *but not before,* connect a wire from 4 to 5. Read each meter and record the readings. Disconnect the wire from 4 to 5.

Decrease the resistance of the rheostat by moving the contact about one half the distance to the opposite position. Again read each meter and record the readings.

Draw a symbolized diagram of your circuit.

Questions

1. What happened to the current as the resistance was decreased?

2. What happened to the voltage drop across the resistor as the resistance of the rheostat was decreased?

3. What happened to the current as the voltage drop across the resistor increased?

Computations

1. (a) What is the voltage drop across the fixed resistor? (b) What is the current in the circuit for the first setting of the rheostat? (c) Compute the ratio of these two values; that is, divide the voltage drop by the current.

2. Compute the ratio of the voltage drop across the fixed resistor to the current in the circuit for the second setting of the rheostat.

3. (a) Compare the ratios obtained in Computations 1 and 2. (b) Give a reason for your answer.

Amperes, Volts, and Ohms

Electric current is measured in **amperes** (AM-peerz). The reading of an ammeter is expressed in amperes. The usual symbol for current in amperes is the letter **I.**

Voltage, or voltage drop, is measured in **volts.** The reading of a voltmeter is expressed in volts. The usual symbol for voltage in volts is the letter **E.**

The resistance of a resistor is expressed in units called **ohms** (OHMS). The usual symbol for resistance in ohms is the letter **R.**

The ratio of E to I represents the resistance.

Example: If the voltage drop across a resistor is 1.5 volts and if the current through the resistor is 0.50 ampere, the resistance of the resistor is

$$R = \frac{E}{I} = \frac{1.5 \text{ volts}}{0.50 \text{ ampere}} = 3.0 \text{ ohms}$$

experiment 8–8

The Relationship between Current and Voltage Drop for a Resistor

Materials

circuit board (or rheostat and 2 resistors, R_1 and R_2); ammeter; voltmeter; dry cell; connecting wires

A. Using the Resistor R_1

Procedure

Arrange the circuit as shown in Figure 8–19, using the resistor R_1 and the rheostat. Attach components to the circuit board as in Experiment 8–7.

Measure and record the voltmeter and ammeter readings for five settings of the rheostat. The ammeter readings should include the largest and the smallest values of the current (obtained from the lowest and highest settings of the rheostat), as well as three values in between. Make a table for the readings as in Figure 8–20.

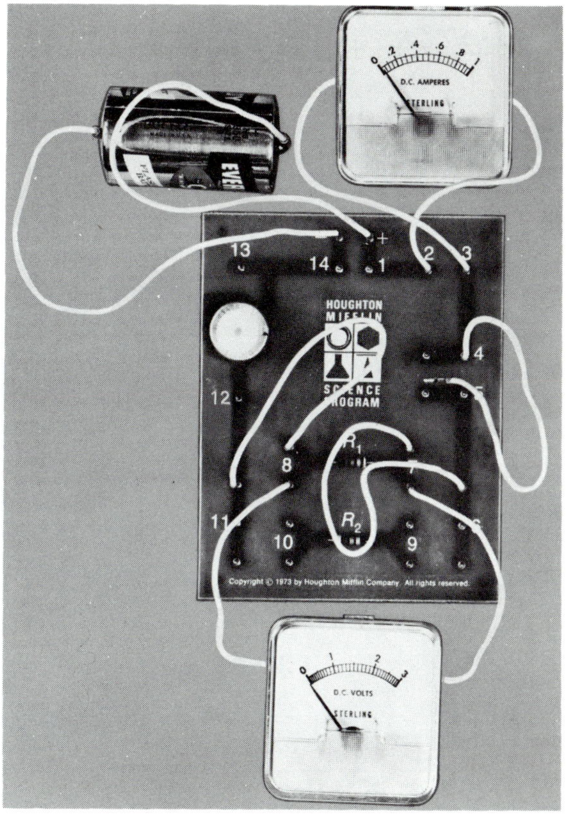

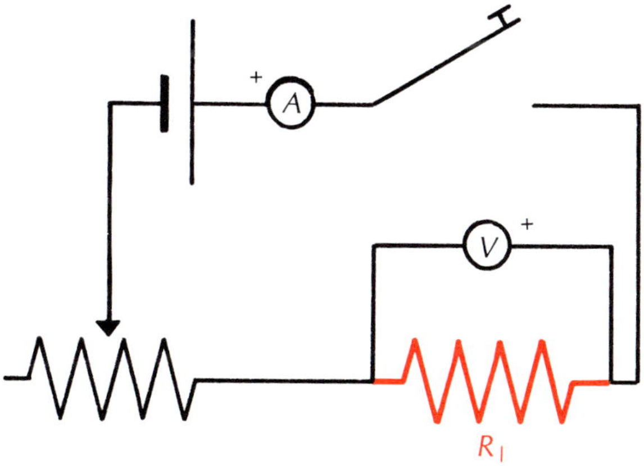

Figure 8–19
Circuit for Experiment 8–8.

B. Using the Resistor R_2

Procedure

In this part of the experiment, you will use R_2 instead of R_1. The only changes in the circuit are the attachment of the connecting wires for the fixed resistor and the voltmeter.

R_2: connecting wire from 6 to 9
connecting wire from 10 to 11

Voltmeter: + terminal to 9, – terminal to 10

Obtain a set of five values of E and I for R_2. Compute the ratio of E to I for each pair of readings and enter the results in the table. Finally, compute the average E/I ratio for each resistor.

Graph

Prepare a full-page graph of the data, plotting current values on the horizontal axis and voltage values on the vertical axis. Select suitable scales so that both sets of values (values for R_1 and for R_2) can be entered on the same graph.

Resistor	Trial	E in volts	I in amperes	E/I in ohms
R_1	1	?	?	?
	2	?	?	?
	3	?	?	?
	4	?	?	?
	5	?	?	?
		Average value of E/I = ?		
R_2	1	?	?	?
	2	?	?	?
	3	?	?	?
	4	?	?	?
	5	?	?	?
		Average value of E/I = ?		

Figure 8–20
Data table for Experiment 8–8.

Questions

1. For resistor R_1, how do the five values of E/I compare?

2. For resistor R_2, how do the five values of E/I compare?

3. How does the average E/I value for R_1 compare with the average E/I value for R_2?

4. Why are R_1 and R_2 called fixed resistors?

5. As the voltage drop across R_1 is increased, what happens to the current?

6. As the voltage drop across R_2 is increased, what happens to the current?

7. State in your own words the relationship between E and I for a fixed resistor. (A scientist named Ohm was the first person to discover and state this relationship.)

8. By reference to the graph, (a) state the current through R_1 if the voltage drop across R_1 is 0.50 volt, and (b) state the voltage drop across R_2 if the current through R_2 is 0.50 ampere.

A Flashlight Bulb

A flashlight bulb contains a short length of resistance wire, called a **filament** (FILL-uh-ment). The filament is connected to terminals outside the airtight bulb. The high resistance of the filament opposes the flow of electrons. As a result, work done by the dry cell in forcing electrons through the wire is changed into heat. That is, the filament gets hot and glows. At first it glows a dull red, then bright red, then orange, then yellow, until eventually it becomes white hot. There is a relationship between the color of a hot wire and its temperature. This relationship is shown in Figure 8–21.

Figure 8–21
The relationship between the temperature and the color of a flashlight bulb filament.

Color of Filament	*Approximate Temperature of Filament*
dull red	500°C
bright red	650°C
orange	900°C
yellow	1000°C
white	1500°C

experiment 8–9

The Relationship between Current and Voltage Drop for a Flashlight Bulb Filament

Materials

circuit board; ammeter; voltmeter; rheostat; dry cell; flashlight bulb and socket; connecting wires

Procedure

Arrange the circuit as shown in Figure 8–22. The circuit is the same as the one in Figure 8–19, except that the fixed resistor is replaced by a flashlight bulb. Attach the components to the circuit board as follows:

Dry cell: + terminal to 1, – terminal to 14
Ammeter: + terminal to 2, – terminal to 3
Flashlight bulb: connecting wire from bulb to 6
connecting wire from bulb to 11
Voltmeter: + terminal to 6, – terminal to 11
Connecting wire from 4 to 5

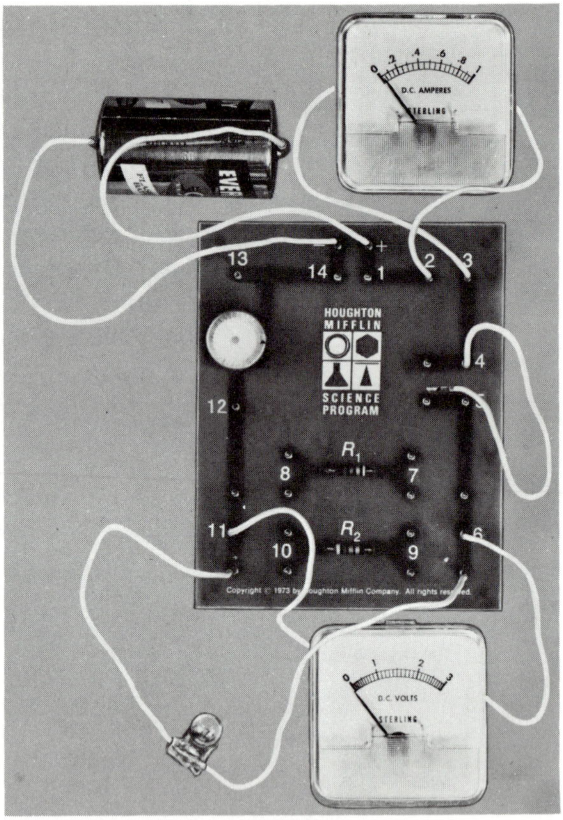

Figure 8–22
Circuit for Experiment 8–9.

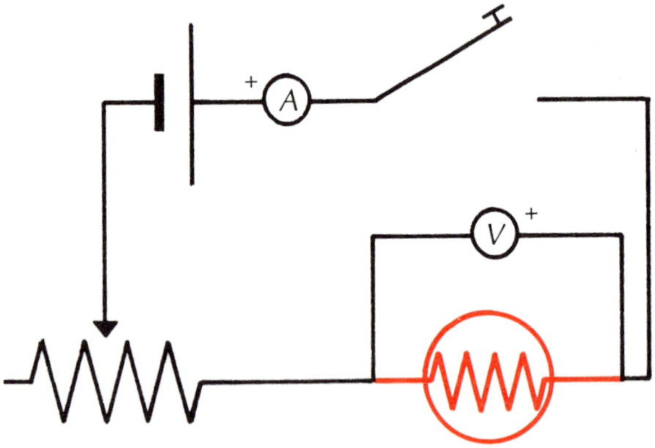

Obtain five pairs of values for voltage drop (E) and current (I). Tabulate the data and then compute the resistance (the E/I ratio) for each pair of values.

Make a graph from your data.

Questions

1. What does the line of the graph mean?

2. What happens to the resistance of the filament of the bulb as its temperature rises?

3. Would you be justified in computing the average value of the resistance of the filament? Give a reason for your answer.

experiment 8–10 The Resistance of Two Resistors in Series and in Parallel

Materials

circuit board or 2 resistors, R_1 and R_2; ammeter; voltmeter; dry cell; connecting wires

Procedure

In this experiment you will compute the resistance of resistors in series and in parallel from readings of the ammeter and the voltmeter.

Arrange the apparatus as shown in Figure 8–23. The resistance to be measured, either a single resistor or a combination of resistors, is connected between points *B* and *C*.

The components should be attached to the circuit board as follows. This basic circuit is used in parts A, B, and C of the experiment.

Dry cell: + terminal to 1, − terminal to 14
Ammeter: + terminal to 2, − terminal to 3
Voltmeter: + terminal to 6, − terminal to 11
Connecting wire from 4 to 5
Connecting wire from 12 to 13

Using the two resistors, R_1 and R_2, proceed as follows. Record the experimental data in your notebook as a table.

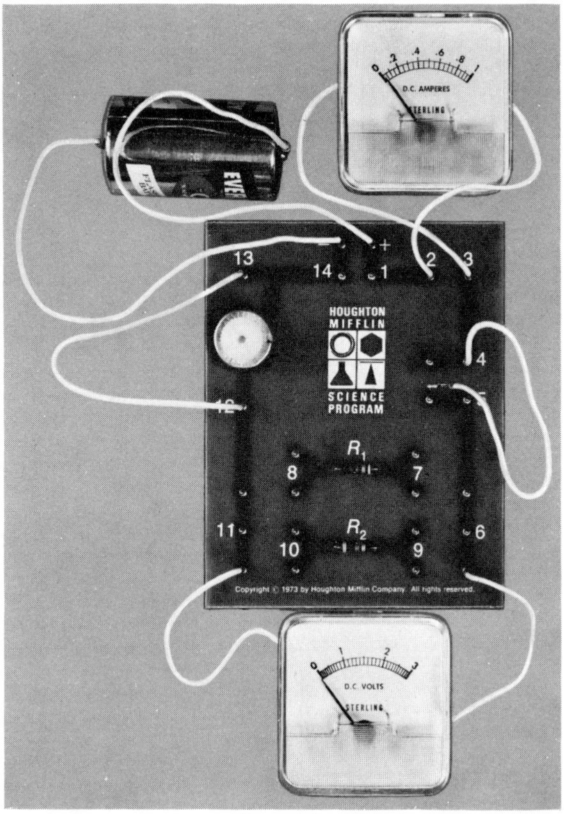

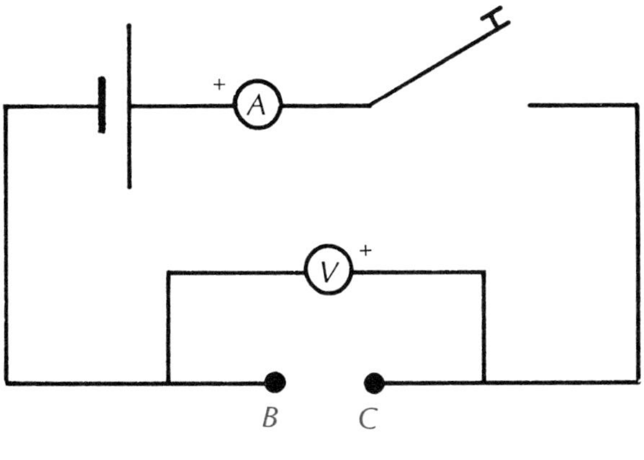

Figure 8–23
The basic circuit for Experiment 8–10. The resistance to be measured should be connected between points *B* and *C* (posts 6 and 11 on the circuit board).

A. The Resistance of Each of the Two Resistors

Attach resistor R_1 between points B and C. The components should be attached to the basic circuit as follows:

R_1: connecting wire from 6 to 7
connecting wire from 8 to 11

From the readings of the voltmeter and ammeter, find the resistance of R_1 in ohms.

Find the resistance of R_2 in the same way.

R_2: connecting wire from 6 to 9
connecting wire from 10 to 11

B. The Resistance of the Two Resistors in Series

Attach resistors R_1 and R_2 between points B and C as shown in Figure 8–24.

The components should be attached to the basic circuit as follows:

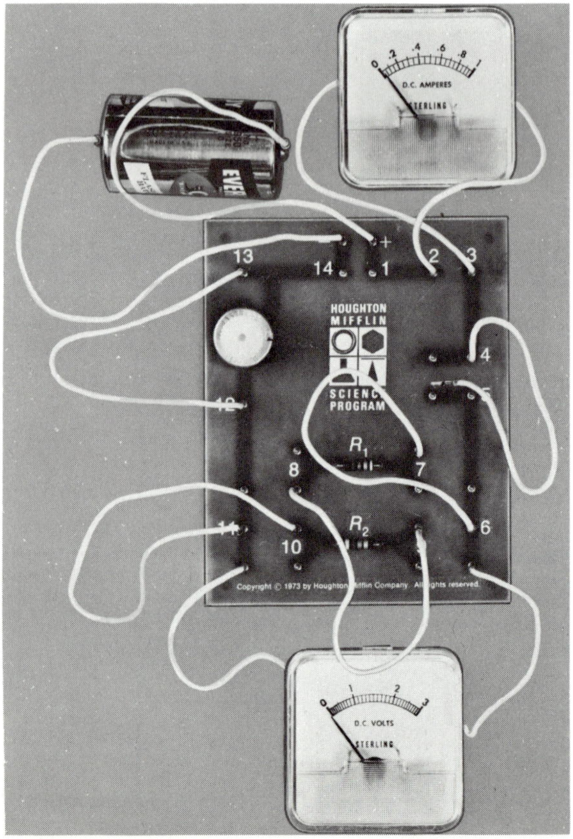

Figure 8–24
Connect the resistors like this for Experiment 8–10B.

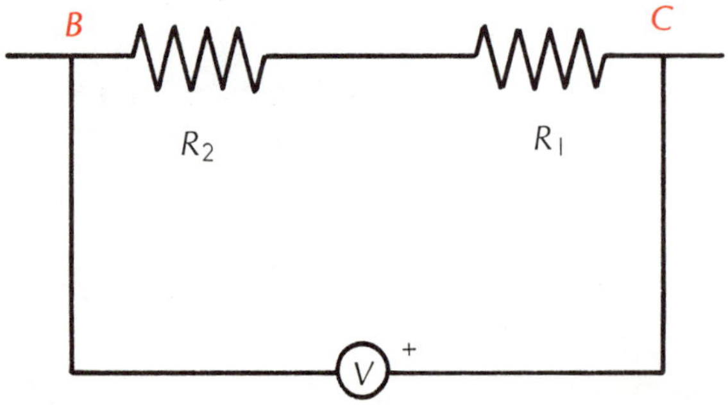

Connecting wire from 6 to 7
Connecting wire from 8 to 9
Connecting wire from 10 to 11

Find the resistance of the combination of the resistors in series. Now compute the resistance of the two resistors in series from the separate resistance values as found in part A.

Compute the percentage difference between the experimental value in part B and the value computed from the values in part A.

Questions

1. Write a general equation to express the resistance of two or more resistors in series.

C. The Resistance of Two Resistors in Parallel

Procedure

Attach resistors R_1 and R_2 between points B and C as shown in Figure 8–25.

Figure 8–25
Connect the resistors like this for Experiment 8–10C.

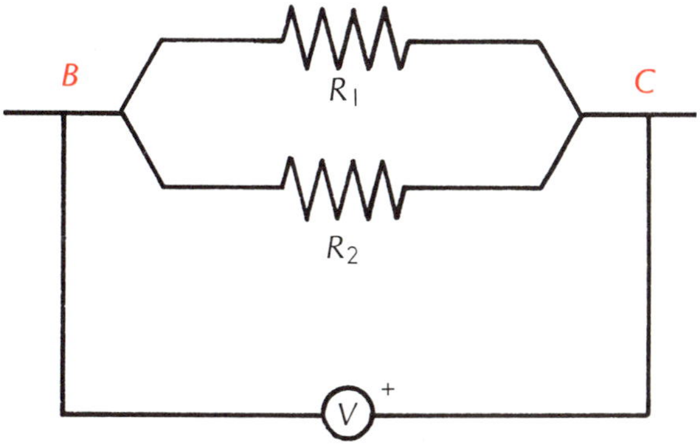

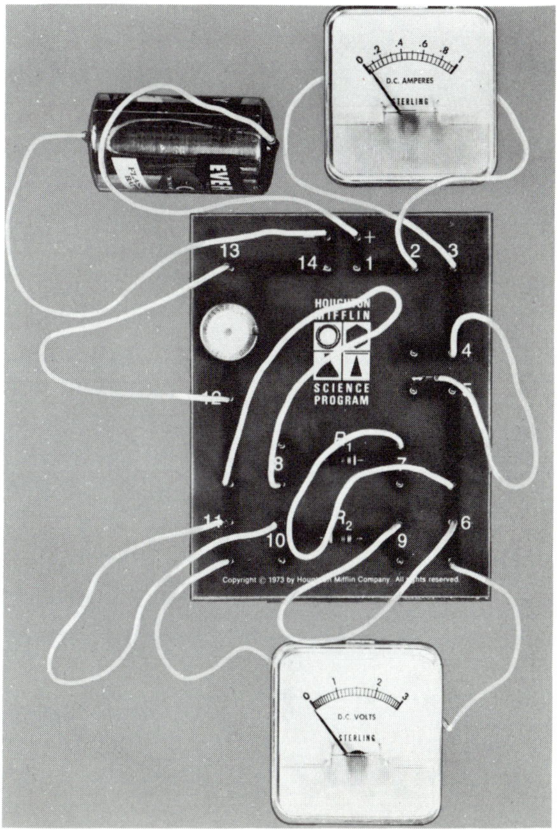

The components should be attached to the basic circuit as follows:
Connecting wire from 6 to 7
Connecting wire from 8 to 11
Connecting wire from 6 to 9
Connecting wire from 10 to 11

From readings of voltmeter and ammeter, compute the resistance of the resistors arranged in parallel.

Questions

1. Compare the resistance of the resistors in series and in parallel.
2. How do you account for the difference in resistance?

RESISTANCE OF A WIRE

Factors Which Might Affect the Resistance of a Wire

There are several factors which might affect the resistance of a wire. These factors include (a) length of the wire, (b) the cross-sectional area, called the **gauge** (GAYJ) of the wire, (c) the material of which the wire is made, and (d) the temperature of the wire.

You studied the effect of temperature change on the resistance of a flashlight bulb filament in Experiment 8–9. What did you conclude about the resistance of the wire as it became hotter?

You will now study the other three factors in Experiments 8–11, 8–12, and 8–13. In each experiment, assemble the circuit as shown in Figure 8–26. You will attach the wires to be studied to points *B* and *C* in the circuit.

experiment 8–11

The Resistances of Wires of Different Lengths

Materials

circuit board; voltmeter; ammeter; dry cell; 2 alligator clips; ruler; connecting wires; gauge 20 nichrome wire, 21 cm in length

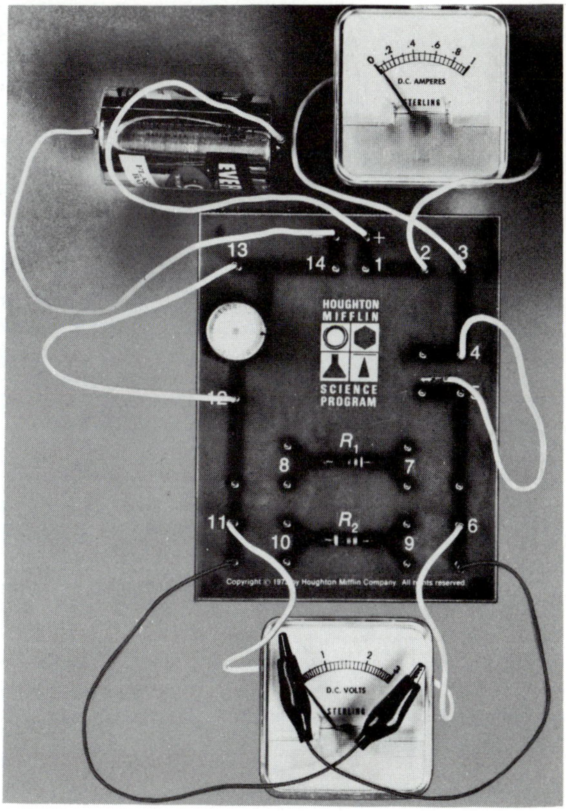

Figure 8–26
The basic circuit for Experiments 8–11, 8–12, and 8–13. The wire to be tested should be connected between points *B* and *C*.

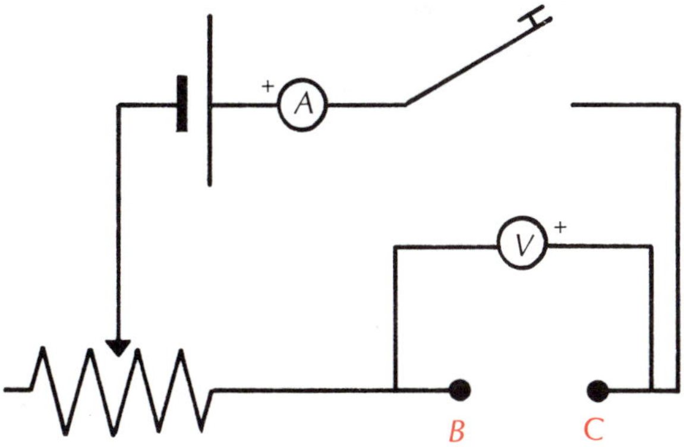

In this experiment you will compare the resistances of two known lengths of **nichrome** (NY-krohm) wire. Nichrome is not an element like copper but is an alloy of nickel and chromium.

Set up the circuit as in Figure 8–26. You will use the same piece of nichrome wire for the two measurements. First, use only 10 cm of the length of the wire in the circuit. To do this, attach one of the alligator clips to the left end of the wire and attach the other clip on the wire at a distance of 10 cm from the first clip. The arrangement is shown in Figure 8–27.

The components should be attached to the circuit board as follows:

Dry cell: + terminal to 1, – terminal to 14
Ammeter: + terminal to 2, – terminal to 3
Voltmeter: + terminal to 6, – terminal to 11
Connecting wire from 12 to 13
Connecting wire from 4 to 5
Nichrome wire: connecting wires and alligator clips to 6 and 11

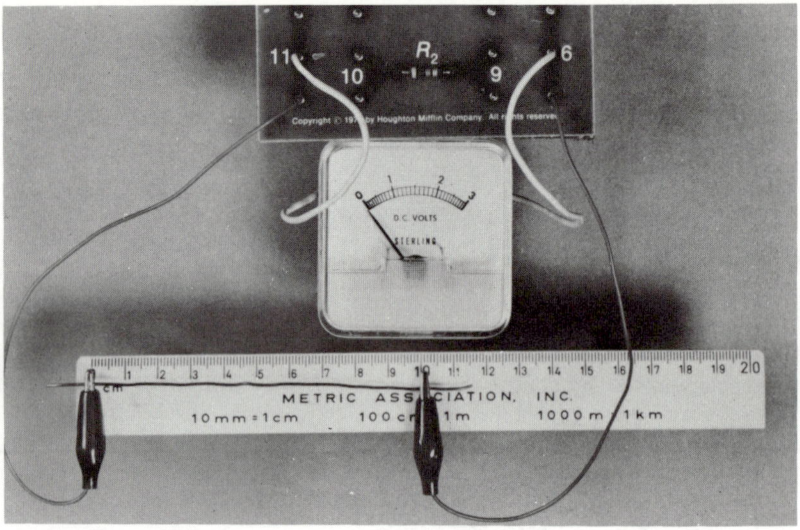

Figure 8–27
Be sure that the wire is straight, and that the alligator clips are exactly 10 cm apart.

Take readings of the voltmeter and ammeter. Compute the resistance of the 10-cm length of wire by finding the *E*/*I* ratio.

Now move the alligator clip on the right until the length of wire between the two clips is 20 cm. Again read the ammeter and the voltmeter. Compute the resistance of the 20-cm length of wire.

Questions

1. Is the length of a wire one of the factors that affects its resistance? Give a reason for your answer.

2. What is the relationship between the length of a wire and its resistance?

3. If the length of a wire is decreased to $\frac{1}{5}$ of its former length, by how much is the resistance changed?

experiment 8–12 The Resistances of Wires with Different Cross-Sectional Areas

Materials

circuit board; voltmeter; ammeter; dry cell; 2 alligator clips; ruler; 11-cm length of gauge 28 nichrome wire; connecting wires

Procedure

In Experiment 8–11 you used nichrome wire of gauge 20. In this experiment you will use nichrome wire of gauge 28. Then you will compute the resistance of nichrome wires of the same length but with different cross-sections. The gauge number of a wire is an indication of its area of cross-section. The higher the number, the smaller is the area of cross-section.

Set up the circuit as in Figure 8–26. The circuit board attachments are given in Experiment 8–11. Use two alligator clips to attach the gauge 28 nichrome wire to points *B* and *C* of the circuit. Adjust the clips so that the length of the wire in the circuit is 10 cm.

Read the ammeter and the voltmeter. Compute the resistance of the 10-cm length of wire from the *E*/*I* ratio. Then calculate the resistance of a 1-cm length of gauge 28 nichrome wire.

Using the data of Experiment 8–11, calculate the resistance of a 1-cm length of gauge 20 nichrome wire.

Questions

1. Which wire has the greater resistance, gauge 20 or 28?

2. Does the cross-sectional area of a wire affect its resistance?

3. What is the relationship between the cross-sectional area of a wire and its resistance?

4. What is the relationship between the gauge number of a wire and its resistance?

5. What is the relationship between the resistance of a wire and the diameter of the wire?

experiment 8–13 The Resistances of Wires of Different Metals

Materials

circuit board; voltmeter; ammeter; dry cell; 2 alligator clips; connecting wires; 15-m coil of gauge 28 copper wire, insulated with plastic or lacquer

Procedure

In Experiment 8–12 you found the resistance of gauge 28 nichrome wire. In this experiment you will find the resistance of a 15-m coil of gauge 28 copper wire. Then you will compare the resistances of the two different wires of the same length and the same cross-sectional area.

Set up the circuit as in Figure 8–26. The circuit board attachments are given in Experiment 8–11. Use alligator clips to attach the ends of the coil of gauge 28 copper wire to points *B* and *C* of the circuit. Obtain ammeter and voltmeter readings, and then compute the resistance of the coil of copper wire. Then calculate the resistance of a 1-cm length of the gauge 28 copper wire. Remember that 15 m = 1500 cm.

Questions

1. Which offers the greater resistance, a 1-cm length of gauge 28 copper wire or a 1-cm length of gauge 28 nichrome wire?

2. Is the material of a wire one of the factors that affects its resistance?

3. A good electric conductor is one with a low resistance. Which is the better electric conductor, nichrome or copper?

experiment 8–14

The Resistance of a Wire

Laboratory Test

Materials

the same as for Experiment 8–13, but substitute a 16-cm length of gauge 24 nichrome wire for the coil of gauge 28 copper wire

Procedure

The circuit board attachments needed are given in Experiment 8–11. Follow the procedure you used in Experiment 8–12. Place a 15-cm length of gauge 24 nichrome wire in the jaws of the alligator clips between points *B* and *C* of the circuit. Obtain voltmeter and ammeter readings, compute the *E*/*I* ratio, and then compute the resistance of a 1-cm length of the wire.

Ask your teacher for the correct value for the resistance of a 1-cm length of gauge 24 nichrome wire and record this value in your notebook. Compute your percentage error.

Questions

1. Compare the resistance of a 1-cm length of the gauge 24 nichrome wire with the resistance of a 1-cm length of gauge 20 nichrome wire as computed in Experiment 8–11. Is this what you expect?

2. Compare the resistance of a 1-cm length of the gauge 24 nichrome wire with the resistance of a 1-cm length of gauge 28 nichrome wire as computed in Experiment 8–12. Is this what you expect?

3. How does the cross-sectional area of the gauge 24 nichrome wire compare with the cross-sectional area of (a) gauge 20 nichrome wire, (b) gauge 28 nichrome wire?

PERSPECTIVES

perspective 8–1

A Voltaic Cell

A voltaic cell usually consists of two different metals, one more active than the other, in a solution that contains certain kinds of ions. The metals are called the terminals of the cell. In a voltaic cell there are two reactions, one at each terminal. In the zinc-copper cell, the reaction at the zinc terminal is

$$Zn = Zn^{2+} + 2e^-$$

The reaction at the copper terminal is

$$2H^+ + 2e^- = H_2$$

The two reactions above can be thought of as the two halves of the whole reaction. The whole reaction is

$$Zn + 2H^+ + 2e^- = Zn^{2+} + 2e^- + H_2$$

Omitting the two electrons, which appear on both sides of the equation, the simplest equation for the reaction is

$$Zn + 2H^+ = Zn^{2+} + H_2$$

Since electric energy is released in this reaction, the equation is better written as

$$Zn + 2H^+ = Zn^{2+} + H_2 + \text{energy}$$

That is, the potential chemical energy of the products (zinc ions and hydrogen) is less than the potential chemical energy of the reactants (zinc and hydrogen ions). The difference in potential energy between reactants and products is transformed into electric energy and released.

The zinc-copper cell is only one of many kinds of voltaic cells. One of the most familiar is the dry cell. You will use dry cells as a source of current in some of your experiments. Still another familiar voltaic cell is the storage battery used in an automobile to start the engine.

Voltaic cells are a ready source of electricity in small amounts. But they cannot provide electricity on a large scale. To provide vast amounts of energy continuously, such as the electricity needed by a city, a generator is used. A **generator** is a machine that changes kinetic energy into electric energy. To do this, a coil of wire is rotated in a magnetic field. The energy needed to rotate the coil is usually supplied by steam or by running water. The magnetic field then changes the kinetic energy of the rotating coil into electric energy.

perspective 8–2

An Electric Circuit

A circuit is the path along which electrons are able to flow. For current to flow, the circuit must be complete or unbroken. In a complete circuit, a voltaic cell operates like an electron pump. For example, a zinc-copper cell pumps electrons continuously along the wire from the zinc terminal to the copper terminal. (Look back at Figure 8–2.)

Since electrons flow out of the zinc terminal, it is called the negative (–) terminal. Since electrons flow into the copper terminal, it is the positive (+) terminal of the cell. Thus the direction of the electron flow is from the negative terminal to the positive terminal along the wire.

It is important that electrons flow through an ammeter in the right direction. The negative terminal of the cell must be connected to the negative terminal of the ammeter. That is, when the ammeter is correctly connected, electrons enter it through the negative terminal.

perspective 8–3

A Dry Cell

The most common voltaic cell is the dry cell. It is used in flashlight batteries, transistor radios, and frequently as a source of current in experimental work.

A dry cell consists of a zinc case and a central rod of carbon. Carbon, like copper, is chemically less active than zinc. However, it is a non-metal, not a metal. Inside the zinc case is a paste of various chemicals. The top of the cell is sealed with wax or pitch to prevent the water in the paste from evaporating. If the paste dried out, the chemical reaction would stop. One terminal of the cell is attached to the zinc case and the other is attached to the carbon rod as shown in Figure 8–28.

When the terminals are connected by a wire, the zinc reacts with the chemicals. Electrons flow from the zinc terminal to the carbon terminal. Therefore the zinc terminal is negative (–) and the carbon terminal is positive (+).

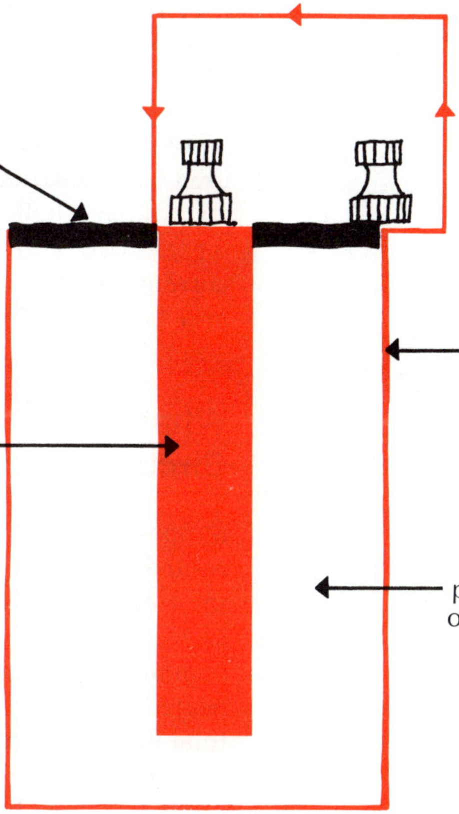

Figure 8–28
A dry cell contains a paste of chemicals. Although it is not really dry, it is sealed to prevent evaporation.

perspective 8–4

Amperes and Volts

An electric pressure is needed to force electrons through a wire. The electric pressure is due to a difference of chemical activity between the terminals of the cell. This difference in activity is called voltage. The voltage of a cell is generated by the chemical reaction within the cell. The voltage forces electrons along the connecting wire and so produces a current.

Voltage is measured by a voltmeter and recorded in volts. The voltage of a zinc-copper cell is about 1 volt. The voltage of a dry cell is about 1.5 volts. An automobile storage battery delivers about 12 volts (about 2 volts per cell). The circuit that provides light and heat in your home operates on about 110 volts.

A stream of electrons flowing in a wire is an electric current. The unit of current is called an ampere. A current of one ampere means that a certain number of electrons (actually 6.25×10^{18} or 6,250,000,000,000,000,000 electrons) flow past a point in the wire every second. An electric current is measured by an instrument called an ammeter. The current through an ordinary light bulb on a house electrical circuit is about 0.5 ampere. An electric toaster takes 5 or 6 amperes. The current supplied by an automobile battery to start the engine may be several hundred amperes.

perspective 8–5

What is a Series Circuit?

Suppose that three light bulbs and three dry cells are connected as shown in Figure 8–29. For each of the three cells the zinc terminal of one cell is connected to the carbon terminal of the next. The three cells are then connected in series. A combination of cells in series is called a **battery.** When connected in series, each cell makes its full contribution to the total voltage of the battery. The total voltage of three dry cells in series is therefore 3×1.5 volts, or 4.5 volts.

The light bulbs are also connected in series. That is, the total current passes through each of them, one after the other. Since an ammeter measures the total current in a circuit, it too must always be connected in series. The positive (+) terminal of the ammeter must be connected to the + terminal of the

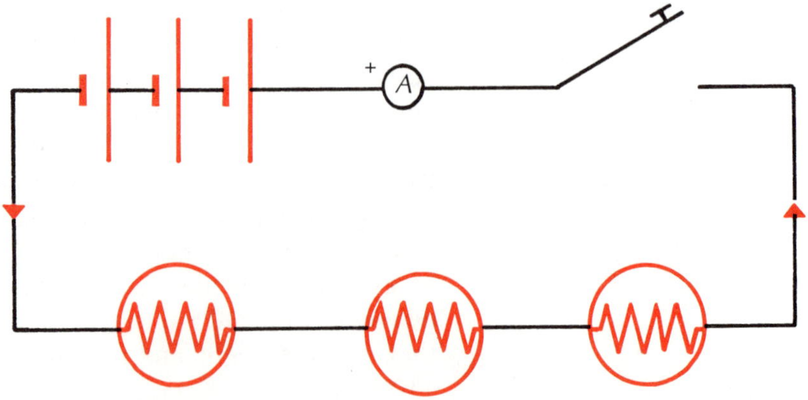

Figure 8–29
These three light bulbs are connected in series. The dry cells are also connected in series.

cell (or battery). If the + terminal of the ammeter is connected to the negative (+) terminal of the cell or battery, the pointer moves backwards off the scale and the ammeter may be damaged.

perspective 8–6

What is a Parallel Circuit?

Suppose that three light bulbs are connected as shown in Figure 8–30. The main current (I) divides into three branches at point A and then re-forms the main current at point B. That part of the main current that flows through the upper light bulb is represented by the symbol I_1. The current that flows through the central bulb is I_2. The current that flows through the lower bulb is I_3. The three light bulbs are said to be arranged in parallel. In other words, the light bulbs in a parallel circuit are so arranged that only a part of the main current flows through each.

In the electric circuit of a house, light bulbs are arranged in parallel, not in series. Why? When a bulb burns out, the filament breaks. If the bulbs were connected in series, a burned-out bulb would break the circuit completely, so no current would flow. No light would be emitted, even by the good bulbs. However, in a parallel circuit a burned-out bulb would break the current only in its own branch. The bulbs in the other branches would still light up.

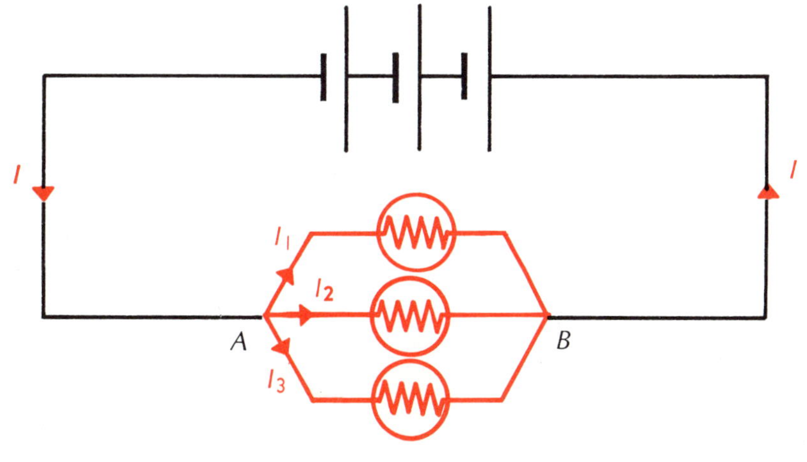

Figure 8–30
These three light bulbs are connected in parallel.

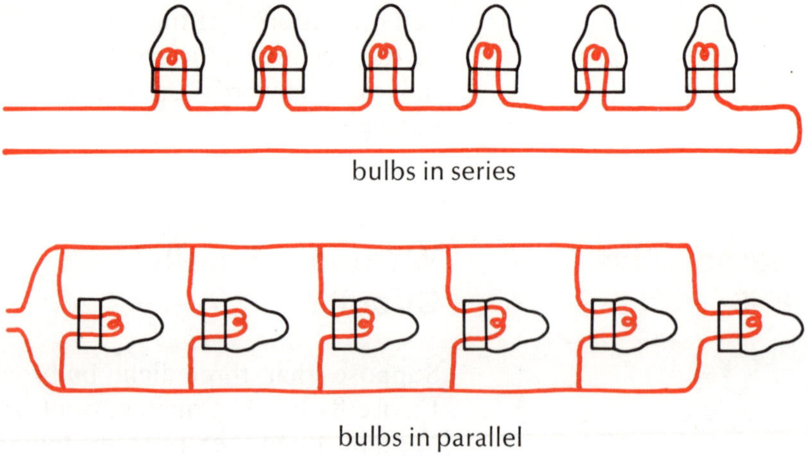

Figure 8–31
Light bulbs can be connected either in series or in parallel. In each arrangement, what happens to the current when a bulb burns out?

Christmas tree lights used to be wired in series, but now they are wired in parallel. Series and parallel connections of Christmas tree lights are shown in Figure 8–31.

perspective 8–7

Resistance

A. Resistors and Conductors

When the terminals of a voltaic cell are connected by a copper wire, electrons escape through the negative terminal and are pushed from atom to atom within the wire. Some atoms in a greatly magnified piece of connecting wire are shown in Figure 8–32. An electron from the cell displaces and replaces an outer electron in a nearby copper atom of the wire. This released electron then displaces an outer electron of another nearby copper atom. In turn, this released electron displaces an outer electron of still another copper atom. And so the process continues through the whole length of the wire. The electrons displaced at the far end of the wire enter the positive terminal.

The ease with which electrons are displaced depends upon the electronic structure of the particular atom. The outer electrons of silver and copper atoms are very easily displaced. That is, silver and copper offer little resistance to the flow of electrons. These metals are called good conductors. On the other hand, atoms of iron, tungsten, and mercury do not readily

Figure 8–32
In a conductor, electrons are transferred from atom to atom. The electrons that come out are not the same ones that went in.

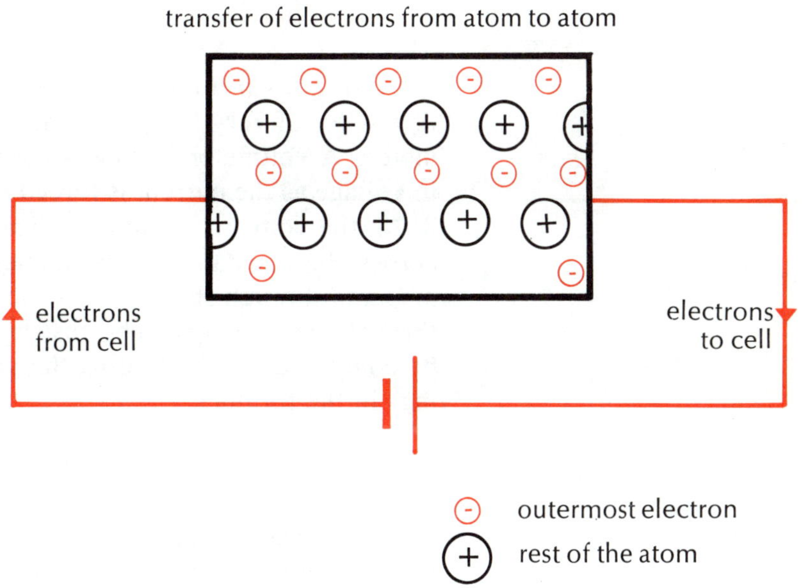

release their outer electrons. These metals offer more resistance to the flow of electrons. The resistance of a wire depends on the material it is made of. Silver has the least resistance of all metals. Expressed another way, silver is the best conductor. If the resistance of silver is represented by 1, the resistance of the other metals can be represented by corresponding numbers. Some of these are listed in Figure 8–33.

Figure 8–33
A comparison of the resistance of some common metals.

Metal	*Relative Resistance*
silver	1.0
copper	1.1
tungsten	3.5
iron	6.3
mercury	60.0

It should be apparent that the resistance of mercury is $\frac{60.0}{1.1}$ or 55 times as great as the resistance of copper, provided that the two metals have the same length, the same cross-section, and the same temperature.

B. Resistance

A fixed (not varying) resistance in a circuit is called a resistor. A resistor of very high resistance is an essential part of a voltmeter. A voltmeter measures the voltage used up or the drop in voltage as the current is forced through a resistor. In Figure 8–34, the voltmeter reading indicates the voltage drop in volts across the resistance. The ammeter indicates the current in amperes through the circuit (which is the same as the current through the resistor). The resistance of the resistor can then be computed by substituting the voltmeter and ammeter readings in the formula

$$R = \frac{E}{I}$$

where R stands for resistance. Resistance is measured in ohms. From the formula it is apparent that 1 ohm is the resistance that allows a current of 1 ampere to flow through the circuit if a voltage of 1 volt is applied to the circuit.

Example: if the voltmeter reads 3.0 volts and the ammeter reads 1.5 amperes, the resistance of the resistor is

$$R = \frac{E}{I} = \frac{3.0 \text{ volts}}{1.5 \text{ amperes}} = 2.0 \text{ ohms}$$

C. Variable Resistors

Sometimes it is desirable to change the current that flows through a circuit. This can be done by using a variable resistor,

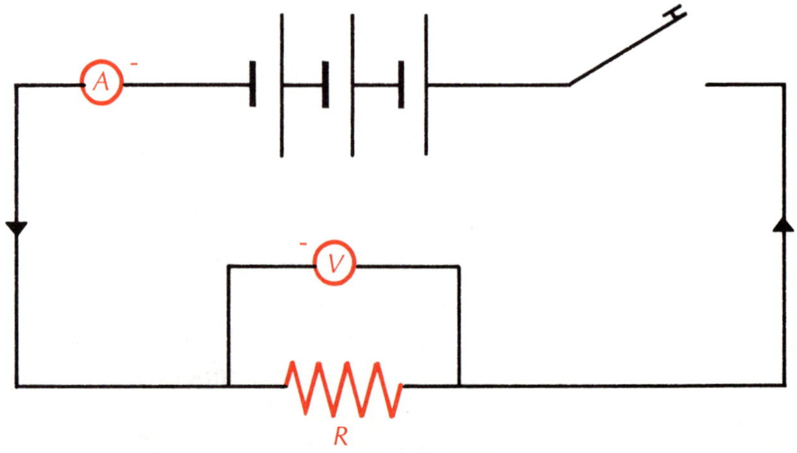

Figure 8–34
The resistance of the resistor can be computed from the voltmeter and ammeter readings.

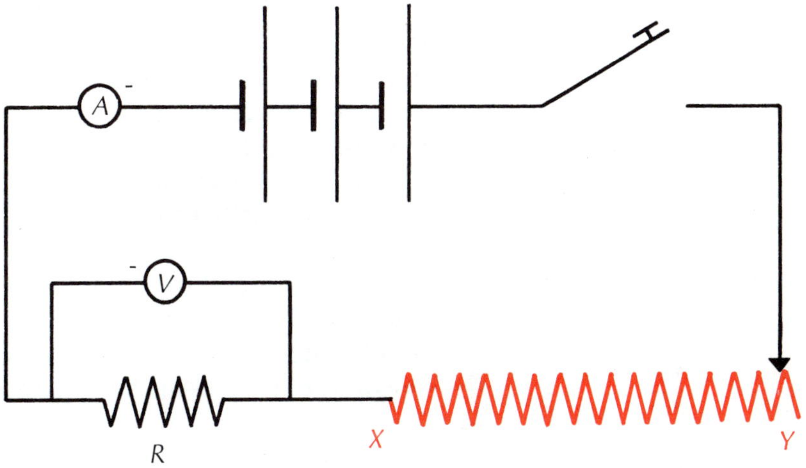

Figure 8–35
In one kind of rheostat, the length of resistance wire in the circuit can be varied.

called a rheostat. A common form of rheostat has a sliding contact as shown in Figure 8–35. If contact is made at *Y*, the rheostat offers its maximum resistance. As a result, the current through the circuit is at a minimum. If contact is made at *X*, the rheostat offers no resistance at all. Then the largest possible current flows through the circuit.

If the contact is moved all the way from *X* to *Y*, the current is steadily decreased from a maximum value at *X* to a minimum value at *Y*. Thus, the current in a circuit can be increased or decreased by a suitable adjustment of the rheostat. The volume-control knobs in many radios and television sets are rheostats.

perspective 8–8 Electricity from Generator to Appliance

It would be difficult to imagine a house without electric appliances. There are electric lights, electric stoves, television sets, percolators, irons, vacuum cleaners, and many more. None of these appliances were possible until the generator was invented in the early part of the 19th century.

As stated in Perspective 8–1, a generator transforms energy from steam or falling water into electric energy. The generator plant may be many miles from the city where the electric energy is used. The electricity is carried by overhead transmission lines to a substation outside the city. From the substation it is carried by wires to factories and houses.

Electric wire must be a good conductor. Silver is too expensive, so wire is made of pure copper. As protection against fire, the wire carrying a current must be well insulated. Cotton, silk, and rubber are usually used for this purpose. The cord that brings electricity to a building contains two copper wires, each insulated from the other. The two wires are then wrapped together with rubber or plastic.

In a modern house there may be five or more separate circuits. Each circuit has a number of wall outlets. The two wires of the circuit pass through the wall and are attached to one of the insulated sections of the outlet. A diagram of an outlet is shown in Figure 8–36. The voltage of a house circuit is usually

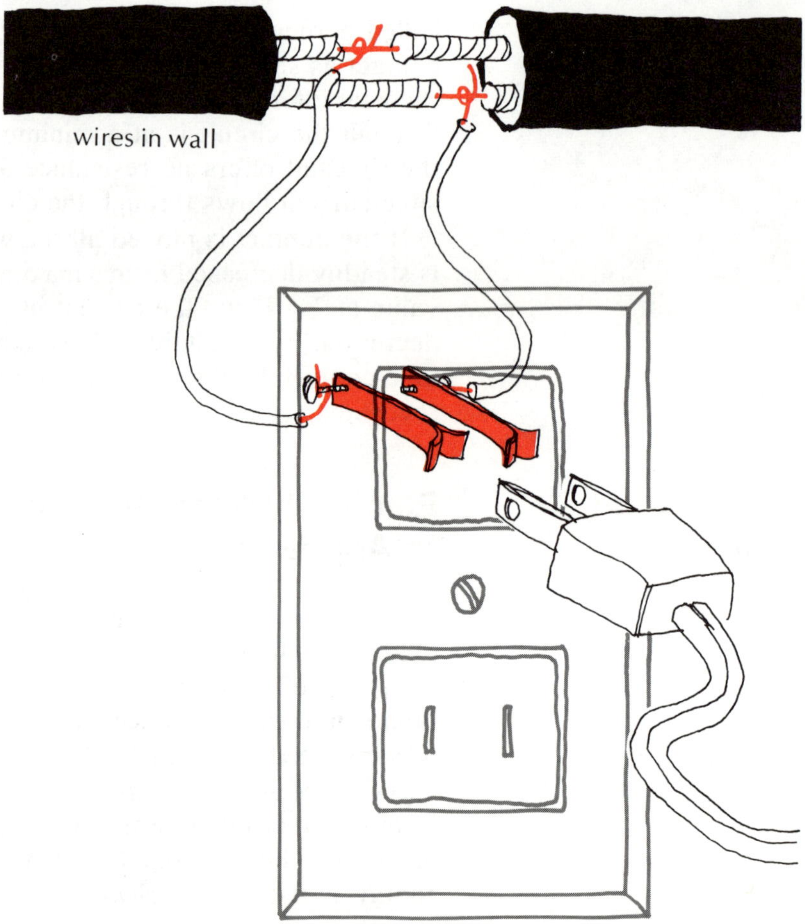

Figure 8–36
An electric outlet is wired like this. What could go wrong to make the outlet not work?

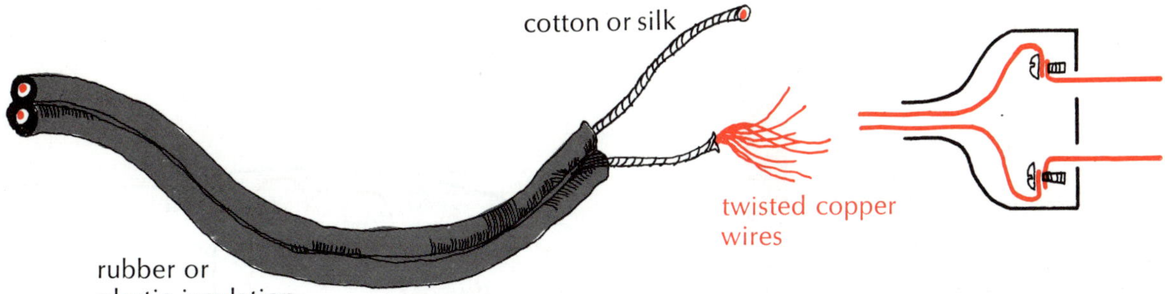

Figure 8–37
An electric cord is insulated to keep the wires from touching. A worn cord should be replaced immediately. An electric plug contains two wires, insulated from each other.

110 volts. If you touched wires at this voltage, you would get a severe shock. As a protection, connection with these wires can be made only by sticking something into the outlet.

In an electric cord, such as an extension cord or the cord attached to a lamp, the wires are made of many thin strands of copper twisted together. Such a wire is shown in Figure 8–37. The twisted strands make the wire flexible. At the end of the cord is a plug like that shown in Figure 8–37. When the prongs of the plugs are pushed into the slits of the wall outlet, the two wires in the cord become connected to the two outlet wires. When they are connected, a current flows.

perspective 8–9

Heat from Electricity

Electricity is a way of getting heat without fire. All electric heating appliances contain high-resistance wire, usually nichrome wire. Remember that the resistance of nichrome is much greater than the resistance of copper. The energy needed to force electrons through a nichrome wire is transformed into heat. The wires in a heating appliance do not get as hot as the filament in a light bulb. You have probably noticed that the nichrome wires in an electric toaster become only red hot, not white hot like a light bulb. In an electric iron the wires are hidden in the base of the iron, as shown in Figure 8–38. Notice that the plug is connected to the iron by two wires.

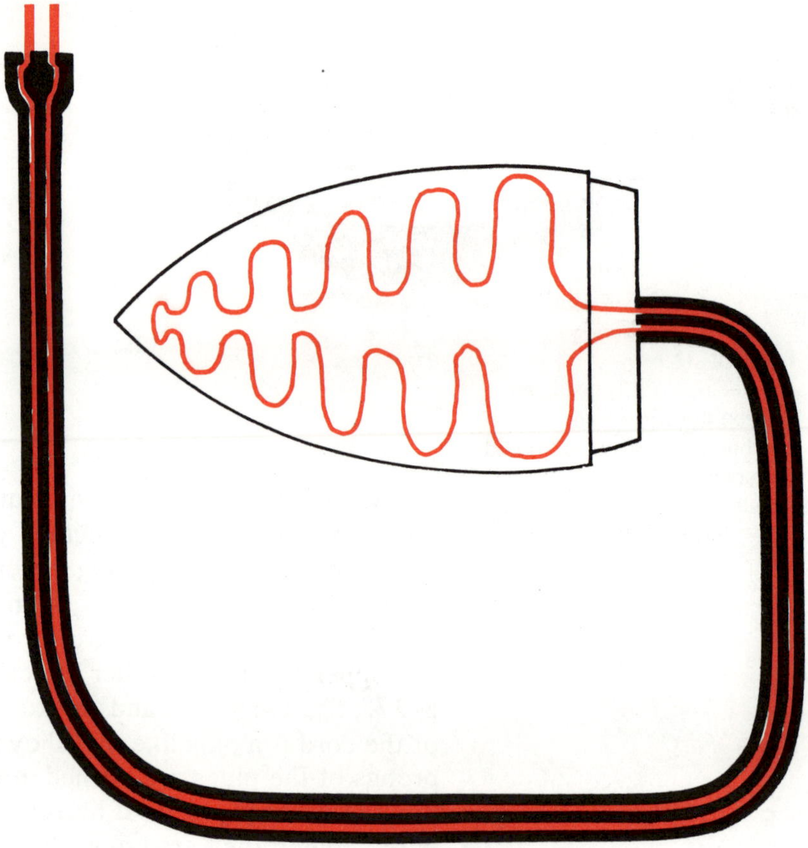

Figure 8–38
The wiring in an electric iron. Follow the path of the current from the wall outlet to the iron and back.

Resistance is not the only factor that determines how much heat is produced in an electric circuit. Another factor is current, sometimes called amperage. If the current is increased by two times, the heat produced in the wire is increased by four times. If the current is increased by three times, the heat produced is increased by nine times. Thus the heat produced is proportional to the square of the current.

The heating effects of a current are used in industry as well as in the home. For example, an electric furnace is used in making steel and glass. The temperature of an electric furnace may be as high as 4000°C, far above the melting points of steel and glass.

What happens when the 110-volt line of the house circuit is attached to a light bulb? The resistance of the bulb is such that about one ampere of current flows through the circuit. An electric iron draws about seven amperes, and a toaster about ten amperes. All these appliances heat up the copper wire in the circuit. If only one appliance is used, the amount of heat released is not great. However, if several light bulbs and several appliances are all plugged into the same circuit at the same time, the wires may become hot enough to cause a fire.

If the cord of an electric iron is worn and frayed, the bare wires may touch, causing a short circuit. A short circuit can be dangerous. Let us see why. As long as the iron is in the circuit, only a small current flows through the wires. However, if the wires touch, the current returns to the source by the easiest path. Since the resistance of the iron is eliminated, the current is greatly increased. To meet the risk of a short circuit and to prevent overloading by plugging in too many appliances, every house circuit has a safety device. One kind of safety device is a circuit breaker, which is an automatic switch that shuts off the current if the circuit is overloaded. Another safety device that does the same job is a fuse.

perspective 8–10

A Fuse

A fuse consists of two parts: a socket and a fuse plug. The socket is inserted in the circuit, and the fuse plug is screwed into the socket. A cross-section of a fuse plug is shown in Figure 8–39.

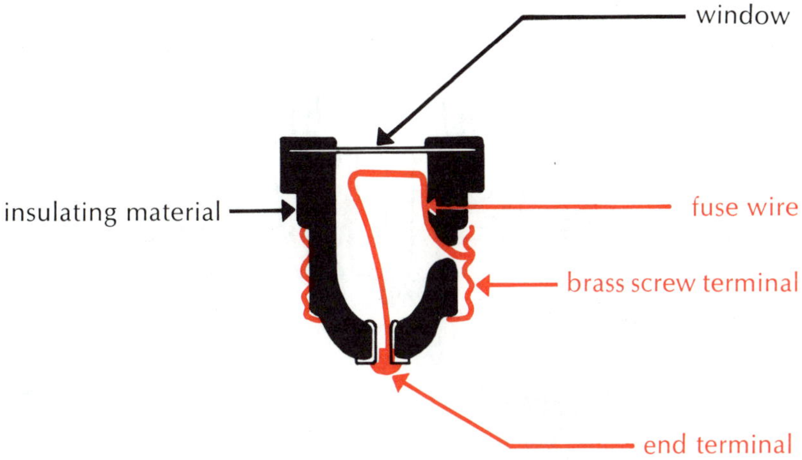

Figure 8–39
A fuse is an important safety device.

A common type of fuse is an insulated container with a window at the top through which the fuse wire can be seen. The fuse wire runs through the container and is attached to two brass terminals on the outside. The fuse wire is a metal strip made of an alloy of lead and tin. This alloy has a low melting point. The melting point is so low that the heat of a burning match is enough to melt fuse wire. The carrying capacity of a

Figure 8–40
Putting too many appliances on the same circuit can overload the circuit and blow a fuse.

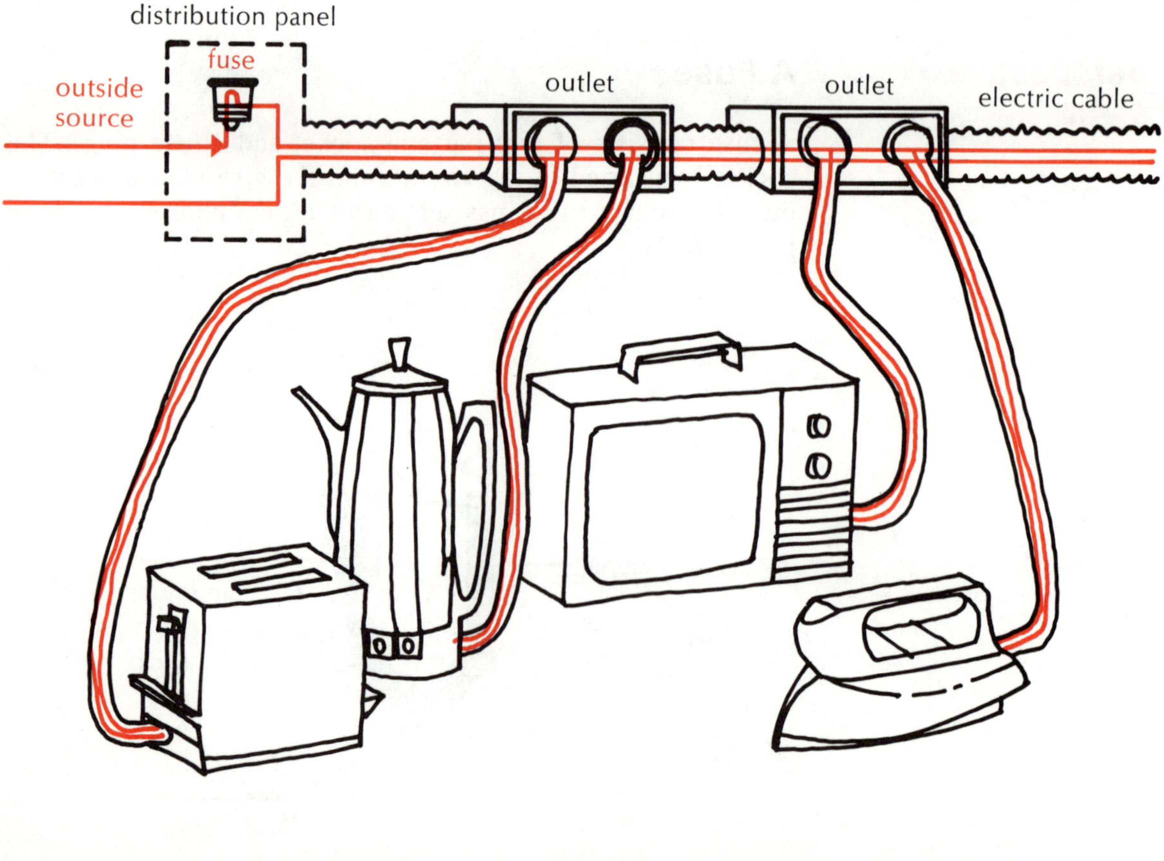

fuse wire is usually stamped on the base of the fuse. For instance, the number 15 means that the fuse can carry a current of 15 amperes without melting. A current greater than 15 amperes will "blow" the fuse wire and break the circuit.

Figure 8–40 shows a number of appliances in a circuit. If all these were turned on at the same time, the circuit would be overloaded and the fuse wire would melt.

perspective 8–11 Electric Motors

So far we have discussed the useful heating effect of an electric current. There is also another useful device which transforms electric energy into kinetic energy. This device is called a motor. A motor is able to do work. Motors are used to run dozens of different kinds of machines.

The scientific principle that led to the motor was discovered in 1830 by two scientists working independently. They were Michael Faraday in England and Joseph Henry in the United States. However, it was not until 1880 that this principle was put to useful purposes and the first motor was built. Today, motors are used to do a wide variety of jobs, both in and out of the home.

In the home, the appliances operated by motors include vacuum cleaners, refrigerators, washing machines, sewing machines, air conditioners, and electric fans. Motor-driven machines are even more numerous outside the home. Elevators are operated by motors, and so are giant cranes and shovels. The wheels of locomotives are turned by electric motors and so are the propellers of ocean liners. Even the gasoline engine of an automobile is started by an electric motor called a starter.

You will learn how a motor works in a later course in science. Actually, a generator and a motor have similar working parts. They are closely related, but operate in opposite directions. A generator transforms kinetic energy into electric energy. A motor transforms electric energy into kinetic energy.

QUESTIONS

A

1. What is a voltaic cell?
2. Write equations for the two chemical reactions in a zinc-copper cell.
3. Is the zinc terminal of a zinc-copper cell positive or negative? Why?
4. What is the main difference between a zinc-copper cell and a dry cell?
5. What is meant by the term *voltage*?
6. What is an ampere?
7. What is (a) a closed circuit, (b) a series circuit?
8. (a) What is meant by the term *resistance?* (b) What is the unit of resistance?
9. How does (a) the length, (b) the area of cross-section affect the resistance of a wire?
10. What is an essential difference in the way a voltmeter and an ammeter are arranged in a circuit?
11. What is (a) a resistor, (b) a variable resistor?
12. Is it better to use copper wire or nichrome wire in wiring a house? Explain.
13. Is it better to use copper wire or nichrome wire in the heating element of an electric toaster? Explain.
14. (a) What is a fuse? (b) What is the use of a fuse in a house circuit?
15. What does the number 15 mean when stamped on a fuse?
16. Both a generator and a motor transform energy. Tell how the energy changes are different.

To solve the "e" puzzle, read the definitions (Questions 17–24) and then think of a word that fits each definition. Check to see if the word has the same number of letters as the number of spaces in the puzzle. Also check to see whether the word has the letter e *in the same position as the* e *in the puzzle. Copy the puzzle in your notebook and print the word on the proper line, one letter per space. Proceed with each definition in the same way.*

Definitions:

17. A charged particle of electricity
18. A source of electricity
19. A variable resistor
20. A device that changes kinetic energy to electric energy
21. Two or more cells connected in series
22. A unit of electrical current
23. A circuit in which there are two or more electrical paths
24. A device that measures voltage drop

Puzzle:

```
17.                 e _ _ _ _ _ _ _
18.               _ e _ _
19.             _ _ e _ _ _ _ _
20.           _ _ _ e _ _ _ _ _
21.         _ _ _ _ e _ _
22.       _ _ _ _ _ e
23.     _ _ _ _ _ _ e _
24.   _ _ _ _ _ _ _ e _
```

B **25.** Would zinc and magnesium be satisfactory metals in a voltaic cell? Why or why not? (Look at your data from Experiment 7–9.)

26. The operation of a dry cell is an example of the transformation of energy. Why?

27. In a parallel circuit of two different resistors, where does the larger current flow? Why?

28. Why is a voltmeter arranged in parallel in a circuit rather than in series?

29. As an extra safety precaution, would it be desirable to insert two fuses in a circuit rather than one? Give a reason for your answer.

30. What would you expect to happen if you replaced a burned out 15-ampere fuse by a 30-ampere fuse?

31. State some of the advantages of electricity as a source of heat as compared with coal or wood.

32. As a result of a short circuit, the resistance of a house circuit was reduced from 20 ohms to 0.2 ohm. If the house voltage was 120 volts, what was the change in current when the short circuit occurred?

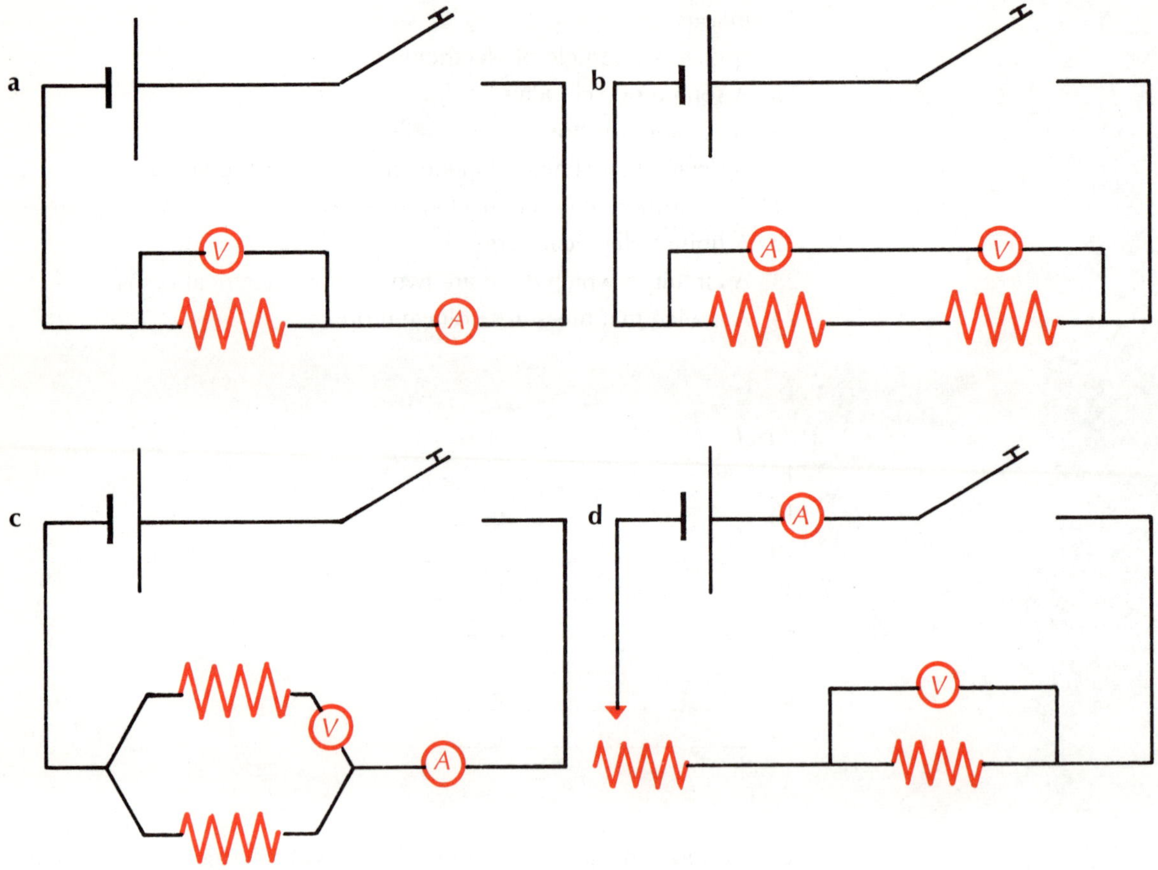

For Questions 33–40, there are four circuit diagrams above. Select the diagram which applies to the circuit described.

33. A circuit with only one resistor.

34. A parallel circuit.

35. A series circuit.

36. The circuit containing a rheostat.

37. A circuit in which the voltmeter is incorrectly placed.

38. A circuit in which the ammeter is incorrectly placed.

39. The circuit which has the greatest resistance, if all the resistors have the same resistance value.

40. The circuit which has the least resistance, if all the resistors have the same resistance value.

Each of Questions 41–45 contains three words that are related to one another, and one word that is not related to the others. For each question, write the three related words and explain how they are related.

41. light bulb, fixed resistor, rheostat, dry cell

42. resistor, ohm, volt, ampere

43. generator, voltmeter, battery, voltaic cell

44. zinc, copper, rubber, carbon

45. ampere, voltage, current, flow of electrons

Chapter 9

LIGHT

The properties of light and the structure of your eye allow you to see scenes like this.

Your eyes are delicate instruments that are sensitive to light. How is your eye constructed and how does it work? Stand in front of a mirror. If possible, select a mirror within reach of a light switch. Examine your eye. Notice the colored circular patch in the middle which may be brown, or blue, or gray, or even green. This circular patch is called the **iris** (EYE-ris). In the middle of the iris there is a smaller dark circle called the **pupil.** The pupil is the opening in the iris through which light passes into the eye.

Observe the size of your pupil in dim light. Then switch on a bright light. What happens to the pupil? Does it become smaller? Switch off the light. Can you see the pupil become larger? Perhaps you have noticed that the pupils of a cat's eyes change in size. They are large in dim light and small in bright light. What causes the pupils to change in size? In some way the iris responds to the amount of light that enters the eye. The iris partially closes the pupil in strong light and widens the pupil in dim light. This is for protection. What part of the eye is protected and why?

The structure of the eye is shown in Figure 9–1. The eye is almost spherical in shape, about an inch in diameter. Covering the front of the eye is a tough "skin" called the **cornea** (KOR-nee-uh). It protects the eye from injury that could be caused by particles in the air. Behind the cornea there is a lens that gathers the light and redirects it. How does the lens work?

At the back of the eye is a sensitive film called the **retina** (RET-ih-nuh). The eye lens directs a picture of an object to the

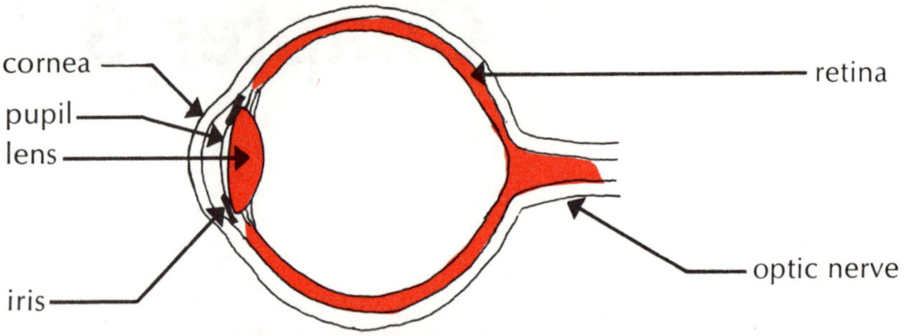

Figure 9–1
The eye has special structures to do its job and to protect itself.

retina. A message is sent to the brain along the optic nerve. Only when the brain receives this message can you see the object. What kind of picture is formed on the retina?

What does the eye do to light? The experiments you will do in this chapter will help you to answer not only these questions but also many others.

ILLUMINATION

Illumination and Sources of Light

You are able to read these words because light is reflected from the printed page to your eyes. That is, light hits the page and bounces back to your eyes. But what is the source of the light? Where does it come from?

If you stand outdoors so that this page is lit by direct sunlight, the source of the light is the sun. There may be so much sunlight reflected into your eyes that the glare is painful. If you stand in the shade of a tree, the source of the light is also the sun. However, in this case the amount of light reflected into your eyes is much less than before. If you read this book indoors at night, the source of light is probably a light bulb or a fluorescent lamp.

The amount of light that falls upon a surface is called the intensity of illumination or just **illumination** (ih-loo-mih-NAY-shun). The illumination of a surface can be measured by a light meter. A light meter is often used to tell whether there is enough light to make a good photograph.

The illumination of the page of the book depends on several factors. One factor is the distance from the source of light to the book. In the following experiment you will study this factor.

experiment 9–1

Intensity of Illumination and Distance from the Source of Light

Materials

slide projector; light meter; screen; meter stick

Procedure

Set up a slide projector so that its bulb is 1 m from the screen. Darken the room and turn on the projector. Turn the "window" of the light meter towards the screen and read the position of the pointer on the scale of the meter.

Now place the bulb of the projector 2 m from the screen and again measure the illumination of the screen. Record both the distance of the bulb to the screen and the reading of the meter. Hold the light meter at the same distance from the screen as you did before.

Finally, place the bulb of the projector 3 m from the screen and record the distance and the meter reading as before.

Record your readings in a data table like that in Figure 9–2.

Figure 9–2
Data table for Experiment 9–1.

Distance from projector to screen	Reading of light meter
1 m	?
2 m	?
3 m	?

Questions

1. (a) What happens to the reading of the light meter as the distance from bulb to screen increases? (b) Is this what you expect? Why or why not?

2. What happens to the meter reading when the distance from bulb to screen is doubled? (Divide the reading at 2 m by the reading at 1 m.)

3. What happens to the meter reading when the distance from source to screen is tripled?

REFLECTION

Angles

In a number of the experiments in this chapter you will be asked to draw angles of a certain size and also to measure angles. A protractor is used for both these purposes.

An **angle** is formed when two lines meet. Angles are measured in degrees. A good way to study angles is to rotate a line or a stick. In Figure 9–3, the line *OP* represents a thin stick. Suppose that one end, *O*, is fixed and that the other end, *P*, is rotated through a quarter of a circle to the position *OQ*. The stick has then traced out the angle *POQ*. Notice that three letters are needed to name an angle. The middle letter in the name marks the point where the two arms of the angle meet.

The angle traced by the rotation of a line through a quarter of a circle is called a **right angle.** There are 90 degrees (90°) in a right angle. Suppose that the rotation of the stick is continued to position *OR*. The angle *POR* is two right angles, or 180°. If the rotation is continued through position *OS* and back to the

Figure 9–3
A line, *OP*, can be rotated to form an angle *POQ*.

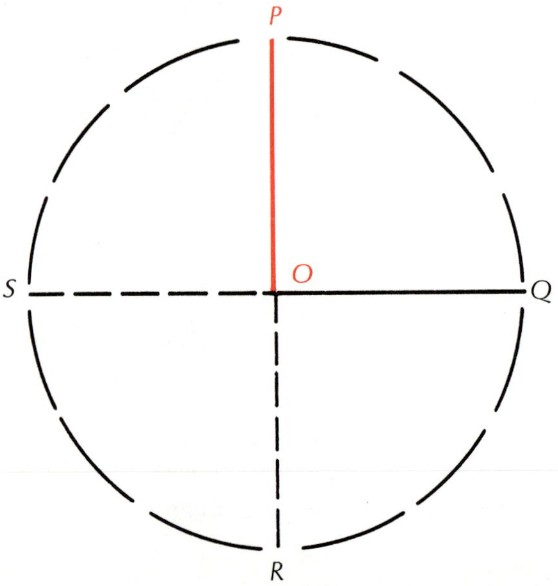

Figure 9–4
The line *XY* is perpendicular to the line *AB*.

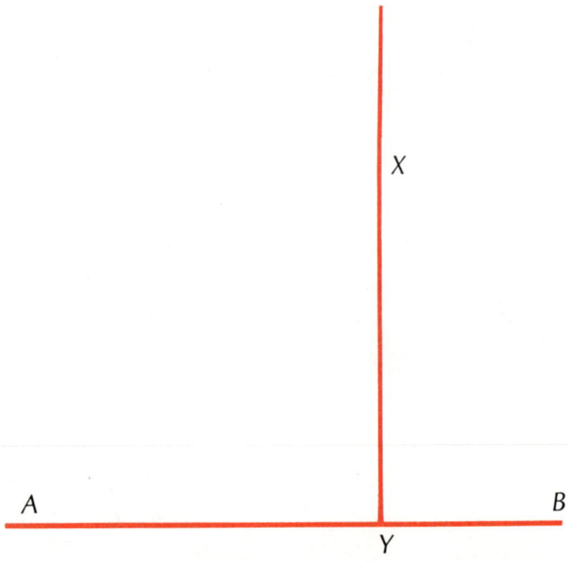

starting point, the point *P* has completed a circular path. It has thus traced out an angle of 360°.

Now examine your protractor. Notice that it is semicircular (half a circle) in shape. The lower edge runs across the middle of the circle. The marked point in the middle of this edge is the center of the circle. The semicircular track is divided into 180 degrees. There are two similar scales on the protractor. The outer scale begins at 0° on the left and ends at 180° on the right. The inner scale begins at 0° on the right and ends at 180° on the left. The two scales enable you to measure angles on either side of a perpendicular line. A **perpendicular** (per-pen-DIK-you-ler) is a line drawn at a right angle to another line. For example, in Figure 9–4 line *XY* is perpendicular to line *AB*.

experiment 9–2

Using a Protractor

Materials

protractor; ruler

A. Drawing a Line Perpendicular to a Given Line, PQ

Procedure

Draw a sloping line *PQ* and mark a point *O* on this line. Place the lower edge of the protractor along this line so that the midpoint of the lower edge is at point *O* as in Figure 9–5. Mark the 90° point (*N*) with a pencil. Remove the protractor. Then, using a ruler, draw a line from *O* through *N*. The line *ON* is perpendicular to line *PQ*.

Figure 9–5
Draw line *PQ*, position the protractor at point *O*, and mark the 90° point *N*.

Figure 9–6
Place the protractor carefully as instructed and mark the 35° point.

Figure 9–7
Since the angle opens to the left, use the outer scale of the protractor. Mark the 60° point.

B. Drawing an Angle BOX *of 35° at Point* O *on Line* AB

Draw line *AB* and mark point *O* somewhere along *AB*. Place the center of the lower edge of the protractor at point *O*. Using the inner scale, make a pencil mark at the 35° mark as in Figure 9–6.

Remove the protractor and draw a line *OX* through the marked point.

The angle named *BOX* opens to the right. There is also an angle *AOX* that opens to the left.

C. Drawing an Angle AOX of 60° at Point O on Line AB

Since the angle is named *AOX* it opens to the left. Draw line *AB* and mark point *O*.

Place the center of the lower edge of the protractor at point *O* as in Figure 9–7. Using the outer scale, make a pencil mark on the outer semicircle at the 60° point. Remove the protractor and draw a line *OX* through the marked point.

D. Measuring an Acute Angle

An **acute angle** is less than a right angle. Draw an acute angle *ABC* as in Figure 9–8. Place the lower edge of the protractor along line *BC* with the midpoint of the lower edge at point *B* of the angle. Read the inner scale along *AB*. If the reading is 42°, the angle *ABC* measures 42°.

Figure 9–8
Draw an acute angle and measure it. Use the inner scale, which reads 0 at *C*.

Figure 9–9
To measure this angle, use the outer scale, which reads 0 at *Y*.

E. Measuring an Obtuse Angle

An **obtuse angle** is greater than a right angle.

Draw an obtuse angle as in Figure 9–9. Place the midpoint of the protractor at point *O* of the angle. Read the outer scale along *OX*. If the scale reading is 133, the angle *XOY* measures 133°.

F. Measuring the Angles XBA and YBC

To measure *XBA* in Figure 9–10, place the base of the protractor along *XY* with the midpoint at *B*, as in Figure 9–10A. Use the inner scale to measure the angle. In the figure, angle *XBA* measures 39°.

To measure *YBC*, place the base of the protractor along *XY* as in Figure 9–10B. Read the inner scale. In the figure, angle *YBC* measures 19°.

Figure 9–10
Draw a figure like this in your notebook. **A** (left) Measure angle *XBA*. **B** (right) Measure angle *YBC*.

Questions

1. In Figure 9–11, measure angles (a) *PQB*, (b) *PQA*, (c) *BQA*.

2. In Figure 9–12, measure angles (a) *XOC*, (b) *DOY*, (c) *XOD*.

3. Copy Figure 9–13 in your notebook. At point *O*, draw (a) an angle *BOD* of 42°, (b) an angle *BOE* or *AOE* of 90°, (c) an angle *BOF* of 142°.

4. In the figure you drew for Question 3, measure angle *AOF*.

5. In Figure 9–14, measure angles (a) *AOC*, (b) *COB*, (c) *BOD*, (d) *DOA*.

6. In Figure 9–14, (a) add together the degrees in angles *AOC* and *COB*. Is your answer what you expect? Why? (b) What is the sum of angles *AOD* and *DOB* in degrees?

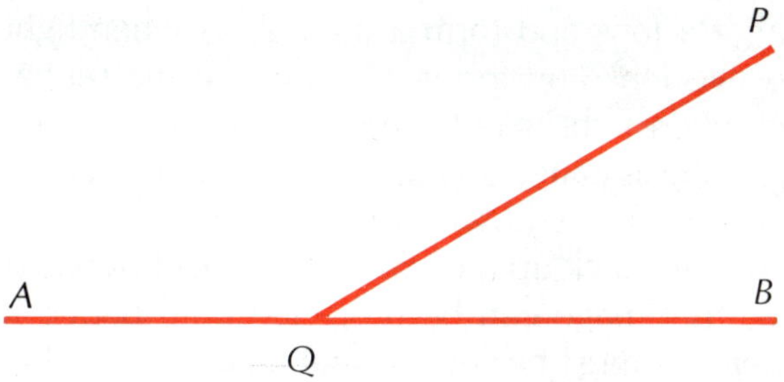

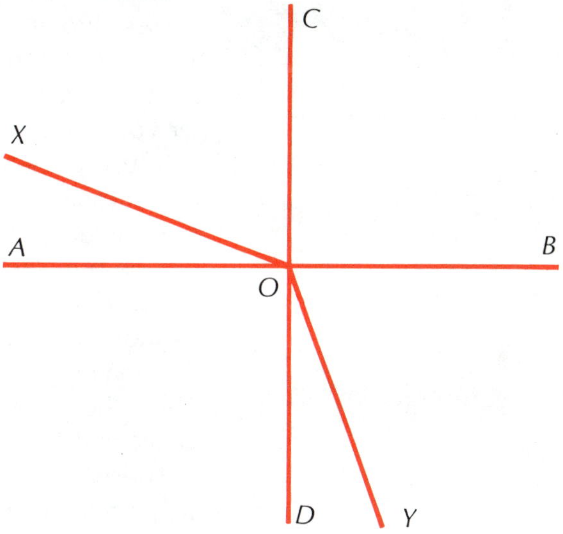

Figure 9–11 (above left)
Measure these angles for Question 1.

Figure 9–12 (above right)
Measure these angles for Question 2.

Figure 9–13 (below left)
Copy this figure for Question 3.

Figure 9–14 (below right)
Measure these angles for Question 5.

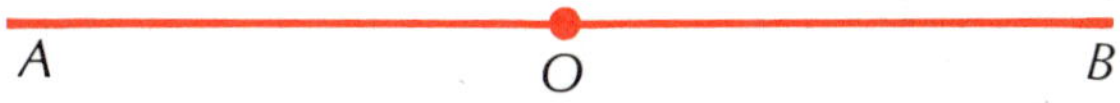

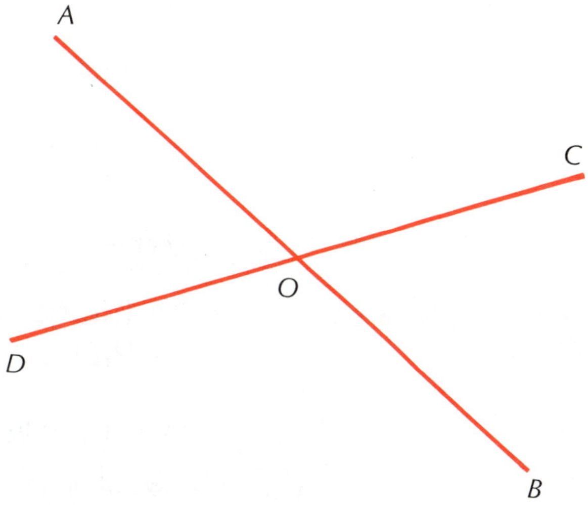

Reflection of Light

When you look into a mirror, you see a picture of yourself. How does a mirror work? What do you actually see in a mirror? A mirror is a reflecting surface. That is, when rays of light strike a mirror, they bounce back or, as we usually say, they are **reflected.** In a lighted room you can see objects because light from the objects enters your eyes. If you stand in front of a mirror, rays of light from you strike the mirror and bounce back. If the reflected rays enter your eye, you see a picture of yourself. Light can also pass through a lens and form a picture, as when light passes through the eye lens and forms a picture on the retina. The picture formed when light is reflected by a mirror or when light passes through a lens is called an **image** (IM-uj). In this chapter you will study images formed by mirrors and lenses.

In using a lens you will pick up the image on a sheet of paper or cardboard. Since the image can be projected onto paper, it must actually be there. It must be real. A **real image** is one that can be projected onto some kind of screen.

The image you see in a mirror is behind the mirror. But light cannot pass through a mirror as it does through a lens. Therefore the image in a mirror cannot be projected onto a sheet of paper. Thus, the image formed by a mirror is not real. An image that is not real is called a **virtual** (VER-chu-al) **image.**

The way you see yourself in a mirror is not the same as the way your friends see you. To see why, look into a mirror and raise your right hand. Notice that, in the image, it is the left hand that is raised. Thus, images formed in a mirror are turned around. They are the reverse of the object in appearance. But this is to be expected when you think about it. When you look into a mirror your image faces the opposite direction to you. Your right eye looks like a left eye when you see it in a mirror.

In this book you will use only **plane** (flat) **mirrors.** In Experiment 9–3, you will find the position of the image of an object in a plane mirror.

experiment 9–3 The Image of a Pin in a Plane Mirror

Materials mirror, supported so that it stands upright; protractor; ruler; pin

Procedure Draw a horizontal line near the middle of a page of your notebook and label it *reflecting surface*. Place the front edge (the reflecting surface) of the mirror on this line, as shown in Figure 9–15. Stick a pin upright in the page, 8 or 10 cm in front of the mirror. Draw a small circle around the pin position and label it *object*.

Bend down so that your head is near the lower right corner of the page. (See Figure 9–16.) Look into the mirror with one eye and observe the image of the pin. Stick two pins on the right side of the object pin, lining them up with the image of the object pin. Indicate the pin positions 1 and 2 by drawing small circles around the pin holes, as shown in Figure 9–17. The line through these two pin positions is called a **line of sight.** The image lies somewhere on the line of sight.

Figure 9–15
The reflecting surface of the mirror should be on the line. Stick a pin in the page and label its position.

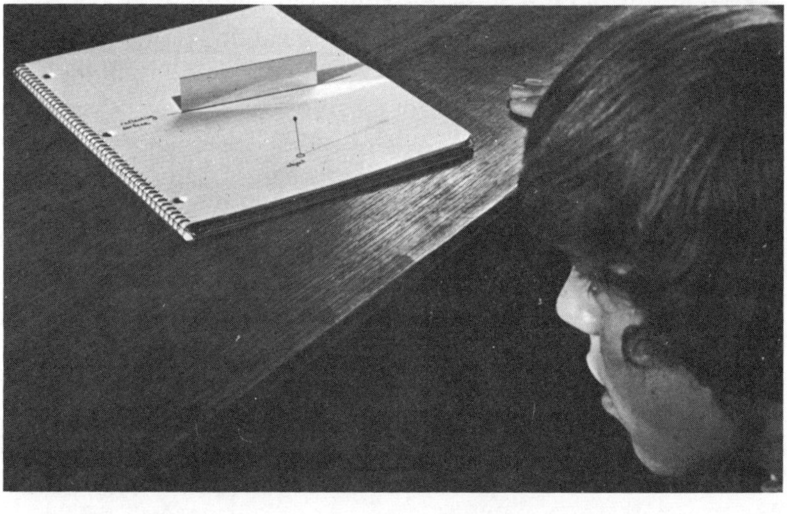

Figure 9–16
Bend down so your head is level with the notebook. Close one eye and observe the image of the pin.

Figure 9–17
Stick pins where they line up with the image of the object pin.

Without moving either the object pin or the mirror, take another line of sight on the object, this time on the left side of it. Again draw circles around the two pins used as markers.

Remove the mirror. Remove the two pins from one of the lines of sight and draw a line through these pin positions as far as the mirror. This line is the path of the **reflected ray.** Label this line *reflected ray path.* Now draw a line from the object pin to the point where the reflected ray path leaves the mirror. This line is the **incident** (IN-sid-ent) **ray** path. Label this line *incident ray path*. Use arrowheads to indicate the directions of ray paths, as shown in Figure 9–18.

Now repeat this procedure with the second line of sight. Using a broken line, extend both the reflected ray paths behind the mirror position until they cross each other. The image lies somewhere along the extension of the lines of sight. Therefore the position of the image of the pin is the point where the broken lines cross. Measure and record the distance from the object to the mirror and the distance from the image position to the mirror.

Using a protractor, draw a perpendicular to the mirror at each of the points where the incident ray hits the mirror. The angle between the perpendicular and the incident ray path is the **angle of incidence.** The angle between the perpendicular and the reflected ray path is the **angle of reflection.** See Figure 9–19.

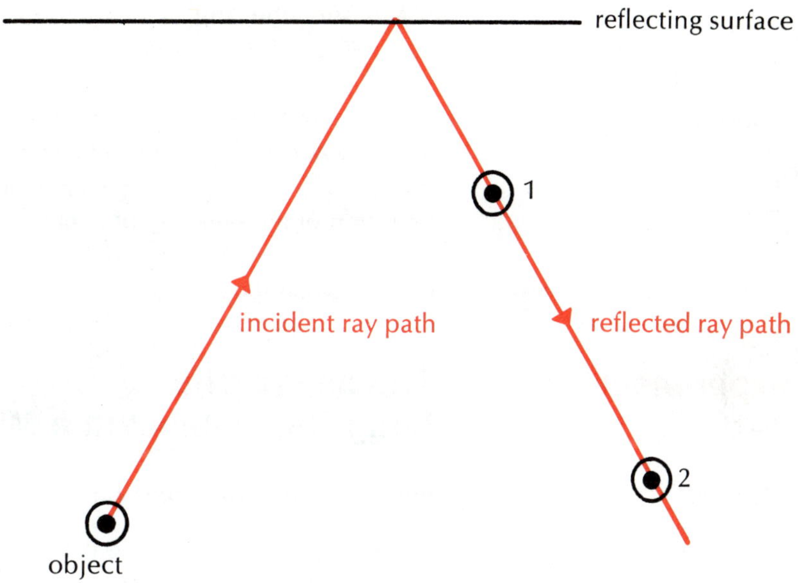

Figure 9–18
Draw the ray paths and label them.

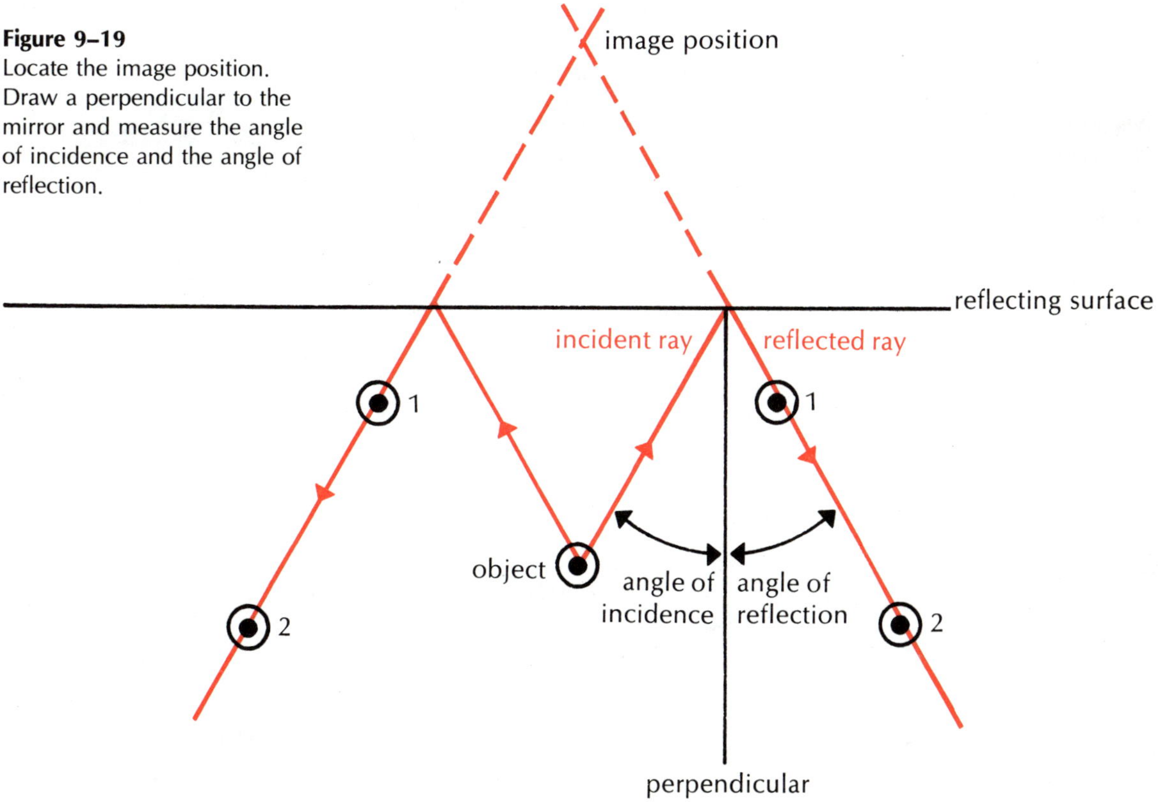

Figure 9–19
Locate the image position. Draw a perpendicular to the mirror and measure the angle of incidence and the angle of reflection.

Measure the angles of incidence and reflection for each ray path. Record these angles.

Questions

1. Compare the distance from object to mirror with the distance from image to mirror. What do you conclude?

2. Compare the angle of incidence of each ray path with the angle of reflection of the same ray of light. What do you conclude?

3. Do light rays from the object actually reach the image position? Give a reason for your answer.

experiment 9–4 The Image of a Long Bent Object in a Mirror

Materials mirror; protractor; ruler; pins

Procedure Draw a horizontal line near the middle of a page of your notebook and label it *reflecting surface*. Place the front edge of the mirror along this line.

Draw a large letter L near the bottom of the page with the long arm pointing toward the upper left-hand corner of the page. The letter L is the object in this experiment. Stick a pin in the

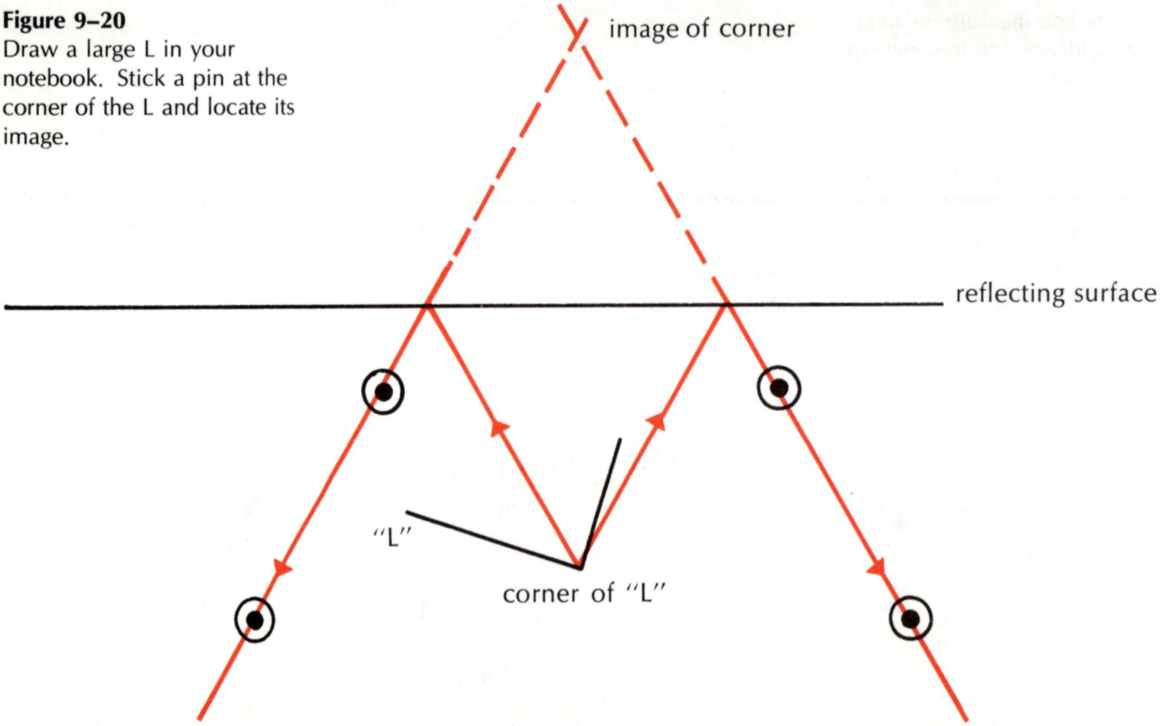

Figure 9–20
Draw a large L in your notebook. Stick a pin at the corner of the L and locate its image.

corner of the L. Locate the position of the image of the corner as in Experiment 9–3 and Figure 9–20.

Next, stick the pin in one end of the L and locate its image position in the same way. Then stick the pin in the other end of the L and also locate its image position.

Finally, remove the pins and draw the image of the object.

Questions

1. Compare the size of the image with the size of the object.

2. Measure the perpendicular distance from the mirror to the corner of the L object and also to the corner of the L image. How does the image distance compare with the object distance?

3. On the basis of your results in Experiments 9–3 and 9–4, state two laws of reflection for a plane mirror.

4. How does the shape of the image compare with the shape of the object?

experiment 9–5

Images in Two Mirrors at a Right Angle to Each Other

Materials 2 mirrors; protractor; ruler; pins

Procedure Arrange the two mirrors at a right angle to each other. Place a pin halfway between the mirrors as in Figure 9–21. Locate the images of the pin.

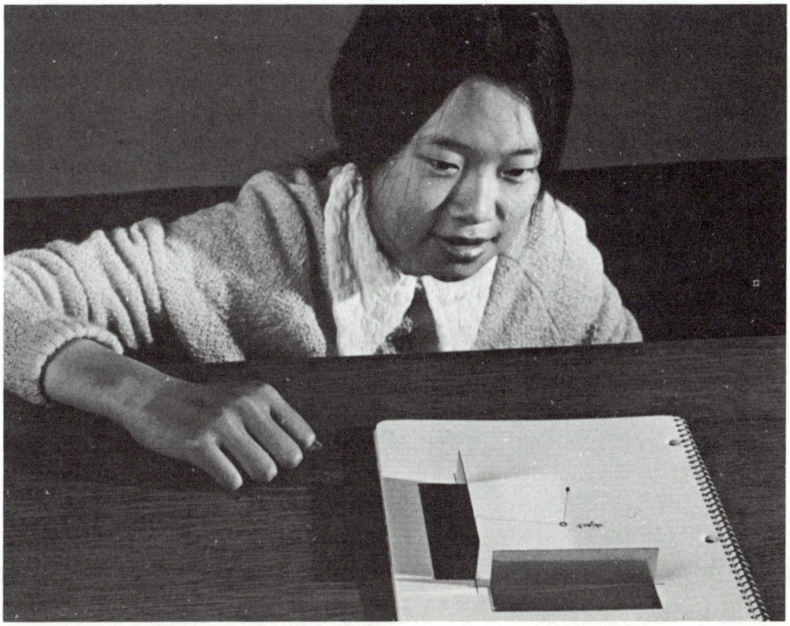

Figure 9–21
Arrange two mirrors at a right angle and locate the images of the pin.

Questions

1. How many images can you see?

2. Are the images where you expect them to be? Give a reason for your answer.

experiment 9–6 Using a Rangefinder to Compare Distances of Object and Image

Materials

large mirror; camera rangefinder

Procedure

Stand about six feet in front of the large mirror. With the camera rangefinder, measure the distance to the frame of the mirror. Then measure the distance to your own image in the mirror. Record these distances in your notebook.

Now stand about 15 feet in front of the mirror. Again, using the rangefinder, measure the distances to the mirror and to your image. Record these distances.

Questions

1. Compare the distance from you to the mirror with the distance from you to your image in each of the two cases.

2. State a general rule for the object-image distances in plane mirrors.

REFRACTION

Light Passing Through Substances

When light strikes a mirror, the light is reflected and you can see images. When light strikes a rough surface, such as a piece of paper, the light is scattered in all directions and no images can be seen.

On the other hand, when light strikes the surface of ordinary glass, some of the light passes right through it. A substance through which light passes is said to be **transparent** (trans-PAIR-ent). What happens to light that passes through a piece of plate glass with parallel sides? You will study this question in Experiment 9–7.

experiment 9–7

How Light Passes Through a Parallel-Sided Glass Plate — I

Materials parallel-sided glass plate; protractor; ruler; pins

Procedure Place a parallel-sided glass plate in the middle of a page of your notebook, with the polished edges of the glass facing the top and bottom of the page. Trace the outline of the plate in ink. Stick two pins, 5 or 6 cm apart, into the paper above the glass plate. Draw small circles around the pins and label the positions 1 and 2 as shown in Figure 9–22. Look through the bottom edge of the glass plate. Move your head until the images of pins 1 and 2 are in line. Stick two pins in line with the images of pins 1 and 2. (See Figure 9–23.) Draw small circles around these pins and label them 3 and 4.

Remove all the pins and the glass plate. Draw a line through pin positions 1 and 2 as far as the top edge of the glass. This line represents the ray of light that goes into the glass. It is the incident ray. Then draw a line through pin positions 3 and 4 up to the lower edge of the glass. This line represents the ray that comes out of the glass. It is called the **emergent** (ee-MER-jent) **ray.** Place arrowheads on the incident and emergent ray paths to indicate their direction. Finally, draw a line connecting the incident ray and the emergent ray at the two edges of the glass plate. This line represents the path of the ray of light in the glass. It is called the **refracted** (ree-FRAK-ted) **ray.** Put an arrowhead

Figure 9–22
Close one eye and find where the images of pins 1 and 2 line up.

Figure 9–23
Stick two pins in line with the images of pins 1 and 2.

on this line to show the direction of the refracted ray. Label the ray paths *incident, refracted,* or *emergent* as in Figure 9–24.

Construct a perpendicular at the point where the incident ray enters the glass. Extend the perpendicular into the region of the glass plate. Construct a perpendicular at the point where the refracted ray emerges from the glass. Extend this perpendicular into the region of the glass plate. Notice that the ray of light is bent when it enters the glass and also when it leaves the glass. The bending of a ray of light as it passes from one material to another is called **refraction** (ree-FRAK-shun).

Measure and record the angle between the incident ray path and the perpendicular. This angle is in the air. It is the angle of incidence. Also measure and record the angle between the refracted ray path and the perpendicular. This angle is in the glass. It is called the angle of refraction. Make a table of your readings as in Figure 9–25.

The word *incidence* comes from a Latin word meaning "to fall upon." At first, the incident ray is the ray in the air that falls upon the glass surface. Then the ray in the glass falls upon the lower glass surface and emerges into the air. Hence, when the ray travels from glass to air, the ray within the glass becomes the incident ray. The emergent ray is the refracted ray.

Measure the angle between the incident ray and the perpendicular. Also measure the angle between the refracted ray and the perpendicular for the ray that passes from glass to air. Make a table of your readings as in Figure 9–26.

Finally, use a broken line to extend the emergent ray backwards through the region of the glass and beyond towards the incident ray.

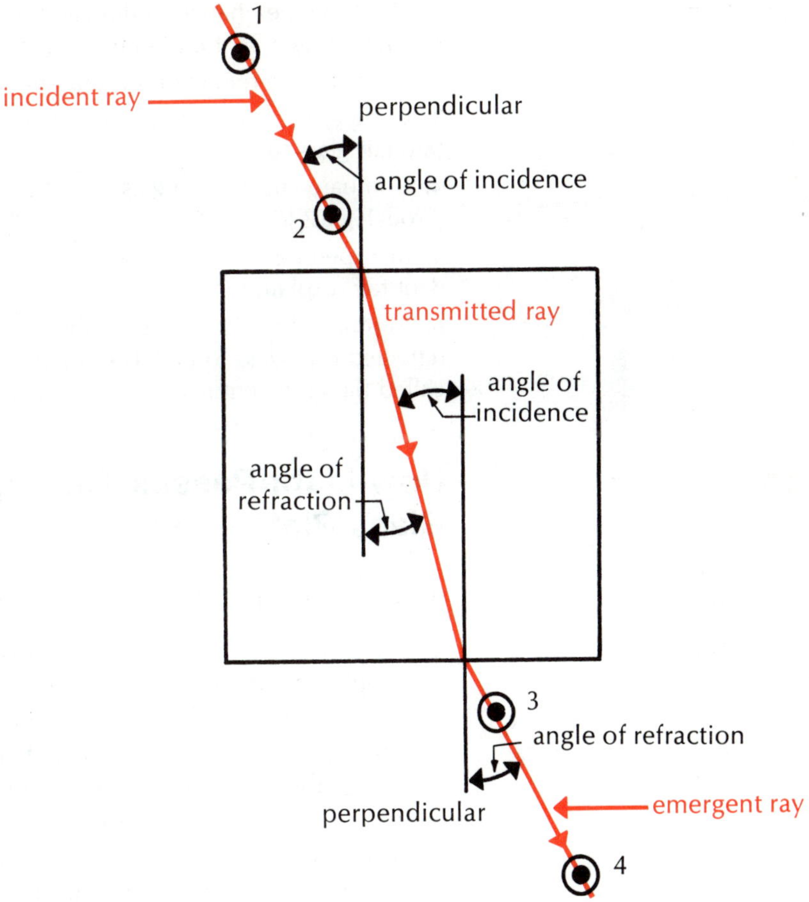

Figure 9–24
Draw the ray paths and label them. Draw perpendiculars and measure the angles.

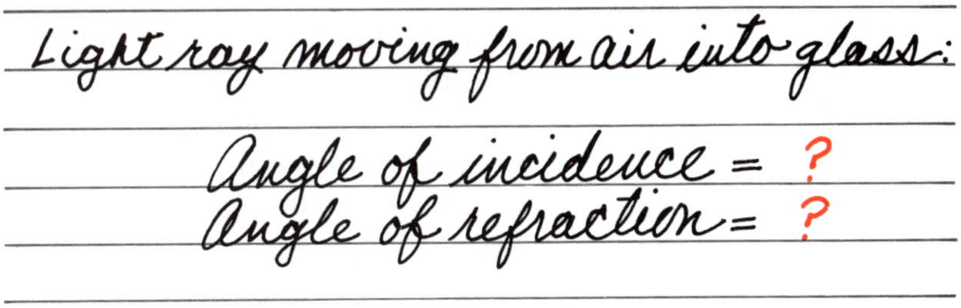

Figure 9–25
Data table for the ray moving from air into glass.

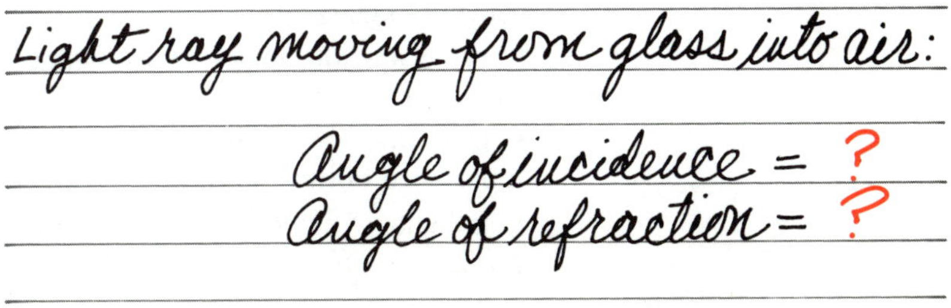

Figure 9–26
Data table for the ray moving from glass into air.

Questions

1. Describe the change in direction, with respect to the perpendicular, of a ray of light as it passes from air to glass.

2. Describe the refraction of a ray of light as it passes from glass to air.

3. Compare the two angles of incidence as the ray passes from air through glass to air.

4. Compare the two angles of refraction as the ray passes from air through glass to air.

5. In Experiment 9–7, the same ray may be a refracted ray or an incident ray. Explain why.

6. Compare the directions of the incident ray (air to glass) and the refracted ray (glass to air). The ray refracted from glass to air is usually called the emergent ray.

experiment 9–8 How Light Passes Through a Parallel-Sided Glass Plate — II

Materials

parallel-sided glass plate; ruler; pins

Procedure

In Experiment 9–7, the incident ray was a sloping line. That is, the angle of incidence was less than 90°. What happens if the incident ray is at right angles to the surface of the glass?

Arrange the glass plate as in Experiment 9–7 and trace the outline of the plate in ink. Draw a line perpendicular to the upper edge and stick two pins on this line at least 5 cm apart. Draw small circles around the pins and label their positions 1 and 2.

Look through the lower edge of the glass plate and stick two pins in line with the images of pins 1 and 2. Circle the lower pin positions and label them 3 and 4. Remove the pins and also the glass plate.

Draw a line through positions 3 and 4 up to the lower edge of the glass. Draw a line to represent the ray of light in glass. Put in arrowheads to show the directions of incident and emergent rays.

Questions

1. What are the angles of incidence and refraction as the ray passes from air to glass?

2. What are the angles of incidence and refraction as the ray passes from glass to air?

3. Does the ray of light change direction as it passes through the upper edge of the glass?

4. Is light refracted if the rays are perpendicular to the glass surface through which they pass?

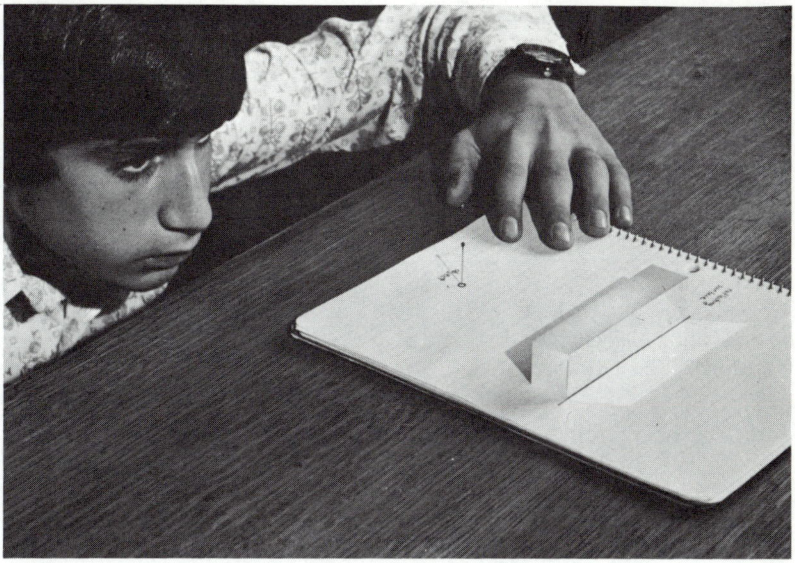

Figure 9–27
Locate the image of the pin in the thick mirror.

experiment 9–9

Locating the Image Formed by a Mirror with a Thick Transparent Coating

Laboratory Test

Materials

mirror with thick transparent coating in front of the reflecting surface; pins; ruler

Procedure

Place the mirror on a page of your notebook and outline the back edge (the reflecting surface) and also the front edge (that is, the front edge of the coating) in ink. Then place a pin about 10 cm in front of the coating and draw a small circle around the position of the pin. The arrangement is shown in Figure 9–27.

Now locate the image of the object (the pin) as in Experiment 9–3.

Questions

1. Is there any evidence of refraction? Give a reason for your answer.

2. Is the distance from object to mirror surface the same as the distance from image to mirror surface? Why or why not?

3. Why are reflecting surfaces usually coated with glass or some other transparent material?

4. The transparent coating that covers a reflecting surface should be thin. Why?

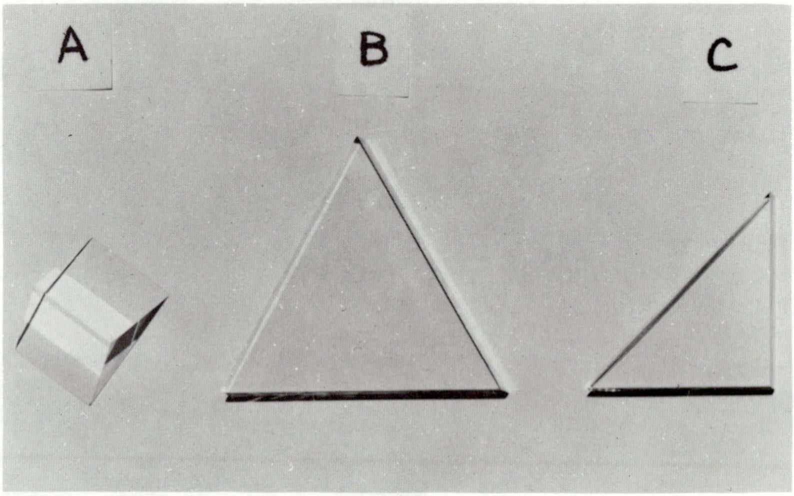

Figure 9–28
A. A triangular prism. **B.** This prism is an equiangular triangle. **C.** This prism is a right-angle triangle.

Refraction through a Prism

A **prism** (PRIZ-um) is a solid which has similar and parallel ends. If the ends are triangular as in Figure 9–28A, the prism is called a triangular prism. In Experiments 9–10 and 9–11 the triangle has equal angles and equal sides as in Figure 9–28B. Such a triangle is called **equiangular.** In Experiment 9–12 the ends are right-angle triangles as in Figure 9–28C. In an equiangular triangle, all the sides are equal. In a right-angle triangle, the side opposite the right angle is longer than the other two sides. In your experiments the glass prisms are thin slices of plate glass. Does light that passes through a glass prism behave in the same way as light that passes through glass with parallel sides? You will find the answer to this question by doing the next three experiments.

experiment 9–10

How Light Passes Through an Equiangular Glass Prism — I

Materials

equiangular glass prism; pins; ruler; protractor

Procedure

In this experiment a ray of light enters the prism at an angle.

Center the prism on a page of your notebook. Make an outline of the prism in ink. Stick two pins in the page on the left side of the prism as shown in Figure 9–29 (positions 1 and 2).

Look through the right-hand side of the prism and move your head until the images of pins 1 and 2 come together. Stick two pins in line with the images of the object pins. Circle all four pin positions, then remove the pins and the prism.

Draw the path of the ray of light (a) as it approaches and hits the prism, (b) as it emerges from the prism, and (c) as it passes through the prism. Use arrowheads to indicate the direction of the incident ray, the refracted ray, and the emergent ray.

Construct a perpendicular to the prism at the point where the ray enters the prism. Construct another perpendicular to the prism at the point where the ray emerges from the prism.

Questions

1. Is the entering ray refracted toward or away from the perpendicular? Why?

2. Is the emergent ray refracted toward or away from the perpendicular? Why?

3. (a) Compare the passage of light through a glass prism with the passage through a block with parallel sides. (b) Is the comparison what you expect? Why?

Figure 9–29
Find where the images of pins 1 and 2 come together.

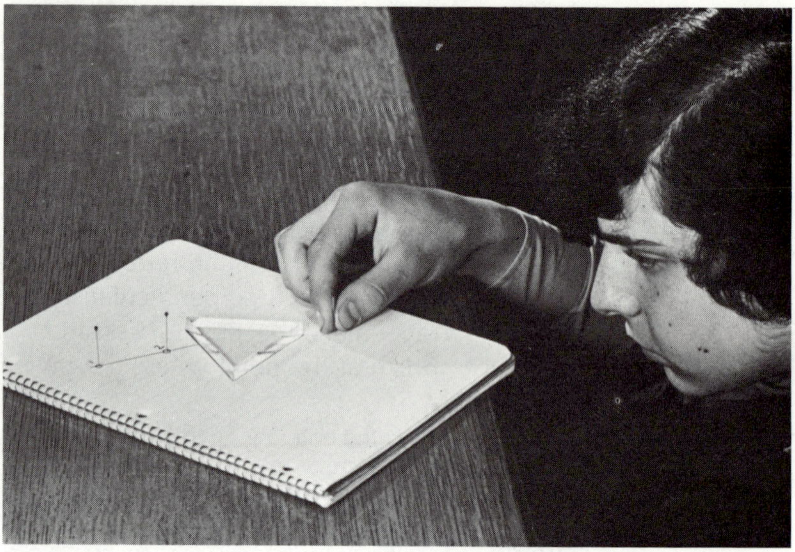

Figure 9–30
Draw a perpendicular to the face of the prism and stick two pins along it. Find where the images of the pins come together.

experiment 9–11 How Light Passes Through an Equiangular Glass Prism — II

Materials equiangular glass prism; pins; ruler; protractor

Procedure Center the equiangular prism on a page of your notebook and outline the prism in ink. Construct a perpendicular above the center of the left face of the prism and stick two pins on this line, as shown in Figure 9–30.

Sight through the lower side of the prism and move your head until the images of the object pins come together. Place two pins in line with the images of the object pins. Circle all four pinholes, then remove the pins and the prism.

Draw the path of the ray of light (a) as it approaches and enters the prism, (b) as it emerges from the prism, and (c) as it passes through the prism. Use arrowheads to indicate the direction of the incident ray, the refracted ray, and the emergent ray.

Questions

1. Is the ray of light bent as it enters the prism? If not, why not?

2. Is the ray of light bent as it emerges from the prism? If not, why not?

3. Describe what happens to the ray in glass when it reaches the glass surface where it emerges.

4. Why can you not observe images of the pins by sighting back through the glass surface opposite the entering surface?

experiment 9–12 How Light Passes Through a Right-Angle Prism

Laboratory Test

Materials

right-angle prism; ruler; protractor; pins

A. Light Entering Perpendicular to a Short Side

Procedure

Place the prism on a page of your notebook and trace its outline in ink. Construct a perpendicular to one of the short sides of the prism and stick two pins on this line as shown in Figure 9–31.

Find where the images of the object pins line up and place two pins in line with the images. Circle all four pinholes, then remove the pins and prism.

Construct the complete path of the ray of light, indicating its direction.

Questions

1. Compare the transmission of light through the prism with reflection by a plane mirror.

Figure 9–31
Draw a perpendicular to a short side of the prism and stick two pins along it.

Figure 9–32
Draw a perpendicular to the long side of the prism and stick two pins along it.

B. Light Entering Perpendicular to the Long Side

Place the prism on a page of your notebook and outline it in ink. Construct a perpendicular to one side of the center of the long side, and stick two pins on this line as shown in Figure 9–32.

Find where the images of the object pins come together. Place two pins in line with the images of the object pins. Circle all four pinholes. Then remove the pins and the prism. Construct the path of the ray of light, indicating its direction.

LENSES

Refraction of Light by a Lens

A **lens** is a piece of transparent material with curved surfaces. Lenses are of two basic kinds, convex and concave. A **convex lens** (Figure 9–33A) is thicker at the center than at the edge. A **concave lens** (Figure 9–33B) is thinner at the center than at the edge. You will investigate only convex lenses.

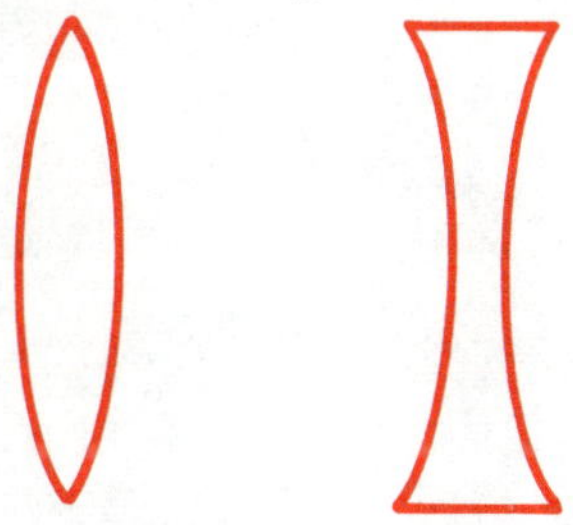

Figure 9–33
A. A convex lens.
B. A concave lens.

A lens, like a prism, refracts light. The refraction of sunlight by a convex lens can be shown very easily. Hold a large convex lens towards the sun and place a sheet of paper just behind the lens. Notice the circular patch of light on the paper. Slowly move the paper away from the lens and observe that the patch of light gets smaller and smaller. Eventually the light passing through the lens is concentrated into a sharp spot. The position of the spot is called the **focal point** of the lens, or the focus. The spot itself is an image of the sun. A sharp, clear image is said to be focused (FOH-kusd). In order to focus the sunlight onto this spot, the light rays from the sun must have been bent toward the spot as they passed through the lens. Notice particularly that the image is projected onto the paper. As you learned earlier, an image that can be picked up on a paper or cardboard screen is a real image. In this respect the image formed by the convex lens is quite different from a virtual image formed by a plane mirror.

The distance from the lens to the point where light from a distant object such as the sun is brought to a focus is called the **focal length** of the lens. In Experiment 9–13 you will find the focal length of a convex lens. Instead of using the sun as the object you will use a house, or a tree, or any other convenient distant object.

experiment 9–13 The Focal Length of a Convex Lens

Materials

convex lens; cardboard; ruler

Procedure

Hold a piece of cardboard upright against the zero end of the centimeter scale of a ruler, and hold a convex lens upright on the ruler, as shown in Figure 9–34.

Point the ruler out the window or at a lamp on the ceiling. Move the lens back and forth until you get the sharpest possible image of the object on the cardboard. Read the distance from the lens to the cardboard and record it in your notebook. This distance is the focal length of the lens.

Figure 9–34
Move the lens until you get a sharp image.

Questions

1. Describe the image of the object.

2. How does the real image formed by a convex lens look different from the virtual image formed by a plane mirror?

experiment 9–14

Using a Mirror to Find the Focal Length of a Convex Lens

Materials

convex lens (the same lens used in Experiment 9–13); mirror; pins; cardboard

Procedure

Place a mirror just behind a convex lens. Stick a pin in a sheet of cardboard placed on the bench in front of the lens-mirror combination. The arrangement is shown in Figure 9–35.

Place your eye on the same level as the pin and lens. Look through the lens and locate the upside-down image of the pin. Move your head from side to side and notice that a gap appears between the pin and the image of the pin, as shown in Figure 9–36A.

Now move the pin either toward the lens or away from it until the gap disappears. The gap should disappear when the

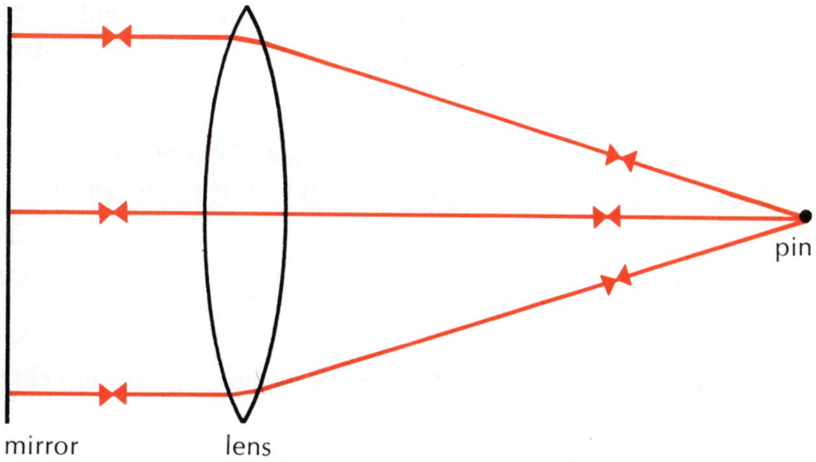

Figure 9–35
Hold the lens in front of the mirror, and place the pin in front of the lens.

heads of the image and object pins appear to remain together as you move your eye from side to side. You can see this in Figure 9–36B. This position of the pin marks the focal point of the lens.

Measure the distance from the pin to the lens and record it as the focal length of the lens.

Questions

1. Explain why the arrows in Figure 9–35 point in opposite directions.

2. Compare the focal length of the lens as measured in Experiment 9–13 with the focal length of the same lens as measured in Experiment 9–14. What do you conclude?

experiment 9–15 The Image of an Object Placed Within the Focal Length of a Convex Lens

Materials

convex lens used in Experiment 9–14; ruler

Procedure

In Experiments 9–13 and 9–14 you examined the image of a distant object formed by a convex lens. You observed that the image formed in this way is real and also upside down. What is the nature of the image if a nearby object is used instead of a distant object?

Hold the convex lens above the page of the textbook at a distance equal to the focal length of the lens. Look through the lens towards the book. Can you see the letters on the page?

Slowly lower the lens until the letters of the printed page are in sharp focus. Measure the distance from lens to page. Compare the size of a letter as seen through the lens with the size of the same letter as seen without the lens. This comparison is called the **magnification** (mag-nih-fih-KAY-shun) of the lens. This relationship can be expressed as

$$\text{magnification} = \frac{\text{size of image}}{\text{size of object}}$$

Questions

1. Compare the image of the object in Experiment 9–14 with the image in Experiment 9–13.

2. Compare the image of the object in Experiment 9–15 with the image in Experiment 9–13.

3. A convex lens can be used as a magnifying glass if the printed page is in a certain position. What is this position?

4. When a convex lens is used as a magnifying glass is the image real or virtual? Give a reason for your answer.

5. The image formed by a convex lens may be (a) upside down or right side up, (b) larger or smaller than the object, (c) real or virtual. Describe how you would get an image fitting each of these six descriptions.

PERSPECTIVES

perspective 9–1

Illumination

Our chief source of light is the sun. For the cave man, the sun was practically the only source of light. Almost all human activities were limited to the period between sunrise and sunset. Only after sources of artificial light were discovered, could people work at night.

Burning sticks (torches) were the first source of artificial light. Later, the Greeks and Romans discovered that oil squeezed out of olives would burn. The first lamps were

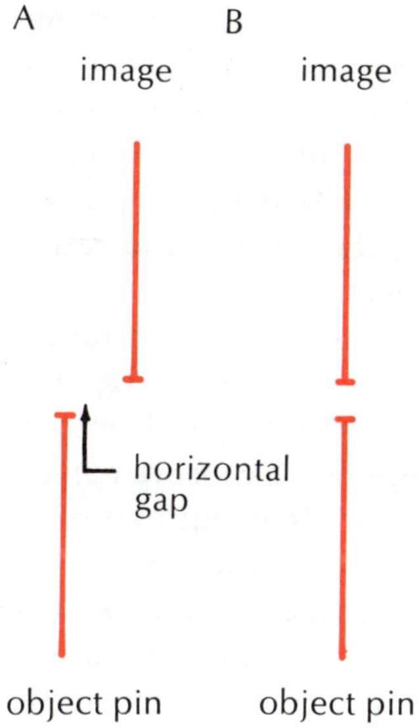

Figure 9–36
A. When the pin is in the wrong position, a horizontal gap appears between the pin and its image when you move your head. **B.** When the pin is in the right position, no gap appears.

merely shallow dishes to hold the oil. The flames from these early lamps were small and smoky, and they had an unpleasant smell.

Next came candles which were made from sheep fat, called tallow. In the 19th century, coal gas and kerosene were used as sources of light. The light from a kerosene lamp was much stronger than the light from a candle. The pioneers in America often suffered from the lack of artificial light. When Abraham Lincoln was a boy, he had neither candles nor lamps. In his log cabin he studied by the light of a wood fire.

At the end of the 19th century, Thomas Edison invented the electric lamp. He passed electric current through a thread of carbon. This fragile thread was later replaced by a thin wire of tungsten. Today electricity is the main source of artificial light.

The sun produces light. So does the white-hot filament of a light bulb. But the moon does not produce light. You see the moon because it reflects the light it receives from the sun. Similarly, books, chairs, and people do not produce light. You see them because their surfaces reflect light.

perspective 9–2

The Eye

All the light that enters your eye passes through a lens. Between the lens and the retina, the eyeball is filled with a transparent jellylike substance. The light passes through this substance before it strikes the retina, the part of the eye that is sensitive to light. What is meant by the phrase "sensitive to light"?

The retina is made up of nerve cells. It is connected to the back part of the brain by the optic nerve. Before you can see an object, light must pass through the lens, an image of the object must fall upon the retina, and a message must pass from the retina to the brain.

When an image is focused on the retina, signals are transmitted along the optic nerve to the brain. These signals are electric. They are due to electric currents which, in turn, are caused by chemical action. The signal of the image is interpreted by the brain. This interpretation is what we know as "sight."

When you see an object, the signals travel from eye to brain. Signals can also travel in the opposite direction, from brain to eye. If too much light falls upon the retina, the sensitive cells could be damaged. However, in strong light, the brain sends signals to the muscles of the iris. The iris responds by partially closing the pupil. Then less light reaches the retina and the cells are not damaged. (See Figure 9–37.)

In some respects your eye is like a camera. For instance, both eye and camera have a lens in front and a sensitive film at the back. However, the eye is a far better optical instrument than the camera. In a camera only one exposure can be made on a given section of the film. If more than one exposure is made, the fiım is spoiled. The exposed film has to be developed in a solution of chemicals. But for the eye, light enters continuously and ever-changing images are formed on the retina. These images are "developed" by the brain instantly, so you see ever-changing pictures. Thus, the eye behaves more like a movie camera than an ordinary camera.

The eye can also focus the light much faster than a camera. In a camera, focusing of images of near and distant objects is done by moving the lens. But in the eye the lens does not move. Focusing is done by changing the shape of the lens. To bring a near object into focus, the lens becomes thicker in the middle. To focus the image of a distant object on the retina, the lens becomes thinner.

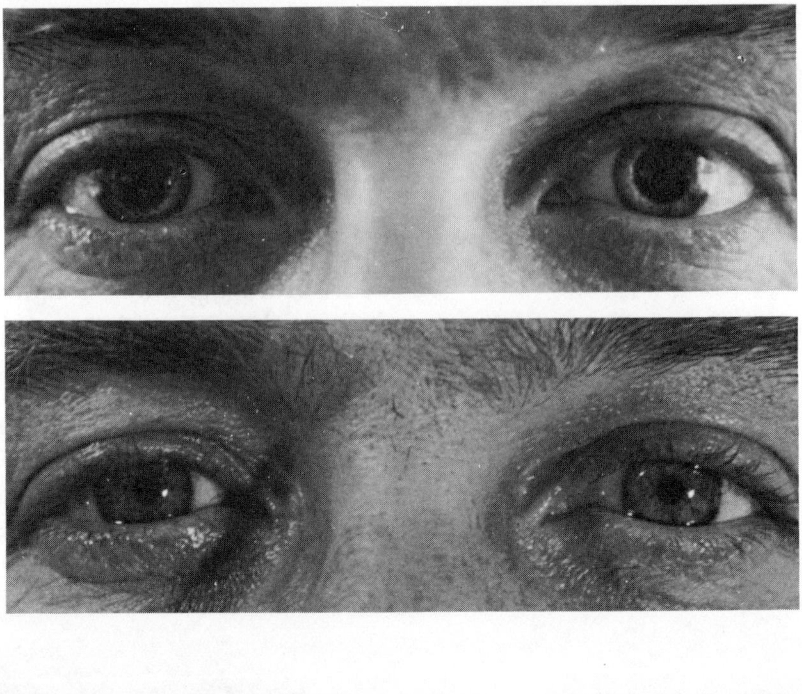

Figure 9–37
In dim light, the pupil is wide open to catch as much light as possible. In strong light, the iris partially closes the pupil to prevent damage to the retina.

perspective 9-3

What Is Light?

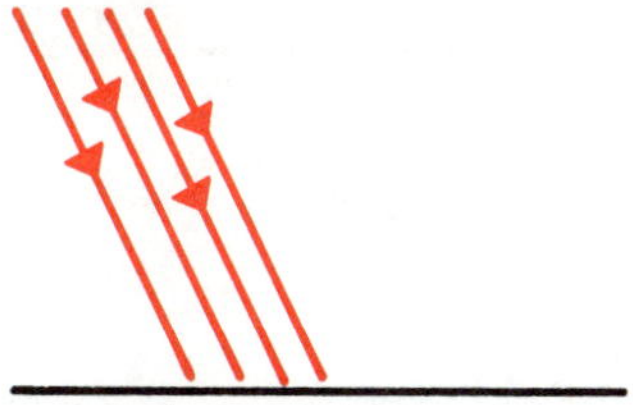

Figure 9–38
Rays from the sun are parallel because the sun is so far away.

Light enables us to see things. Isaac Newton, of gravitational law fame, was the first scientist to discuss the nature of light. According to Newton, light consists of tiny weightless particles shot out of a luminous body in all directions. These "bullets" of light travel at high speed in straight lines. He called the paths of these bullets **rays,** and said that a person could see an object only if particles of light from the object entered his eye. Newton's theory of light is called the particle theory. Using the particle theory, Newton could explain reflection and refraction.

This book uses the particle theory of the nature of light. But there are other theories. One theory is that light is a kind of wave motion. The wave theory can explain reflection and refraction, and some other effects of light.

perspective 9-4

Reflection

The sun emits rays of light into space in all directions. However, the sun is so far from the earth (93 million miles, or 150 million kilometers) that we think of the sunlight that strikes the earth as made up of parallel rays. A bundle of such parallel rays is shown in Figure 9–38. What happens to a bundle of rays of sunlight (called a beam) when it strikes the earth? The answer to this question depends on whether the surface of the earth is smooth or rough. The polished surface of a metal such as copper or aluminum is smooth, and so is the water surface of a still pond. The most familiar object with a smooth surface is a mirror. A mirror may be a piece of highly polished metal or a piece of glass with a thin layer of silver on one side. To keep the silver highly polished, it must be protected from the air. This can be done by covering the silver with a thin coat of varnish or paint.

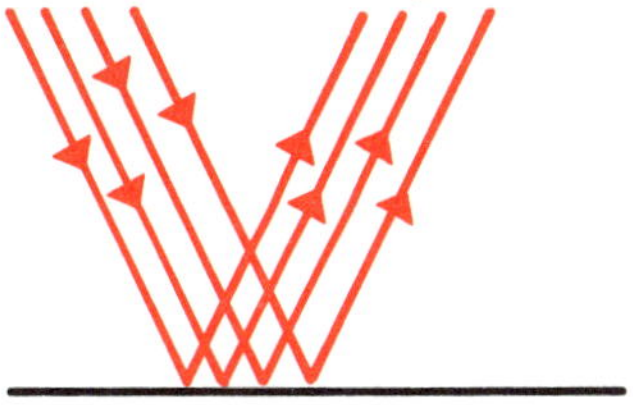

Figure 9–39
Parallel rays striking a smooth surface are reflected as parallel rays.

If a bundle of parallel rays strikes a smooth surface, it is reflected as a bundle of parallel rays, as shown in Figure 9–39. If the smooth surface is a mirror on a table, you see a patch of light only where the reflected rays strike the mirror. The rest of the room stays almost dark.

If you look directly into the patch of light, your eye receives the reflected rays as shown in Figure 9–40. Your eye cannot look in two directions. So, to your eye, the ray *OB* appears to

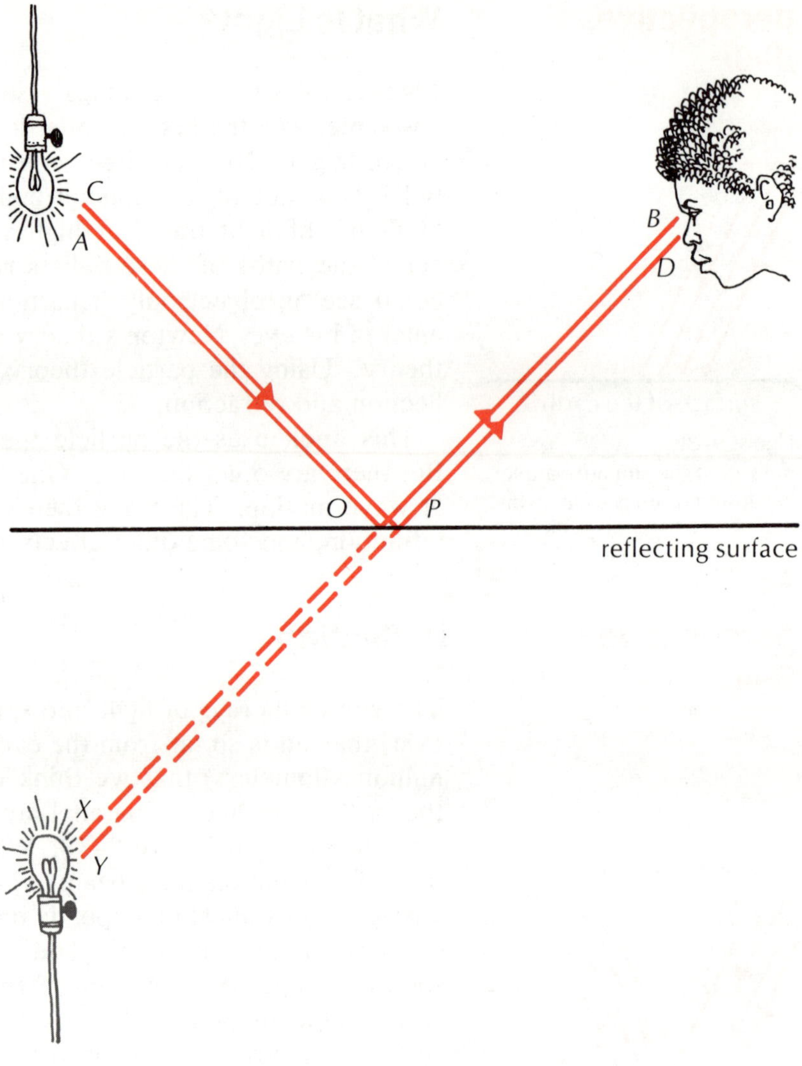

Figure 9–40
An object reflected in a mirror produces a virtual image.

come from direction OX, not from OA. Similarly, the ray PD appears to come from direction PY, not from PC. Thus, although the real object (the source of the light) is in the direction of OA and PC, your eye cannot see it. Instead, it sees an image of the object along the direction of OX and PY. Since there are no rays of light on the far side of the reflecting surface, the image is an optical illusion. An image of this kind is called a virtual image.

perspective 9–5

Reflection from a Rough Surface

If a beam of sunlight reflected from a mirror enters your eye, it is so bright as to be almost blinding. If the same beam is reflected from the page of a book there is little or no glare. Why the difference?

Although the surface of the paper appears to be smooth, it is made up of thousands of tiny humps and hollows, and is rough. As a result of this rough surface, the parallel incident rays are reflected in all directions. This is shown in Figure 9–41. The scattered rays reach all parts of the room, and the book can thus be seen from any part of the room.

Because the surfaces of most objects, such as walls, chairs, books, and even dust particles, are rough, scattering of light is a common effect. Without the scattering effect we would not be able to see most common objects. Suppose that sunlight passing through a window into a room was always reflected from objects in a regular way. That is, suppose the reflected light was made up of only parallel rays. The areas struck by direct sunlight would be dazzlingly bright. There would be some bright patches of reflected light on the walls, but the part of the room outside the direct paths of the rays would be almost dark. (See Figure 9–42.) You would not be able to see objects in the room, even in daylight.

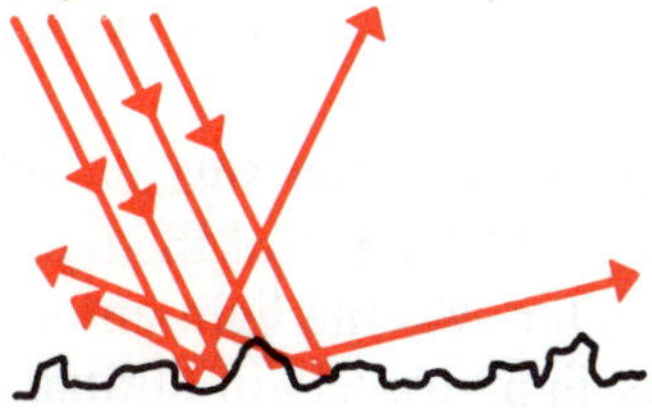

Figure 9–41
Parallel rays striking a rough surface are reflected in all directions.

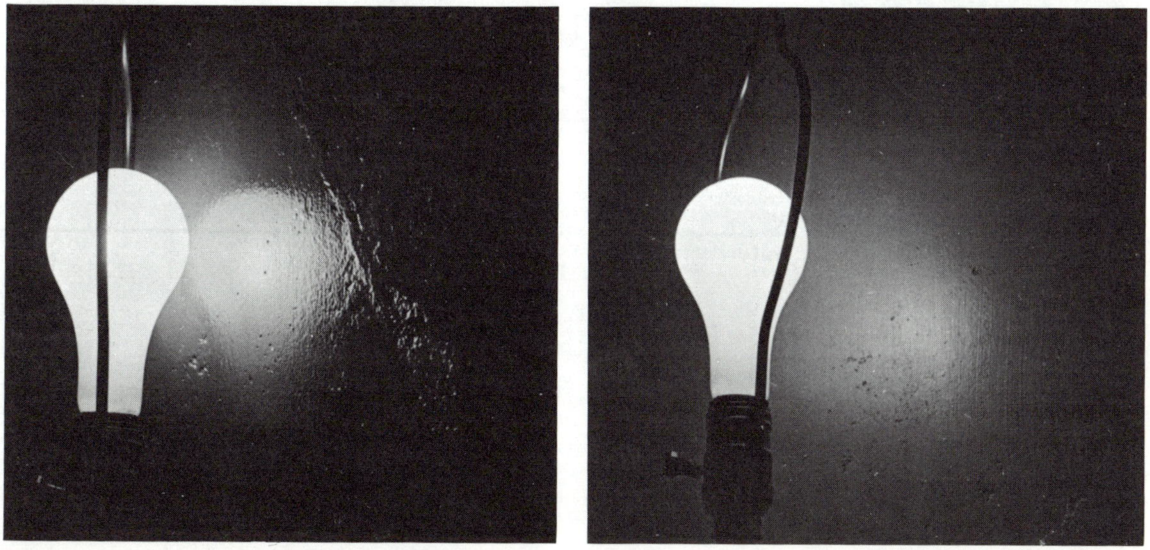

Figure 9–42
When light is reflected in a regular way, as from a wall painted with enamel paint, there is a bright patch of light but the rest of the wall stays dark. When light is reflected from a rough surface like ordinary paint, it is spread out more smoothly.

perspective 9–6

Refraction

You have probably observed that when part of a stick is submerged in water it appears to be bent at the water surface. This is shown in Figure 9–43. The stick does not actually bend. It only appears to bend. Thus you do not see the submerged part of the stick. Instead you see an image of it. How can this be explained?

First, consider a beam of light that strikes the surface of water at a point *P* as shown in Figure 9–44. Some of the light is reflected, but most of it enters the water. As the light enters the water it changes direction. Thus light travels in straight lines only if it moves through the same substance such as air or water. If the substance changes, the direction changes. The change in direction is called refraction, and the ray of light in the water is called the refracted ray. When light travels from a lighter to a denser medium, the rays are refracted toward the perpendicular.

Why does refraction occur? Water is a denser material than air. The speed of light is slowed down as it enters a denser material. The speed of light in air is about 300,000 kilometers

Figure 9–43
The stick appears to be bent because of refraction. Also, the water appears to be shallower than it really is.

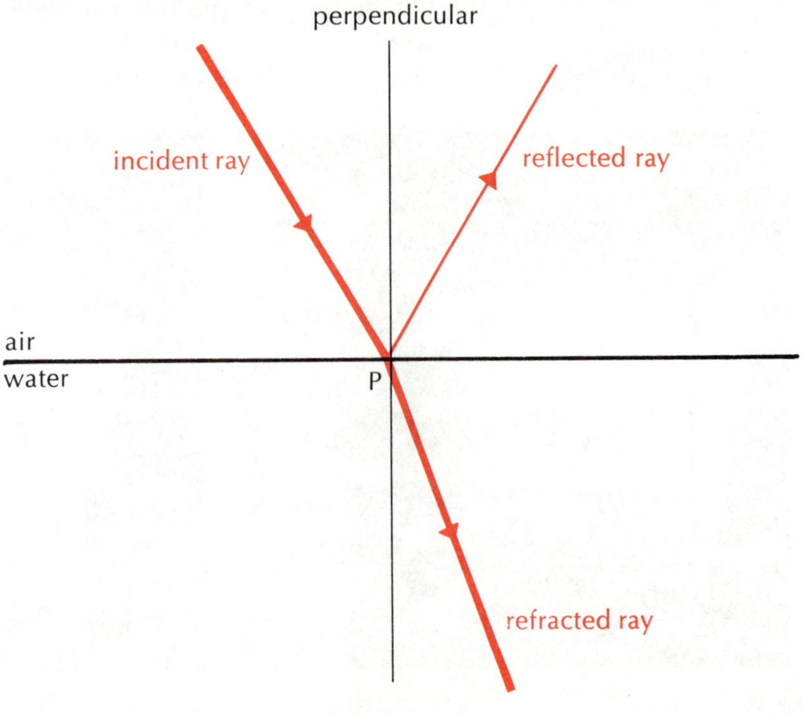

Figure 9–44
A ray of light moving from air into water is refracted.

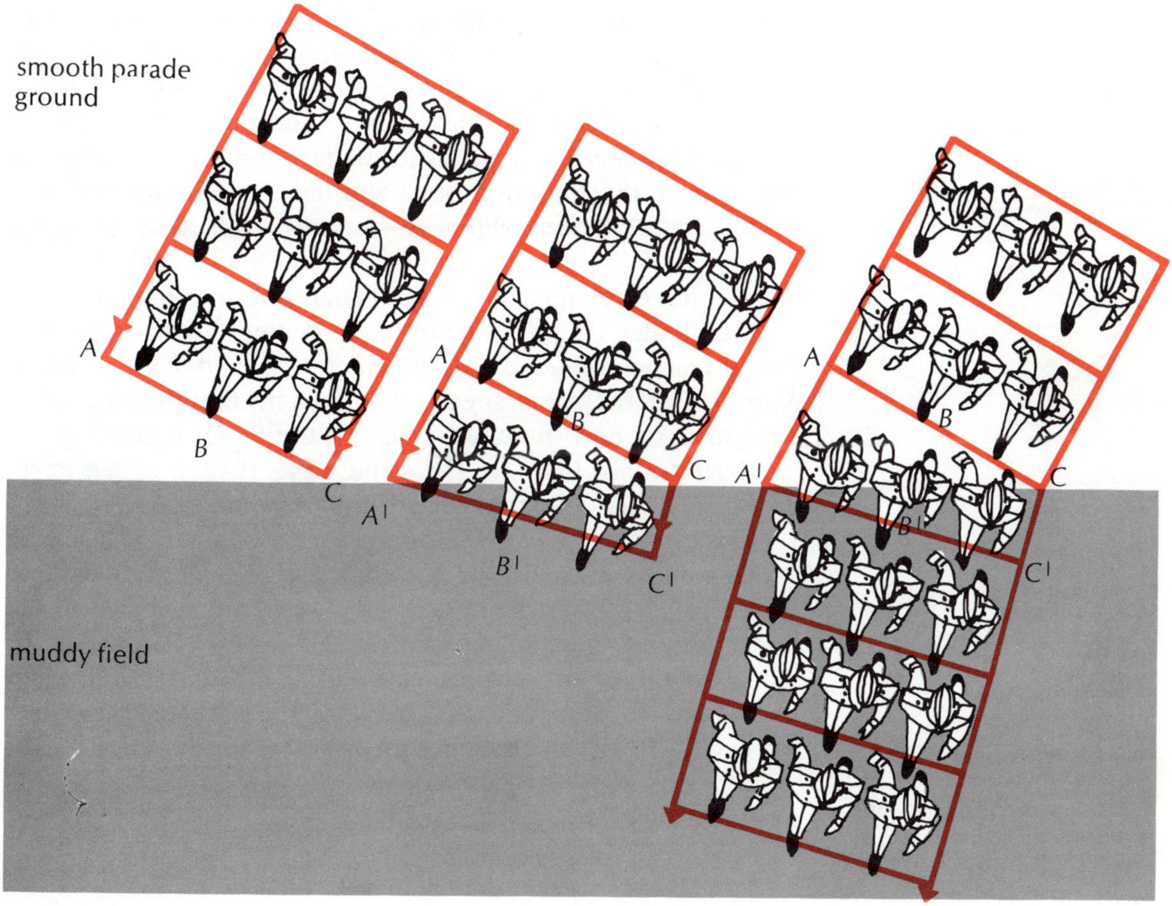

Figure 9–45
When soldiers march at an angle from a smooth parade ground to a muddy field, their line of march is bent.

per second (186,000 miles per second). In water it is about 225,000 kilometers per second. If the speed of light is reduced, the light is refracted towards the perpendicular. On the other hand, if light is refracted away from the perpendicular, it is because the speed of light is increased.

But why does a change in speed cause the light to be refracted? To answer this question, consider the path of a column of marching soldiers, three men across. Suppose the column starts on a smooth parade ground and then marches onto a muddy field. The situation is shown in Figure 9–45. The front of the column changes direction from ABC to $A'B'C'$. Can you see why?

The front of the column approaches the muddy field at an angle. Soldier C reaches the mud while soldiers A and B are still on the smooth ground. (See Figure 9–45, left.) Soldiers A and B are still marching quickly, but soldier C is slowed down by the mud. Suppose the soldiers can march four miles per hour on the smooth ground and only three miles per hour on the mud. While soldier A marches from A to A' at the boundary of the smooth ground and the mud, soldier C can march only three-quarters as far. Therefore the column is bent toward soldier C. (See Figure 9–45, center.) When soldier A finally reaches the mud a A', soldiers A, B, and C are again all marching at the same speed. The column therefore marches straight in its new direction. (See Figure 9–45, right.)

"Particles" of light act the same way. If their speed is reduced, they are refracted toward the perpendicular.

How can the appearance of a partially submerged stick be explained? In Figure 9–46 the stick is represented by the line STK. Rays of light leave the submerged tip K in all directions. One of these rays, KF, leaves the water surface at F. Since the speed of light increases in air, the ray is refracted away from the perpendicular. Suppose the refracted ray enters the eye at E. The eye receives the ray from the direction FE. It

Figure 9–46
Because of refraction, the part of the stick that is under water appears to be higher than it really is. Therefore the stick appears to be bent.

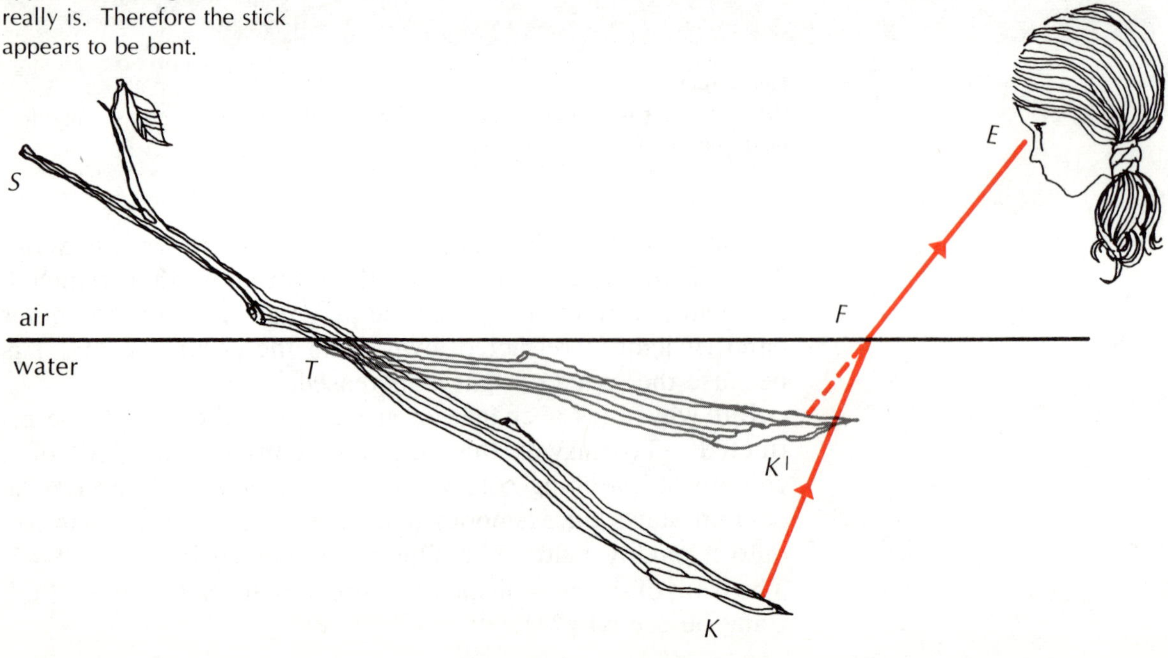

therefore sees the image along this same line, that is, along an extension of EF. The image of the tip K appears to be at K'. The image of the submerged part of the stick is shown by the broken line TK'.

When early American Indians aimed at a fish under water with bow and arrow, they were faced with this same problem of refraction. The fish, which they could not see, was below the image, which they could see. They learned by experience that they could hit the fish by aiming below the image.

Glass is an essential part of most optical instruments. Light is refracted by glass as it is by water. However, the amounts of refraction are different. The speed of light in glass is about 200,000 kilometers per second. This is a little slower than the speed of light through water. The amount of refraction is therefore greater for glass than for water.

perspective 9–7

Refraction through a Lens

If a beam of light strikes a lens, it is refracted as it enters the lens and again as it leaves the lens. If the lens is convex, parallel incident rays are refracted so that the emerging rays pass through the focal point as shown in Figure 9–47. The emerging rays **converge** (come together and pass through a point). A convex lens is sometimes called a **converging lens.**

The distance from focal point to lens is called the focal length. Light passing through a convex lens is focused, whether it is received from the left or the right. The lens therefore has a focal point on the left as well as on the right.

On the other hand, if a beam of parallel rays strikes a concave lens, the emergent rays get farther apart or **diverge,** as shown in Figure 9–48. Thus a concave lens is sometimes called a **diverging lens.** You will use only convex lenses in your experiments, not concave lenses.

A convex lens can give a real image, but a concave lens cannot. The image of a convex lens is real and upside down if the object is beyond the focal point of the lens. But if the object is between the lens and the focal point of the lens, the image is virtual, right side up, and larger than the object. A convex lens is often used in this way as a magnifying glass or, as it is sometimes called, a reading glass.

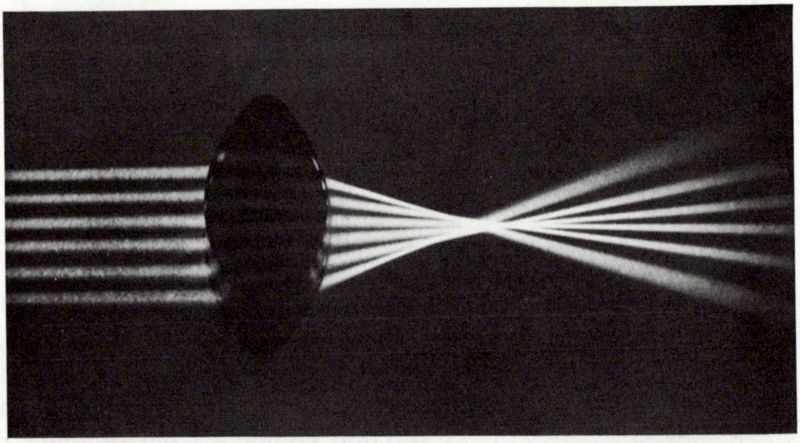

Figure 9–47
Light rays passing through a convex lens converge.

Figure 9–48
Light rays passing through a concave lens diverge. Notice what happens to the light reflected from the left side of the lens.

The lens of your eye is a convex lens. In order for you to see an object, light from the object must pass through the lens and come to a focus on the retina. The image formed on the retina is real and upside down. You do not see an upside-down image, however. This is because your brain interprets the signal so that you see the image right side up.

perspective 9–8

Lenses in Optical Instruments

Convex lenses are essential parts of the eye and of cameras. Lenses are also used to enlarge the field of vision. In ordinary eyeglasses lenses make it possible to see objects more clearly.

In microscopes they make it possible to see tiny particles, such as living cells, too small to be seen with the naked eye. Lenses in a telescope enable us to see thousands of stars which are too far away to be seen by the eye alone.

A. Telescope

The word *telescope* comes from a Greek word meaning "to see far." As you know, telescopes are used to see distant objects such as stars. The first telescope was constructed by an Italian scientist named Galileo in the early part of the 17th century. (Galileo died in the same year that Newton was born, 1642.) Galileo ground and shaped two lenses out of ordinary glass and fitted them into the ends of a tube. This was a small telescope, less than 60 cm in length. The lenses were only 6 cm across. However, with his telescope Galileo could see hundreds of stars that were invisible to the naked eye. When he pointed his telescope at the planet Jupiter, he was amazed to see four moons around the planet. Today we know that Jupiter has twelve moons. Galileo was the first person to see the four largest ones.

The way a telescope works is shown in Figure 9–49. The instrument consists of two convex lenses at the ends of a long tube. The lens that points to the object is called the **objective lens.** The lens you look through is called the **eyepiece.** The objective lens has a long focal length. Light from a distant

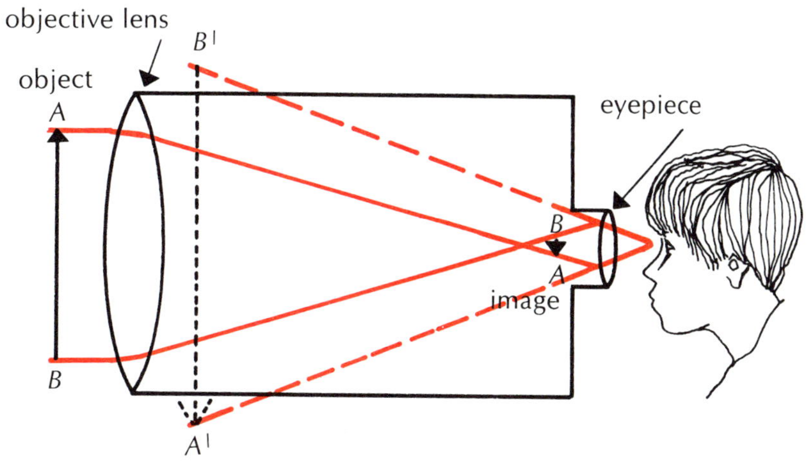

Figure 9–49
Through a refracting telescope you see a virtual image that is upside down and larger than the object.

object is brought to a focus at the focal point. You recall that the image formed at the focal point of a convex lens is real and upside down. This image lies just inside the focal length of the eyepiece. The eye sees a virtual enlarged image of this real image. The virtual image is shown by the dotted arrow $B'A'$.

This kind of telescope is called a **refracting telescope** because light is refracted by both the objective lens and the eyepiece. A refracting telescope is shown in Figure 9–50. There are also reflecting telescopes.

Since light from a distant star is seen through a telescope, this same light would also enter the unaided eye even though it is not seen. The unaided eye cannot see a distant star because the light is much too faint for the retina to respond. A telescope must not only magnify. It must also gather enough light from a star for the eye to see it. For this reason, the area of the objective lens is made very large. The diameter of the objective lens of the largest refracting telescope is 40 inches, just over 1 m. This telescope is at Yerkes Observatory in Wisconsin. The Yerkes telescope gathers almost 30,000 times as much light from a distant star as does the unaided eye. (Look back at Figure 1–1.)

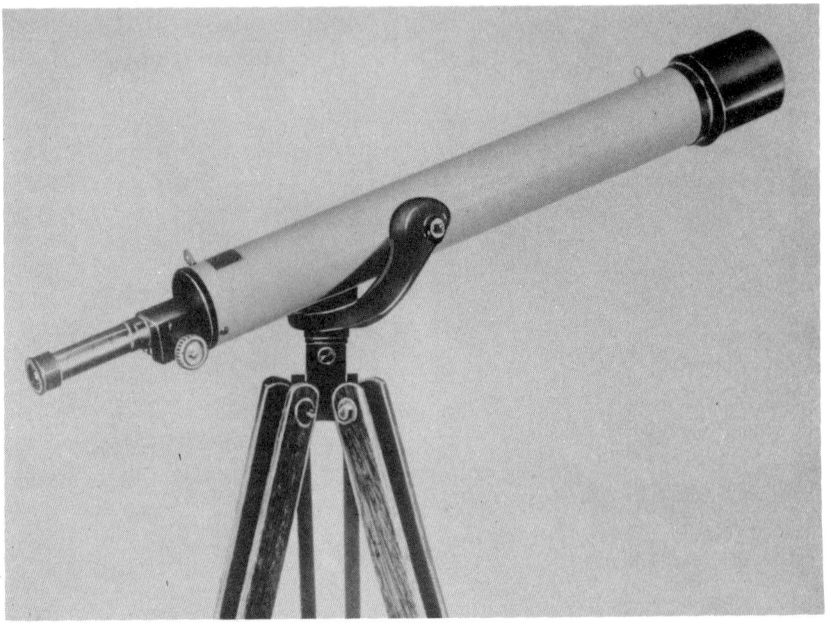

Figure 9–50
A refracting telescope.

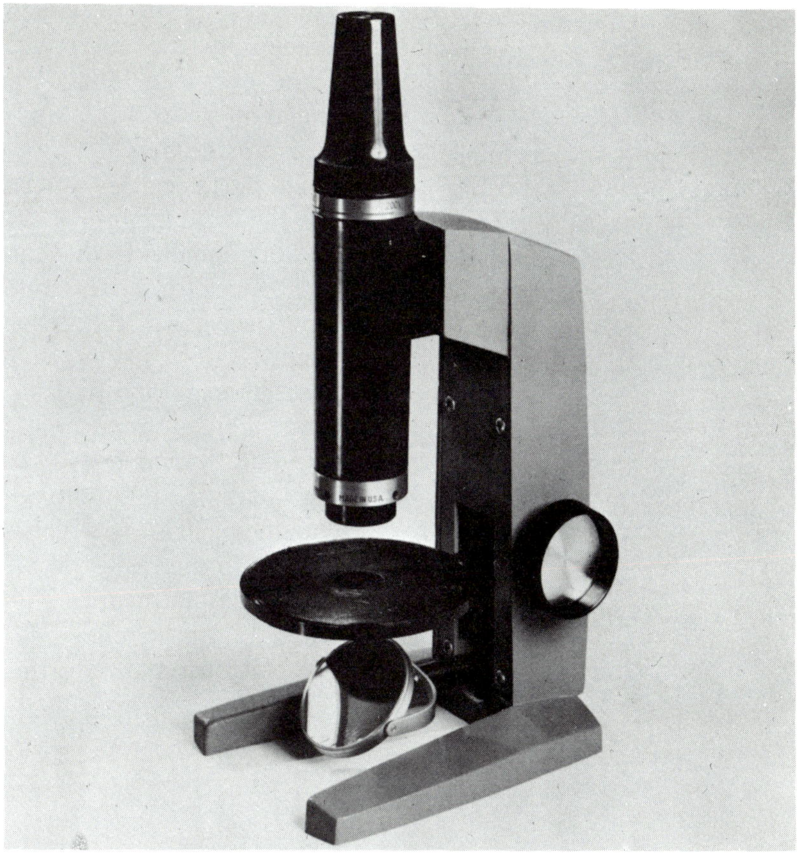

Figure 9–51
A microscope has two convex lenses, the objective lens and the eyepiece. A mirror below the platform reflects light onto the object being examined.

B. Microscope

An ordinary reading glass magnifies six or seven times. But this is not nearly enough to examine objects too small to be seen by the eye. A microscope is needed for this purpose. The word *microscope* comes from a Latin word meaning "to see tiny particles."

A microscope, like a refracting telescope, has two convex lenses. One is the objective lens and the other is the eyepiece. A microscope is shown in Figure 9–51. The object to be examined is mounted on a glass slide and the slide is placed on the platform of the microscope. In this position the tiny object is at the focal point of the objective lens. The eyepiece, at the other end of the tube, can be adjusted so that the real image formed by the objective lens is just inside the focal point of the eyepiece lens. The eye then sees a magnified virtual image of the object. Also, as in the refracting telescope, the image is upside down.

The main purpose of a microscope is magnification. For this reason the focal lengths of both objective lens and eyepiece are made very small. If they are small enough, a magnification of 2000 times (or more) can be obtained. That is, in a good microscope the size of the image can be 2000 times the size of the object.

However, this great increase in size raises a problem in illumination. This is because light from a small object becomes very dim when spread over a large area. So, to get a bright image, the object must be strongly illuminated. This is usually done by artificial lighting. Light from a light bulb is directed through a hole in the center of the platform. Usually there is a mirror below the platform as shown in Figure 9–51. The mirror is turned so that it reflects light from the bulb through the hole in the platform and onto the object being examined.

C. Eyeglasses

You know from experience that you can see near and distant objects almost at the same instant. For example, you can see a tree on a distant hill and then, by changing your field of vision, you can see a book on a nearby table. Both tree and book are seen clearly and their images are in focus. If you picked up the images of these objects with a convex lens, you would have to move the lens. But the eye lens cannot be moved. As you learned in Perspective 9–2, the eye muscles can change the shape of the lens. If you look at a near object, the eye lens is made rounder (thicker) to focus a clear image of the object on the retina. If you look at a distant object the eye lens is made flatter (thinner).

In a normal eye, light from an object a few feet away is focused on the retina without any strain on the eye muscles.

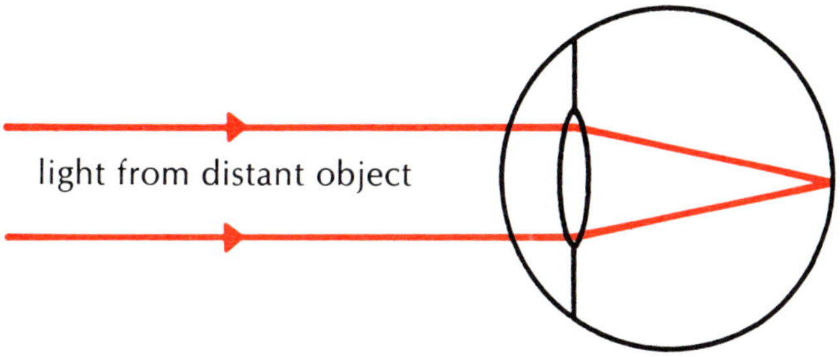

Figure 9–52
A normal eye can focus light from a fairly distant object onto the retina with no strain on the eye muscles.

This situation is shown in Figure 9–52. To look at an object that is closer or farther away, the lens is made thicker or thinner.

The eye is a complex instrument of many parts. Some of the parts may be slightly defective. For instance, you may not be able to see a near object clearly. Or you may not be able to see a distant object clearly. In either case you have defective vision.

Some people can see near objects clearly, but distant objects are blurred, as in Figure 9–53. This condition is called **nearsightedness.** In a nearsighted eye the eyeball is too long or the focal length of the lens is too short. The lens can change enough to focus a near object, but not a distant object. The sharp image of a distant object forms in front of the retina, as shown in Figure 9–54, so the image on the retina is blurred. Nearsightedness can be corrected by a concave or diverging lens as shown in Figure 9–54. The concave lens causes the rays of light to diverge. When these diverging rays pass through the eye lens, they converge to form a clear image on the retina.

You can make a simple test to see if you are nearsighted. When you read a book, the printed page should be about 25 cm (10 inches) from your eye. To bring the words into focus, your eye lens has been made thicker by the eye muscles. This amount of muscular effort causes no eye strain for the average person. However, if you have to hold the book much nearer to your eye than 25 cm, you are nearsighted. That is, although your eye muscles have thickened the eye lens, you are unable to see clearly at 25 cm, so you must move the book closer. If a nearsighted person wears the correct glasses, he can read at a distance of 25 cm.

Figure 9–53
A nearsighted person can see near objects clearly, but distant objects are blurred.

Figure 9–54
In a nearsighted eye, the image of a distant object forms in front of the retina. A concave lens can correct nearsightedness.

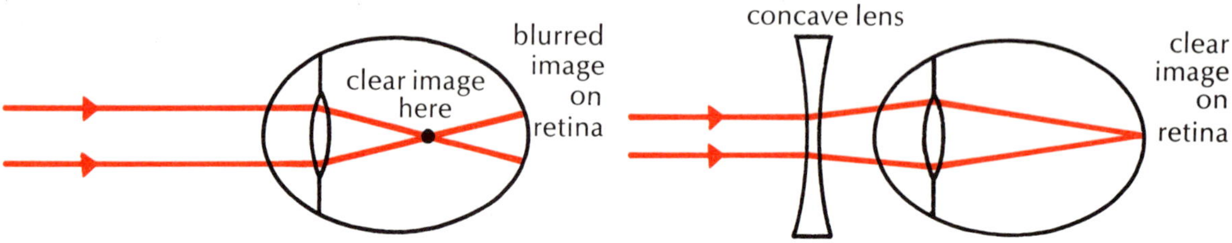

If you can see distant objects clearly but near objects are blurred, as in Figure 9–55, you have a condition called **farsighted.** A farsighted eye is too short or the focal length of its lens is too long, as shown in Figure 9–56. Farsightedness can be corrected by using converging (convex) lenses as eyeglasses. As shown in Figure 9–56, the light that enters the eye is converged by both the convex eyeglass and the eye lens. As a result, a clear image of the object is formed on the retina.

Figure 9–55
A farsighted person can see distant objects clearly, but near objects are blurred.

Figure 9–56
In a farsighted eye, the image would be formed behind the retina. A convex lens can correct farsightedness.

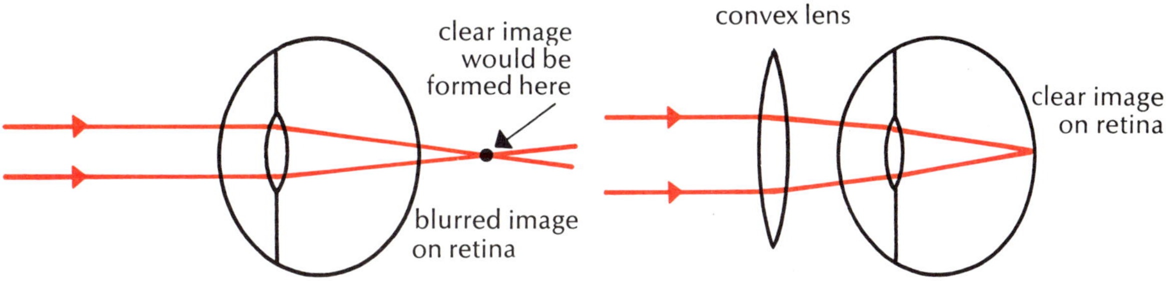

QUESTIONS

A

1. What is illumination?
2. What is the iris of the eye, and what is its purpose?
3. What is the retina and what is its purpose?
4. What was Newton's theory of light?
5. What is the difference between a real image and a virtual image?

6. Using a diagram, explain why an object under water does not appear to be where it really is.

7. What is (a) a convex lens, (b) a concave lens?

8. What is (a) the focal point, (b) the focal length of a convex lens?

9. What is (a) a telescope, (b) a refracting telescope?

10. A microscope is a double lens system. What does this mean?

11. Name two common eye defects and state what the names mean.

12. Copy the crossword puzzle. Study the definitions and fill in the correct words.

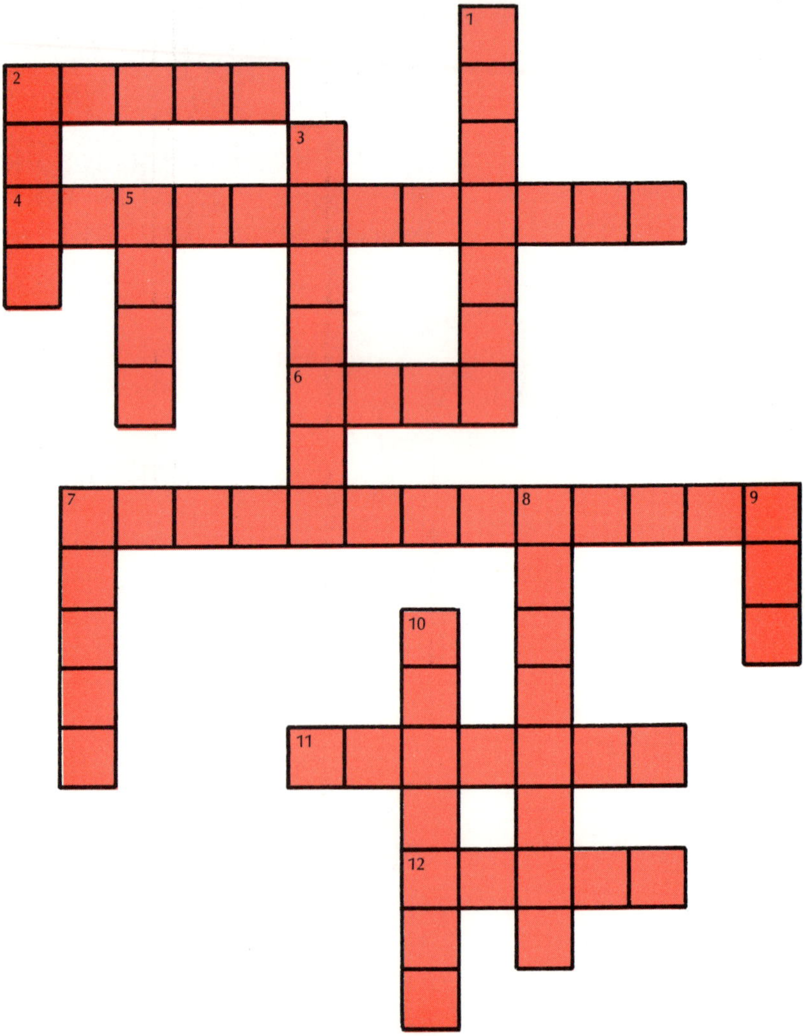

Definitions:

Across:

2. the picture of an object
4. the amount of light that falls upon a surface
6. an image that can be projected on a screen
7. a line drawn 90° to another line
11. to bounce back
12. two lines which meet form a(n) ________

Down:

1. an image that is not real
2. the colored circular part of the eye
3. to get farther apart
5. a piece of glass with a curved surface
7. a triangular piece of glass
8. to come together
9. a beam of light
10. bend

13. What is the difference between the sun and the moon as sources of light?

14. What would the word WHY read if looked at in a plane mirror? Give a reason for your answer.

15. How does the eye (a) resemble a camera, (b) differ from a camera?

16. What causes refraction?

17. The image formed by a convex lens may be either upside down, or right side up. Why?

18. Is the image in a magnifying glass real or virtual? Give a reason for your answer.

19. Why is the objective lens of an astronomical telescope very wide?

20. Draw a diagram of the lenses of a microscope and show where the images are formed.

21. What is the essential difference between a refracting telescope and a microscope?

22. What kind of lenses are used in eyeglasses to correct nearsightedness? Why?

23. A man reads a newspaper at arm's length. What kind of eye defect do you think he has? Why?

24. In the diagram of the telescope (Figure 9–49), the length of the virtual image is greater than the width of the tube of the telescope. That is, part of the virtual image is outside the tube. Does the tube interfere with the sighting of the image? Explain.

B *For Questions 25–34, write in your notebook the letter that completes or answers each statement.*

25. A boy 5 feet tall stands 4 feet in front of a 6-foot-high plane mirror. The boy's image is

a 6 feet tall measured at the mirror and 5 feet tall measured 4 feet in back of the mirror
b 4 feet tall
c located 4 feet behind the mirror
d 4 feet tall and located 6 feet behind the mirror

26. You can see the page of this book because

a the angle of incidence equals the angle of reflection
b it refracts the light that falls upon it
c it diffuses and reflects the light that falls upon it
d light is absorbed by the pages of the book

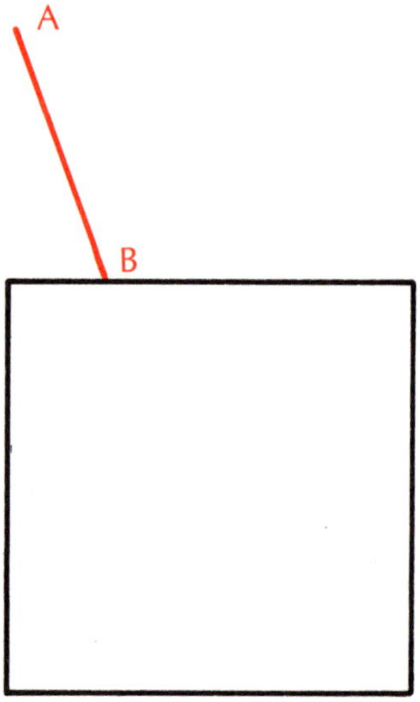

27. In the figure at the left, the incident ray *AB*

a will be reflected through the plate and emerge from the bottom
b will continue straight through the plate on the same path it entered the plate
c will emerge from the bottom of the plate on a path that is parallel to ray *AB*
d will be totally reflected by the plate at the bottom edge

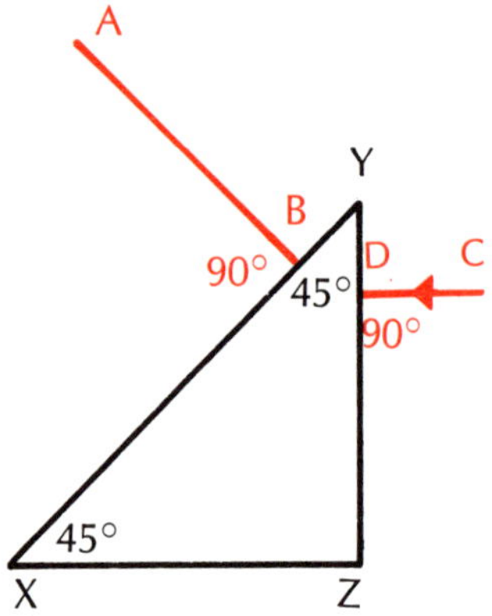

28. In the figure at the left, the incident ray *AB*

a will be reflected twice within the prism and emerge from the same side where it entered
b will be refracted by the prism
c will emerge through side *YZ*
d will be refracted and emerge through side *YZ*

29. In the figure at the left, the incident ray *CD*

a will be reflected internally by side *XY*
b will emerge through side *XY*
c will be refracted through side *XY*
d will be refracted and emerge through side *YZ*

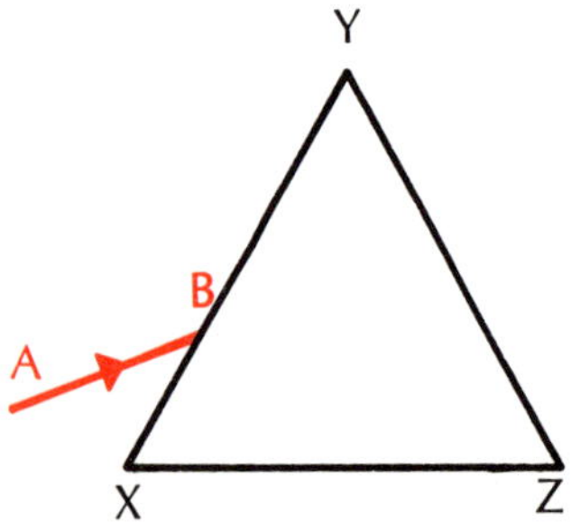

30. In the figure at the left, the incident ray *AB*

a will be reflected internally by side *YZ*
b will emerge through side *YZ*
c will be reflected by the prism, emerging through side *YZ*
d will be refracted away from the base of the prism

31. Your eye, observing an object through a lens with focal points F and F′ as in the figure below, would see the image of the object

a larger and right side up
b upside down
c smaller and upside down
d smaller than the object

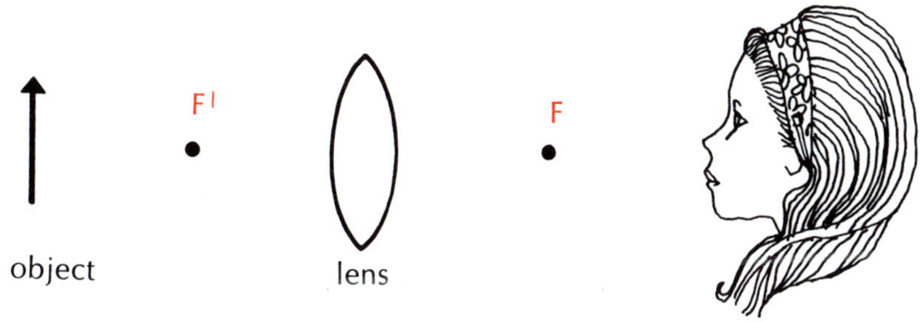

32. The right triangle , if viewed in a plane mirror, would look like

a 
b 
c 
d 

33. If you are 3 m from a source of light, and move 3 m farther from the source, you will receive

a $\frac{1}{2}$ as much illumination
b $\frac{1}{3}$ as much illumination
c $\frac{1}{4}$ as much illumination
d $\frac{1}{9}$ as much illumination

34. If you are 3 m from a source of light, and move to a position $1\frac{1}{2}$ m from the source, you will receive

a $1\frac{1}{2}$ times as much illumination
b twice as much illumination
c three times as much illumination
d four times as much illumination

For each of questions 35–40, there are three words in parentheses to choose from. Choose a word to fill in the blank so that the two words on the right relate to each other in the same way as the two words on the left.

35. iris and pupil — circle and ________ (square, center, perpendicular)

36. real and virtual — converge and ________ (diverge, concave, convex)

37. reflect and mirror — refract and ________ (perpendicular, lens, protractor)

38. eye and retina — camera and ________ (lens, object, film)

39. principal focus and focal point — perpendicular and ________ (angle, protractor, right angle)

40. convex lens and transparent — real image and ________ (virtual image, upside down, glass)

Appendix 1

THE SLIDE RULE

Some slide rules are circular and others are straight. For most purposes, the most convenient is a straight slide rule, about 10 inches long. A 10-inch slide rule is shown in Figure 1. The first thing to notice about a slide rule is the many scales on both sides of it. All the scales are lettered. For example, on one side of the slide rule there are scales lettered K, A, B, C, and D. The scales C and D are used in the processes of multiplication and division. So these are the scales you should learn to use.

Notice that scale D is on the lower frame of the slide rule. It is a fixed scale. Scale C is on the central section of the slide rule, which is movable. This movable section is usually called the **slide.** Another movable part is the glass or plastic **runner** which slides along the outer edges of the slide rule. A fine line, called a **hairline,** crosses the runner. The hairline is used to locate numbers on the scales.

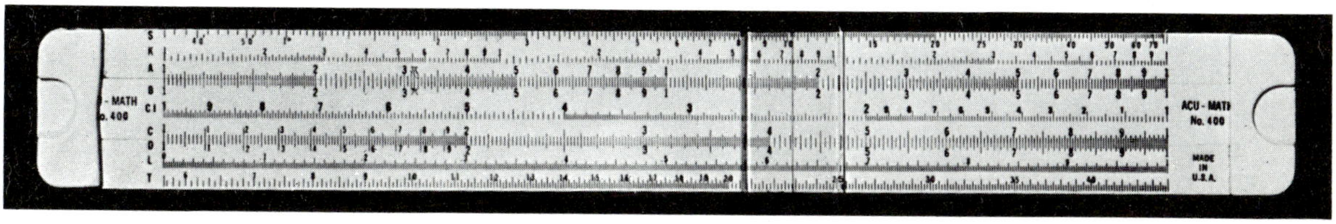

Figure 1
A 10-inch slide rule. Notice the scales lettered K, A, B, C, and D.

1. The C and D Scales

A. The Main Divisions

Move the slide until the C and D scales line up with each other. Notice that each of the scales is divided into 10 **main divisions.** These main divisions are numbered with large numbers as shown in Figure 2. However, the divisions are not equal. They become shorter as we go from left to right on the scale. The mark on the left is numbered 1, the next on the right is 2, then 3 and so on up to 9. The mark on the extreme right is numbered 1, not 10. The numbers of these main divisions are usually called **marks.**

For practice, move the runner to the left until the hairline is over mark 1 on the two scales. Then move the runner to the right until the hairline is over mark 2. Now move the runner to the right and place the hairline in turn over marks 3, 4, 5, 6, 7, 8, 9, and 1.

B. The Secondary Divisions and Subdivisions

On a 10-inch slide rule, the main divisions are divided into 10 parts, with unequal lengths, called **secondary divisions.** For instance, the 10 parts between marks 1 and 2 are labeled with the small numerals 1, 2, 3, 4, etc., representing the secondary divisions. Each of these secondary divisions is also divided into ten parts called **subdivisions.**

Now examine the ten secondary divisions between marks 2 and 3. Because of lack of space these secondary divisions are not numbered as they are between marks 1 and 2. Because the lengths between marks become smaller towards the right, it is not possible to make the same number of clear subdivisions as between marks 1 and 2. For this reason the ten secondary divisions between marks 2 and 3 are subdivided into only five parts. And the same is true of the ten secondary divisions between marks 3 and 4.

The ten secondary divisions between all the other marks are subdivided into only two smaller parts. For example, between marks 4 and 5 there are ten secondary divisions, and each secondary division has only two subdivisions. Similarly, between

Figure 2
The main divisions of the C and D scales.

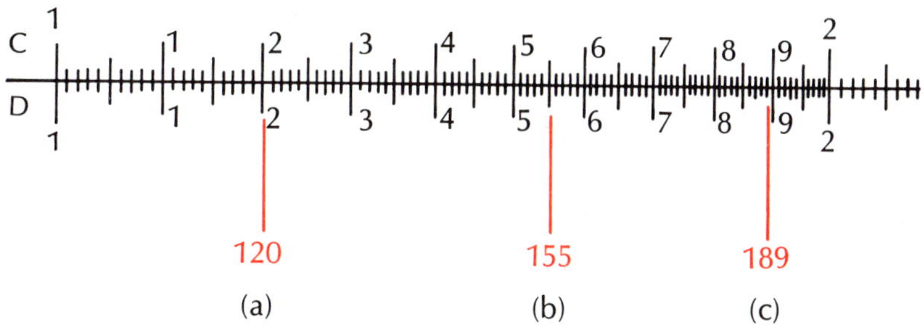

Figure 3
Scale readings between marks 1 and 2 on the C and D scales.

marks 8 and 9 there are ten secondary divisions, each of which is subdivided into two smaller parts. Thus, in reading a scale on a slide rule, there is a main division, a secondary division, and a subdivision. As we shall see, any scale reading can be read to three figures.

2. To Read the Scale

A. Scale Readings Between Marks 1 and 2

(1) Move the runner until the position of the hairline is the same as in Figure 3 (a). The first figure of the scale reading is 1. The second figure of the scale reading is 2. Therefore the reading on both C and D scales is 12 to two figures or 120 to three figures.

(2) Move the runner until the hairline is the same as in Figure 3 (b). The first figure is 1, the second is 5, and the third is 5. Therefore the scale reading is 155.

(3) Move the runner until the hairline is the same as in position (c). The second figure is 8 and the third is 9. So the scale reading is 189.

B. Scale Readings Between Marks 2 and 4

In this section of the slide rule, the secondary divisions are divided into five parts. Therefore each subdivision has a value of 2 in the third figure of the scale reading.

(1) Move the runner until the hairline is the same as in Figure 4 (a). The main division number is 2, the secondary division number is 3, and the subdivision number is 0. So the scale reading is 230 to three figures, or 23 to two figures.

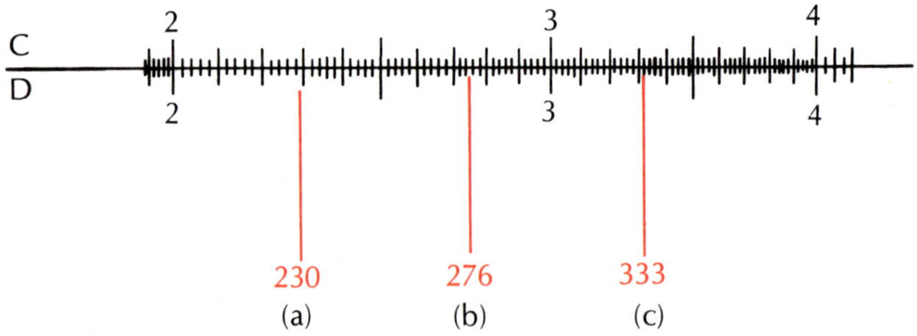

Figure 4
Scale readings between marks 2 and 4 on the C and D scales.

(2) Move the hairline to the position shown in Figure 4 (b). The main division number is 2, and the secondary division number is 7. What is the subdivision number? The subdivision number is the third line between secondary divisions 7 and 8. The space between secondary numbers 7 and 8 is divided into five parts, not ten. Therefore each subdivision has a value of 2. Since the hairline is over the third line of the subdivision, its scale value is 6. Hence the scale reading is 276.

(3) Move the runner to position (c). The main division number is 3 and the secondary division number is 3. The arrow points midway between the subdivision numbers 2 and 4. So the third figure is 3 and the scale reading is 333.

C. Scale Readings Between Marks 4 and 1

In this section of the slide rule, the secondary divisions are divided into only two parts (not five or ten). Therefore each subdivision has a value of 5 in the third figure of the scale reading.

(1) Move the hairline of the runner to position (a) of Figure 5. The main division number is 4, the secondary division number is 5, and the subdivision number is 5. Therefore the scale reading is 455.

(2) Move the hairline to position (b). The main division number is 5, the secondary division number is 0, and the subdivision number is 5. Therefore the scale reading is 505.

(3) Move the runner to position (c). The main division number is 6, the secondary number is 8, and the subdivision number is 0. So the scale reading is 680 to three figures, or 68 to two figures.

(4) Move the runner to position (d). The scale reading is 740 to 3 figures, or 74 to 2 figures.

(5) Move the runner to position (e). The scale reading is 995.

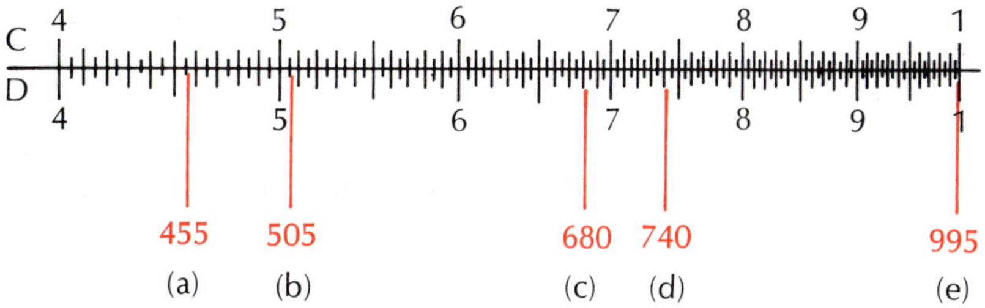

Figure 5
Scale readings between marks 4 and 1 on the C and D scales.

3. Multiplication of Numbers by Slide Rule

The movable slide carries the numbers on the C scale. So by moving the slide, you can "add" numbers on the C scale to numbers on the D scale. The addition of the two scales is in effect the process of multiplication. You can see how by multiplying 2 × 3.

Move the slide so that mark 1 on scale C (or main division number 1) is above mark 2 on scale D. Move the runner until the hairline is over mark 3 of scale C. The hairline now reads 6 (the main division number) on scale D. You know that $2 \times 3 = 6$. We also see that a reading of 2 on scale D plus a reading of 3 on scale C gives a reading of 6 on scale D. Hence the rule for multiplying two numbers is as follows: *Set mark 1 of scale C over the first of the numbers. The scale reading of the product is on scale D below the second of the numbers.*

Thus to multiply 2 by 3, proceed as follows. Move the slide until mark 1 on scale C is above mark 2 on scale D as shown in Figure 6. Move the runner until the hairline is over mark 3 on scale C. The answer (6) is the reading of scale D under the hairline.

To multiply 2 × 4: Leave the rule with the same setting of scales C and D. Move the runner so that the hairline is above mark 4 on scale C. The answer (8) is the number under the hairline on scale D.

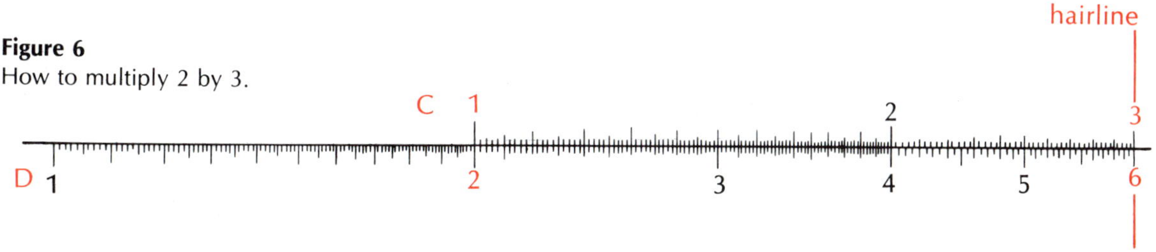

Figure 6
How to multiply 2 by 3.

To multiply 2×5*:* With the same setting of the rule, move the hairline of the runner to mark 5 on scale C. The mark under the hairline on scale D is 1, which indicates the number 10. Thus, by slide rule, $2 \times 5 = 10$.

To multiply 2×6*:* With the same setting of the rule, mark 6 of scale C is too far to the right. It is beyond the right end of scale D. When this happens, start your measurements from the right end of scale D instead of the left end. That is, move the slide to the left so that mark 1 at the right end of scale C coincides with mark 2 on scale D. Move the runner until the hairline is over mark 6 on scale C. Read the answer on scale D. Notice that the runner is over small number 2 on scale D. This small number 2 is a secondary division between marks 1 and 2. Hence the answer is 12.

To multiply 2×7 *and* 2×9*:* With the same setting of the rule, move the runner until the hairline is over mark 7 on scale C. Read the answer on scale D. It is 14. Hence, by slide rule, $2 \times 7 = 14$. Now move the runner until the hairline is over mark 9. The answer is read on scale D, and it is 18.

4. Multiplication of Decimal Numbers

There are no decimal points on a slide rule, only numbers. Therefore by slide rule the answer to 2×0.3 is exactly the same as 2×3 or 20×300. In each case the answer by slide rule is 6. You must put in the decimal point from an inspection of the problem. Thus, $2 \times 0.3 = 0.6$, and $20 \times 300 = 6000$.

Example 1: 0.3×25. Place mark 1 of scale C on mark 3 of scale D. Move the hairline of the runner to the position over 25 on scale C. The answer, read on scale D, is 75. However, you know by inspection that the correct answer is 7.5, not 75.

Example 2: 70×0.45. Place mark 1 on the right end of scale C on mark 7 of scale D. Move the runner until the hairline is over number 45 (between marks 4 and 5) on scale C. The scale reading on scale D is 315. The decimal point is found by a rough calculation: 0.45 is nearly 0.5, and 70×0.5 is 35. Hence, the answer is nearly 35. Therefore the correct answer is 31.5.

Example 3: 7.1×5.75. To calculate roughly, substitute 7 for 7.1 and 6 for 5.75. This gives 42 as an approximate answer. By slide rule, place the right end of scale C over 71 on scale D. Place the hairline over 575 on scale C. The reading on scale D is between 405 and 410, but nearer 410. The best slide rule reading is 408. The approximate answer gives us the correct position of the decimal point. So the correct answer is 40.8.

Example 4: 3.14×283. By rough calculation, $3 \times 300 = 900$. By slide rule: Notice that 3.14 is the constant *pi*. Because it is used in many computations, this value is especially marked on both the D and C scales. Place mark 1 (left end) on scale C over 314 on scale D. Move the hairline to 283 on scale C. Notice that there is no mark for 283. It lies between the marks for 282 and 284. The reading on the D scale is 890. The approximation tells us that the correct answer is 890, not 89 or 8900.

5. Multiplication of Three Numbers

Example 5: $25.4 \times 0.0314 \times 7.82$. By rough calculation

$$25 \times 0.03 \times 8 = 200 \times 0.03 = 6$$

By slide rule: Set mark 1 (left end) of C on 254 on scale D. Move the hairline to 314 on C. We have now multiplied 254 by 314 and the reading on D is 797. But this reading is not the answer. You still have to multiply it by 782. Therefore, move the runner until the right end (mark 1) of C is under the hairline. Then move the hairline to 782 on C. The reading on D under the hairline is 624. Check the position of the decimal point in the rough calculation. The correct answer must be 6.24, not 624, or 62.4, or 0.624.

6. Multiplication of Four Numbers

Example 6: $0.645 \times 2.86 \times 0.0373 \times 7.32$. By rough calculation

$$0.6 \times 3 \times 0.04 \times 7 = 1.8 \times 0.28 = 0.6$$

By slide rule: Set mark 1 (right end) of C on 645 of D. Move the hairline to the left to 286 on C. Move the slide to the right until mark 1 (left) of C is under the hairline. Move the hairline to 373 on C. Move the slide until mark 1 (right) of C is under the hairline. Move the hairline to 732 on C. The reading on D under the hairline is 505. Therefore the correct answer is 0.505.

7. To Compare Multiplication by Slide Rule and Arithmetic

In Experiment 2–2 you computed the volume of a regular object from its length, width, and thickness. This can be done either by slide rule or by arithmetic. In the following example we will compute the volume by both methods.

Example 7: Find the volume of a regular object whose dimensions are: length 11.3 cm, width 5.7 cm, and thickness 3.2 cm. By rough calculation, $11 \times 6 \times 3 = 198$. The answer by slide rule is 206 cm^3. By arithmetic

$$\begin{array}{r} 11.3 \\ \times\ 5.7 \\ \hline 791 \\ 565 \\ \hline 64.41 \end{array} \qquad\qquad \begin{array}{r} 64.41 \\ \times\ 3.2 \\ \hline 12882 \\ 19323 \\ \hline 206.112 \end{array}$$

That is, the answer by slide rule is 206 cm^3 and the answer by arithmetic is 206.112 cm^3. In other words, the slide rule gives an answer to 3 figures and ordinary multiplication gives an answer to 6 figures. Which is the better answer? At first it looks as though the answer with the larger number of figures is the more accurate. But this is not true. Let us see why.

The length of the object is given to 3 figures and the width and thickness are given to 2 figures. When data are accurate to 3 figures, the product of these data is also accurate to 3 figures, and not more than 3 figures. If the factors are accurate to only 2 figures, then the product is accurate to only 2 figures. Thus, in the above problem the slide rule answer to 3 figures is the limit of accuracy we can expect in this problem.

So, if you solve this problem only by arithmetic, you know that the first two figures 2 and 0 are accurate. The third figure 6 may be accurate. But you know nothing about the accuracy of the last three figures 1, 1, and 2. So these figures should be discarded. In short, if your multiplication is done by arithmetic, your answer should be rounded off to three figures. Thus an answer of 12.725 would be rounded off to 12.7. An answer of 678.84 would become 679, and an answer of 0.85695 would become 0.857.

Computations

1. 2.9×2.9	Answer 8.4
2. 2.9×9.2	Answer 26.6
3. 16×4.8	Answer 77
4. 6.6×4.8	Answer 31.7
5. 7.23×0.372	Answer 2.69
6. $4.1 \times 2.2 \times 7.4$	Answer 67
7. $4.1 \times 7.4 \times 0.22$	Answer 6.7
8. $7.4 \times 0.22 \times 0.41$	Answer 0.67
9. $16.3 \times 121 \times 3.65$	Answer 7200
10. $6.31 \times 22.1 \times 0.631$	Answer 88.0
11. 173×0.0444	Answer 7.68
12. $8.25 \times 4.12 \times 1.07 \times 0.036$	Answer 1.32

8. Division

You know that 36 divided by 12 equals 3.

$$\frac{36}{12} = 3$$

The number above the line in the fraction is called the *numerator,* and the number below the line is called the *denominator*. The number that results from the division of one number by another is called the *quotient*. Or, using these terms,

$$\frac{36 \text{ (numerator)}}{12 \text{ (denominator)}} = 3 \text{ (quotient)}$$

How is 36 divided by 12 on a slide rule? Set the hairline on line 36 on scale D. Move the slide until line 12 on scale C is under the hairline. The quotient (3) is then read on scale D under mark 1 of scale C as shown in Figure 7.

Figure 7
How to divide 36 by 12.

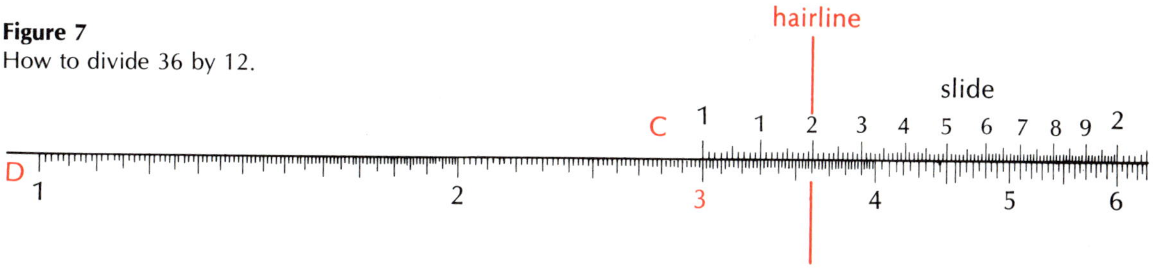

Thus the rule for the division of two numbers is as follows: *Move the slide so that the number of the denominator on scale C is above the number of the numerator on scale D. Then read the quotient under mark 1 of scale C.*

There are three things to remember when you are dividing two numbers: (1) The upper number of the fraction (the numerator) is located on the lower scale (scale D). (2) In division the slide is moved (in multiplication the runner is moved). (3) In both division and multiplication the answer is read on the D scale.

Example 1. Divide 87 by 35.

$$\frac{87 \text{ (scale D)}}{35 \text{ (scale C)}}$$

As in multiplication, the position of the decimal point must be found by rough calculation (by inspection). For example, 87 is about 90 and 35 is about 30, and $90/30 = 3$. Hence the answer (the quotient) has one figure before the decimal point.

Set the hairline on 87 on scale D. Move the slide until 35 on scale C is under the hairline. The reading on scale D under mark 1 of scale C is 248. Therefore the quotient is 2.48.

Example 2. Find the value of

$$\frac{31 \text{ (scale D)}}{87 \text{ (scale C)}}$$

By rough calculation, $30/90 = 0.3$. By slide rule: Place hairline over 31 on D. Move slide until 87 on C is under hairline. The reading on D under mark 1 on C is 356. Therefore the quotient is 0.36.

Example 3. Find the value of

$$\frac{50.8}{765}$$

By rough calculation, $50/800 = 0.06$. Move the runner so that the hairline is over 508 on D. Move the slide until 765 on C is under the hairline as shown in Figure 8. The reading on scale D under mark 1 of C is 666. The answer is 0.0666.

Example 4. Find the value of

$$\frac{765}{508}$$

By rough calculation, $800/500 = 1.6$. By slide rule: Remember that 765 (the numerator) is on scale D and 508 (the denominator)

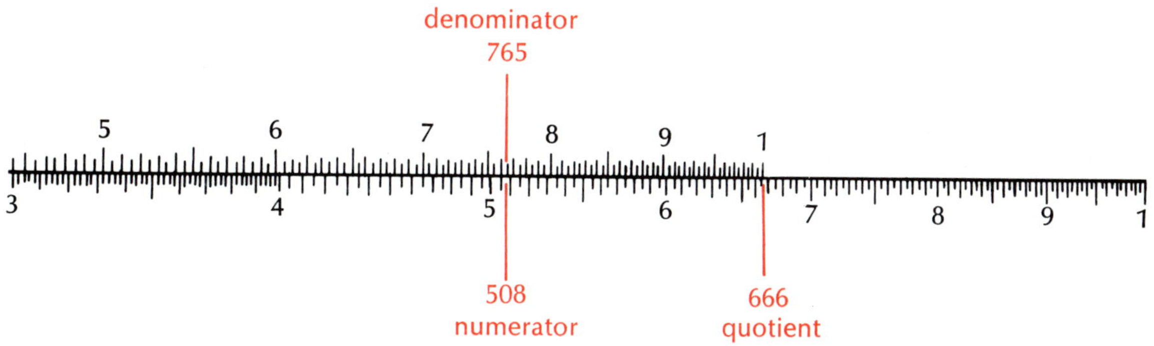

Figure 8
How to divide 50.8 by 765.

is on scale C and over 765 on D. Mark 1 on C is then above 151 on D. The slide rule reading is 151, and the quotient is 1.51.

Computations

Find the value of each of the following:

1. $\frac{200}{3.14}$ Answer 63.7
2. $\frac{1.000}{3.14}$ Answer 0.318
3. $\frac{31.4}{7.24}$ Answer 4.34
4. $\frac{385}{732}$ Answer 0.526
5. $\frac{1.24}{57.5}$ Answer 0.0216
6. $\frac{95.6}{73.8}$ Answer 1.29
7. $\frac{0.475}{65}$ Answer 0.0073
8. $\frac{3060}{360}$ Answer 8.50

9. The volume of 46.8 cm^3 of platinum has a mass of 1000 g. Compute the density of the metal.

$$D = \frac{M}{V} = \frac{1000 \text{ g}}{46.8 \text{ cm}^3} = 21.4 \text{ g/cm}^3$$

10. The density of a liquid is 0.79 g/ml. Find the volume of 250 g of this liquid.

$$V = \frac{M}{D} = \frac{250 \text{ g}}{0.79 \text{ g/ml}} = 320 \text{ ml}$$

11. The density of copper is 8.9 g/cm^3. Find the volume of 1 kg of copper.

$$V = \frac{M}{D} = \frac{1000 \text{ g}}{8.9 \text{ g/cm}^3} = 110 \text{ cm}^3$$

12. The mass of a cube of copper is 123 g. The length of each side of the cube is 2.4 cm. Compute the density of copper.

Volume of copper = $(2.4 \text{ cm})^3 = 13.8 \text{ cm}^3$

$$D = \frac{M}{V} = \frac{123 \text{ g}}{13.8 \text{ cm}^3} = 8.9 \text{ g/cm}^3$$

9. Combined Multiplication and Division

The usual way to solve this kind of problem is to divide and multiply alternately. That is, first divide, second multiply, third divide, fourth multiply, and so on to the end.

Example 1:

$$\frac{820 \times 650 \times 450}{780 \times 540}$$

By inspection the approximate answer is 600.
Step 1 (division): 820/780 = 1.05
Step 2 (multiplication): 1.05 × 650 = 684
Step 3 (division): 684/540 = 1.27
Step 4 (multiplication): 1.27 × 450 = 570
Actually it is unnecessary to read the numbers after the various steps. Simply move the slide and runner. Use the hairline instead of reading the numbers after Steps 1, 2, and 3. Then read the answer on scale D after Step 4.

Example 2:

$$\frac{13.5 \times 326}{61.8}$$

By rough calculation, the approximate answer is 60. By slide rule: Remember: To divide, use the slide; to multiply, use the runner. Read the answer on scale D. Move the hairline to 135 on D. Move the slide until 618 on C is under the hairline. Place the hairline on mark 1 (right) of C. Move the runner until mark 1 (left) of C is under the hairline. Move the hairline to 326 on C. Read the answer under the hairline. It is 716. Therefore the correct answer is 71.6.

Example 3:

$$\frac{61.8}{0.555 \times 0.0286}$$

By rough calculation, the answer is about 3000. To simplify the computation, there should be as many numbers in the numerator as in the denominator. So rewrite the fraction as

$$\frac{61.8 \times 1}{0.555 \times 0.0286}$$

With practice, you can imagine this step. It is not necessary to actually write "× 1" in the numerator. Place the hairline on 618 of scale D. Move the slide until 555 on C is under the hairline. Move the runner to the left until mark 1 (left) on C is under the hairline. Move the slide until the hairline is over 286 on C. Read the answer on D, under mark 1 (right) of C. It is 388. Hence the correct answer is 3880.

Example 4:

$$\frac{1}{2.24 \times 18.5 \times 0.055}$$

By rough calculation, the answer is approximately 0.5. In your imagination, rewrite the fraction as

$$\frac{1 \times 1 \times 1}{2.24 \times 18.5 \times 0.055}$$

Place the hairline over mark 1 on scale D. Move the slide until 224 is under the hairline. Move the runner until the hairline is over mark 1 (right) on C. Move the slide until 185 is under the hairline. Move the runner until the hairline is over mark 1 (left) on C. Move the slide until 55 is under the hairline. Move the runner until the hairline is over mark 1 (right) on C. Read the answer on scale D under mark 1 on C. It is 438. Therefore the correct answer is 0.438.

Computations

Compute the answers to the following problems.

1. $\frac{1.40 \times 532}{48}$ Answer 15.6
2. $\frac{1.37 \times 3.14}{62}$ Answer 0.069

3. $\frac{2.66 \times 7.28}{0.52}$ Answer 37.2

4. $\frac{82.5 \times 0.105}{7.42}$ Answer 1.17

5. $\frac{25.3 \times 43.5}{2760}$ Answer 0.399

6. $\frac{500 \times 298}{273}$ Answer 546

7. $\frac{273}{500 \times 298}$ Answer 0.00183

8. $\frac{7.54}{3.56 \times 0.0277}$ Answer 76.5

9. $\frac{4.40 \times 0.313}{2.2 \times 2.73}$ Answer 0.229

10. $\frac{14.6 \times 273 \times 70}{303 \times 76}$ Answer 12.1

11. $\frac{1}{26.6 \times 0.085 \times 2.54}$ Answer 0.174

12. $\frac{30.5 \times 506}{3.64 \times 380 \times 42.5}$ Answer 0.262

10. Square Root

In some cases it is necessary to find the square root of a number. It is much easier to find a square root by slide rule than by arithmetic. To find a square root by slide rule, use the A and D scales.

Notice that scale A is divided into two identical halves. Each half is numbered from 1 to 10. The left half of scale A is used if the number has an odd number of digits on the left side of the decimal point. The right half of scale A is used if the number has an even number of digits on the left side of the decimal point.

To find the square root of a number, set the hairline on the number on scale A and then read the square root on scale D. For example, to find the square root of 4 ($\sqrt{4}$) set the hairline on mark 4 on the left half of scale A and read the answer 2 on scale D. (See Figure 9.)

To find the square root of 40 ($\sqrt{40}$) place the hairline on mark 4 on the right half of scale A. Read the answer 6.3 on scale D. (See Figure 10.) Similarly, $\sqrt{7} = 2.64$, and $\sqrt{70} = 8.35$.

Example 1. Find $\sqrt{2.54}$. Place the hairline on 2.54 on the left half of scale A. Read the square root under the hairline on scale D. $\sqrt{2.54} = 1.59$.

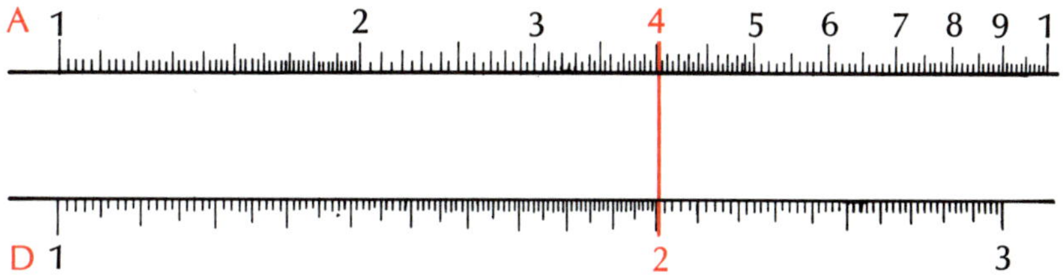

Figure 9
The square root of 4.

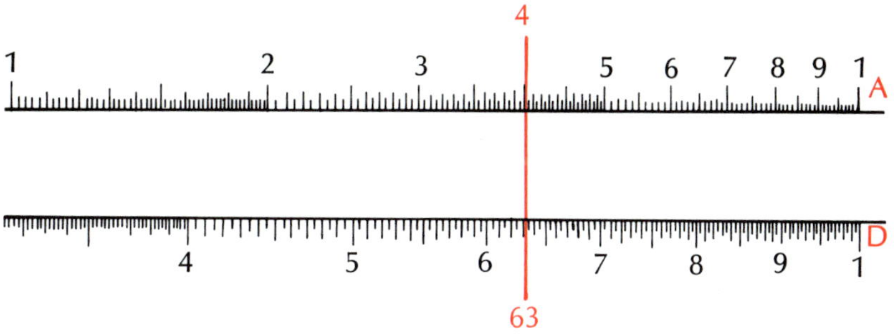

Figure 10
The square root of 40.

Example 2. Find $\sqrt{25.4}$. Use the right half of scale A. $\sqrt{25.4} = 5.03$.

Example 3. Find $\sqrt{625}$. Since the number has an odd number of digits, use the left half of scale A. $\sqrt{625} = 25.0$.

Example 4. Find $\sqrt{6250}$. Since the number has an even number of digits, use the right half of scale A. $\sqrt{6250} = 79.1$.

Example 5. Find $\sqrt{0.365}$. To obtain a number between 1 and 100, move the decimal point 2 places to the right. That is,

$$\sqrt{0.365} = \sqrt{36.5} \times \sqrt{0.01}$$

Since there is an even number of whole digits in 36.5, use the right half of scale A.

$$\sqrt{36.5} = 6.04$$

$$\sqrt{0.365} = 6.04 \times 0.1 = 0.604$$

Computations

1. $\sqrt{36}$ Answer 6.0
2. $\sqrt{360}$ Answer 18.9
3. $\sqrt{3600}$ Answer 60.0
4. $\sqrt{3.6}$ Answer 1.89
5. $\sqrt{0.36}$ Answer 0.60
6. $\sqrt{0.036}$ Answer 0.189
7. $\sqrt{0.0036}$ Answer 0.060

Appendix 2

THE PERIODIC TABLE

KEY

Atomic Number → 6 | Average Atomic Mass → 12.0 | Symbol of Element → **C** | Element Name → Carbon

ROWS ↓ / COLUMNS →	I	II							
1	1 1.0 **H** Hydrogen								
2	3 6.9 **Li** Lithium	4 9.0 **Be** Beryllium							
3	11 23.0 **Na** Sodium	12 24.3 **Mg** Magnesium							
4	19 39.1 **K** Potassium	20 40.1 **Ca** Calcium	21 45.0 **Sc** Scandium	22 47.9 **Ti** Titanium	23 50.9 **V** Vanadium	24 52.0 **Cr** Chromium	25 54.9 **Mn** Manganese	26 55.8 **Fe** Iron	27 58.9 **Co** Cobalt
5	37 85.5 **Rb** Rubidium	38 87.6 **Sr** Strontium	39 88.9 **Y** Yttrium	40 91.2 **Zr** Zirconium	41 92.9 **Nb** Niobium	42 95.9 **Mo** Molybdenum	43 99 **Tc** Technetium	44 101.1 **Ru** Ruthenium	45 102.9 **Rh** Rhodium
6	55 132.9 **Cs** Cesium	56 137.3 **Ba** Barium	57–71* see below	72 178.5 **Hf** Hafnium	73 180.9 **Ta** Tantalum	74 183.9 **W** Tungsten	75 186.2 **Re** Rhenium	76 190.2 **Os** Osmium	77 192.2 **Ir** Iridium
7	87 223 **Fr** Francium	88 226 **Ra** Radium	89–103** see below	104 (unnamed)	105 (unnamed)				

*Lanthanide Series	57 138.9 **La** Lanthanum	58 140.1 **Ce** Cerium	59 140.9 **Pr** Praseodymium	60 144.2 **Nd** Neodymium	61 147 **Pm** Promethium	62 150.4 **Sm** Samarium	63 152 **Eu** Europium
**Actinide Series	89 227 **Ac** Actinium	90 232.0 **Th** Thorium	91 231 **Pa** Protactinium	92 238.0 **U** Uranium	93 237 **Np** Neptunium	94 242 **Pu** Plutonium	95 243 **Am** Americium

			III	IV	V	VI	VII	VIII ←COLUMNS	ROWS ↓
								2 4.0 **He** Helium	1
			5 10.8 **B** Boron	6 12.0 **C** Carbon	7 14.0 **N** Nitrogen	8 16.0 **O** Oxygen	9 19.0 **F** Fluorine	10 20.0 **Ne** Neon	2
			13 27.0 **Al** Aluminum	14 28.1 **Si** Silicon	15 31.0 **P** Phosphorus	16 32.1 **S** Sulfur	17 35.5 **Cl** Chlorine	18 39.9 **Ar** Argon	3
28 58.7 **Ni** Nickel	29 63.5 **Cu** Copper	30 65.4 **Zn** Zinc	31 69.7 **Ga** Gallium	32 72.6 **Ge** Germanium	33 74.9 **As** Arsenic	34 79.0 **Se** Selenium	35 79.9 **Br** Bromine	36 83.8 **Kr** Krypton	4
46 106.4 **Pd** Palladium	47 107.9 **Ag** Silver	48 112.4 **Cd** Cadmium	49 114.8 **In** Indium	50 118.7 **Sn** Tin	51 121.8 **Sb** Antimony	52 127.6 **Te** Tellurium	53 126.9 **I** Iodine	54 131.3 **Xe** Xenon	5
78 195.1 **Pt** Platinum	79 197.0 **Au** Gold	80 200.6 **Hg** Mercury	81 204.4 **Tl** Thallium	82 207.2 **Pb** Lead	83 209.0 **Bi** Bismuth	84 210 **Po** Polonium	85 210 **At** Astatine	86 222 **Rn** Radon	6

64 157.3 **Gd** Gadolinium	65 158.9 **Tb** Terbium	66 162.5 **Dy** Dysprosium	67 164.9 **Ho** Holmium	68 167.3 **Er** Erbium	69 168.9 **Tm** Thulium	70 173.0 **Yb** Ytterbium	71 175.0 **Lu** Lutetium
96 247 **Cm** Curium	97 247 **Bk** Berkelium	98 251 **Cf** Californium	99 254 **Es** Einsteinium	100 253 **Fm** Fermium	101 256 **Md** Mendelevium	102 255 **No** Nobelium	103 256 **Lr** Lawrencium

Glossary

absolute zero: The theoretical temperature at which all molecular motion is reduced to its minimum level. If a substance could be a gas at this temperature, there would be no gas volume at all.

acute angle: An angle less than 90°.

alloy: A solid mixture of two or more metals.

ammeter: An instrument that measures an electric current.

ampere: A unit of measurement of electric current.

analysis: Decomposition of a compound in order to find what elements it is made of.

angle: Two lines that meet.

angle of incidence: The angle between the perpendicular and an incident ray path.

angle of reflection: The angle between the perpendicular and a reflected ray path.

angle of refraction: The angle between the perpendicular and a refracted ray path.

anhydrous: Without water.

Archimedes' Principle: A law of buoyancy.

atom: A building block of matter. The smallest particle of an element that still retains the identity of the element.

atomic mass: The sum of the numbers of protons and neutrons in the nucleus of an atom.

atomic number: The number of protons in the nucleus of an atom.

axes (singular, axis): The horizontal and vertical lines on a graph along which scales are made for the quantities being graphed.

balanced equation: A chemical equation in which the number of atoms of each element is the same on both sides.

barometer: An instrument that measures the pressure of the atmosphere.

boiling point: The temperature at which a substance changes from a liquid to a gas.

Boyle's Law: A law relating the pressure and the volume of a gas.

buoyancy: An upward force exerted by a fluid upon an object submerged in the fluid.

calorie: The amount of heat energy required to raise the temperature of one gram of water through one Celsius degree.

Celsius scale: A temperature scale in which the freezing point of water is zero degrees (0°C) and the boiling point of water is 100 degrees (100°C).

center of gravity: The point where all the weight of a body appears to be concentrated.

centimeter: A unit of length in the metric system. One hundredth of a meter.

Charles' Law: A law relating the temperature and the volume of a gas.

chemical activity: The rate at which a substance reacts with another substance.

chemical change: A change that alters the number or the arrangement of electrons in atoms.

chemical reaction: A chemical change. In a chemical reaction, a substance can be changed into another substance.

circuit: A path followed by an electric current.

cm: The abbreviation for centimeter.

combining capacity: The number of electrons lost or gained by an atom in a reaction.

combustion: Burning. A reaction in which a flame is produced.

components: The parts of an electric circuit.

compound: A substance that contains only one kind of molecule, made of atoms of two or more different elements.

concave lens: A lens thinner at the center than at the edge.

condensation: A process in which a substance changes from a gas to a liquid.

conductor: A substance through which electrons flow easily.

converge: To come together.

converging lens: A convex lens.

convex lens: A lens thicker at the center than at the edge.

current: A flow of electrons.

data: Measurements taken and recorded in an experiment.

decomposition: The breaking down of a compound into its elements.

density: The mass of a unit volume of a substance.

diatomic: Containing two atoms.

diverge: To get farther apart.

diverging lens: A concave lens.

dry cell: A voltaic cell containing a paste of chemicals.

drying agent: A substance that absorbs water vapor from the air.

E: The symbol for voltage in volts.

efficiency: The ratio of output work to input work.

elasticity: A force that acts upon a stretched object, such as a wire, in a direction opposite to the applied force.

electric charge: An accumulation of electricity on an object.

electric force: The force between charged particles. Particles with like charges repel each other, and particles with unlike charges attract each other.

electron: One of the particles making up an atom. It has a negative charge.

electron cloud: The space occupied by the electrons of an atom.

element: A substance that contains only one kind of atom.

emergent ray: A ray that comes out of a transparent object.

energy: The ability to do work.

equation: A representation of a chemical reaction in which symbols and formulas are used.

equiangular: Having equal angles.

equilibrium: A state in which two opposing forces are in balance.

evaporation: A process in which a substance changes from a liquid to a gas.

fluid: Matter that can flow. Both liquids and gases are fluids.

focal length: The distance from a lens to its focal point.

focal point: A position at which light passing through a lens is brought to a focus.

focus: To make a clear image. The focal point of a convex lens.

force: A push or a pull.

formula: An abbreviation for the name of a compound.

freezing point: The temperature at which a substance changes from a liquid to a solid.

friction: A force that opposes motion.

fulcrum: The point of support of a lever.

g: The abbreviation for gram.

gas: Matter that has neither a fixed volume nor a fixed shape.

gas pressure: The force exerted by gas molecules on a surface.

generator: A machine that changes kinetic energy into electric energy.

gram: A unit of mass in the metric system. One thousandth of a kilogram.

gravity: A force that attracts any two masses to each other.

heat: A form of energy.

heat capacity: The number of calories of heat required to raise the temperature of one gram of a substance through one Celsius degree.

heat of combustion: The amount of heat released per gram of substance burned.

heat of fusion: The number of calories of heat that must be supplied to one gram of a solid substance at its melting point to change it to a liquid at the same temperature.

heat of solution: The energy needed to dissolve a solid.

humidity: Moisture in the air.

hydrocarbon: A compound of carbon and hydrogen.

I: The symbol for current in amperes.

illumination: The amount of light that falls upon a surface.

image: The picture of an object formed by a reflecting surface or a lens.

Incident ray: A ray of light falling on an object.

inclined plane: A simple machine consisting of a flat surface that is inclined or slanted.

inference: A judgment based on an observation.

input work: Work that is put into a machine.

insoluble: Of very low solubility.

insulator: A substance through which heat or electrons do not flow easily.

internal energy: The sum of the kinetic and potential energies of the atoms or molecules of a substance.

ion: An atom with an electric charge.

joule: A unit of work in the metric system. Another name for a newton-meter.

kg.: The abbreviation for kilogram.

kilogram: The standard unit of mass in the metric system. One thousand grams.

kilometer: A unit of length in the metric system. One thousand meters.

kinetic energy: Energy due to speed.

km: The abbreviation for kilometer.

l: The abbreviation for liter.

lens: A piece of transparent material with curved surfaces.

lever: A bar that rotates about a point of support. A simple machine.

liquid: Matter with a fixed volume but not a fixed shape.

liter: A unit of liquid volume in the metric system, slightly larger than a quart.

m: The abbreviation for meter.

machine: A device that makes it easier to do work.

magnification: The ratio of the size of an image to the size of the object that produced the image.

mass: A property of matter. The amount of matter in an object.

melting point: The temperature at which a substance changes from a solid to a liquid.

meniscus: The curved free surface of a liquid.

metal: A substance whose atoms usually lose electrons in a chemical reaction.

meter: The standard unit of length in the metric system, slightly longer than a yard.

metric system: A system of measurement based on the meter as a unit length and the kilogram as a unit mass.

milliliter: One thousandth of a liter.

ml: The abbreviation for milliliter.

molecule: A group of atoms bonded together. The smallest particle of a gas.

N: The abbreviation for newton.

neutral: Having no electric charge.

neutron: One of the particles making up an atom. It has no charge.

newton: A unit of weight or force in the metric system.

newton-meter: A unit of work or torque in the metric system.

N·m: The abbreviation for newton-meter.

non-metal: A substance whose atoms usually gain electrons in a chemical reaction.

nuclear force: The force that binds together the particles in the nucleus.

nucleus: The center of an atom, containing all the protons and neutrons of the atom.

observation: Something you see, or hear, or feel.

obtuse angle: An angle greater than 90°.

ohm: A unit of electric resistance.

origin: The point where the two axes of a graph cross.

output work: Work that is accomplished by a machine.

oxide: A compound of oxygen and another element.

oxidation: A chemical process in which an oxide is formed.

parallel circuit: A circuit arrangement in which the total current is divided into two or more paths.

periodic table: An arrangement of the elements according to their atomic numbers.

perpendicular: A line drawn at a right angle to another line.

photosynthesis: A complicated reaction that occurs in green plants in the presence of sunlight. Carbon dioxide and water are changed into starch (or sugars) and oxygen.

potential energy: Energy of position or condition.

precipitate: A solid substance that is formed by the reaction between two solutions.

pressure: Force per unit area.

principle of work: In a frictionless machine, the output work equals the input work.

prism: A solid that has similar and parallel ends.

product: A substance that is produced in a reaction.

property: A feature that identifies something. Two properties of matter are mass and volume.

proton: One of the particles making up an atom. It has a positive charge.

protractor: An instrument used to measure angles.

R: The symbol for resistance in ohms.

radical: A group of atoms that usually remain together in a chemical reaction.

ray: A narrow beam of light.

reactant: A substance that enters into a reaction.

real image: An image that can be projected on a screen.

reflect: To bounce back.

refraction: The bending of a ray of light as it passes from one material to another.

regular solid: A solid that has sides that can be measured.

resistor: A device that slows down the flow of electrons in a circuit.

respiration: A process in which food is burned in the cells of an animal body. Carbon dioxide is formed in this process.

rheostat: A resistor whose resistance can be varied.

right angle: An angle of 90°.

saturated: A condition in which the maximum amount of a solid is dissolved in a liquid.

series circuit: A circuit arrangement in which there is only one path for the flow of electrons through two or more resistors.

solid: Matter with a fixed volume and a fixed shape.

solubility: The maximum mass of a solid that will dissolve in 100 g (100 ml) of a liquid to form a saturated solution.

soluble: Capable of being dissolved.

solution: A mixture of dissolving substance and dissolved substance.

symbol: An abbreviation for the name of an element.

synthesis: Putting together; the building up of a compound from its elements.

temperature: The hotness or coldness of a body. It is related to the speed of the atoms or molecules in the body.

terminal: A place where a wire can be attached to an ammeter, voltmeter, or voltaic cell.

theory: A suggested reason why something happens.

thermometer: An instrument that measures temperature.

torque: The turning effect of a force.

transparent: Allowing light to pass through.

valence: Combining capacity.

vapor: The gaseous state of a substance that is normally a liquid at room temperature.

vaporization: The process in which a substance changes from a liquid to a vapor.

virtual image: An image that is not real; that is, an image that cannot be projected on a screen.

volt: A unit of measurement of voltage or voltage drop.

voltage: The difference in chemical activity between the terminals of a voltaic cell.

voltage drop: The difference in electric pressure at the two ends of a resistor.

voltaic cell: A device that changes chemical energy into electric energy.

voltmeter: An instrument that measures the voltage between two points in a circuit.

volume: The amount of space occupied by an object.

weight: The gravitational force that acts upon a body.

work: A force exerted through a distance. Work is expressed as the product of force and distance.

Index

BCDEFGHIJ–RM–79876543